Advanced Research on Heat Exchangers Networks and Heat Recovery

Advanced Research on Heat Exchangers Networks and Heat Recovery

Guest Editors

Stanislav Boldyryev
Bohong Wang
Timothy Gordon Walmsley

MDPI

Basel • Beijing • Wuhan • Barcelona • Belgrade • Novi Sad • Cluj • Manchester

Guest Editors

Stanislav Boldyryev
Faculty of Mechanical Engineering and Naval Architecture
University of Zagreb
Zagreb
Croatia

Bohong Wang
National & Local Joint Engineering Research Center of Harbour Oil & Gas Storage and Transportation Technology
Zhejiang Ocean University
Zhoushan
China

Timothy Gordon Walmsley
School of Engineering
The University of Waikato
Hamilton
New Zealand

Editorial Office
MDPI AG
Grosspeteranlage 5
4052 Basel, Switzerland

This is a reprint of the Special Issue, published open access by the journal *Energies* (ISSN 1996-1073), freely accessible at: https://www.mdpi.com/journal/energies/special_issues/HH44A95D11.

For citation purposes, cite each article independently as indicated on the article page online and as indicated below:

Lastname, A.A.; Lastname, B.B. Article Title. *Journal Name* **Year**, *Volume Number*, Page Range.

ISBN 978-3-7258-4989-5 (Hbk)
ISBN 978-3-7258-4990-1 (PDF)
https://doi.org/10.3390/books978-3-7258-4990-1

Contents

About the Editors

Stanislav Boldyryev

Stanislav Boldyryev is a Senior Researcher at the University of Zagreb, Faculty of Mechanical Engineering and Naval Architecture. He graduated from the National Technical University "Kharkiv Polytechnic Institute" (Kharkiv, Ukraine) in 1999 and defended his PhD thesis in 2008. His PhD study was in the field of Process Integration. During his academic career, he worked at several universities as an Assistant and Associate Professor. He developed new curricula for academia and industry in process integration, energy efficiency, and process simulation. During his teaching career, he has created several teaching guides and supervised more than 40 BSC, MSc, and PhD students. Since 2000, he has been involved in different capacities in 20+ research and development projects funded by the EU, EIB, Danish Energy Agency, and other bodies. He was awarded an individual grant for experienced researchers from the EU Marie Curie Fund and the Croatian Government and has many years of experience working in the industry, engineering, consulting, and teaching. He has completed over 40 projects in various sectors, with a focus on energy efficiency and debottlenecking, also working at a chemical bromine plant as Deputy General Director of Development. In 2017, he established a Process Integration Department at Neva-TeploTechnika Engineering Company, where he collaborated with CALGAVIN, HTRI. and Process Systems Enterprise to share knowledge and experience with the industry. In 2025, he co-founded a renewable energy and sustainability start-up. He is a Subject Editor and Editorial Board Member of international scientific journals and a Scientific Advisory Board Member of several international conferences. His research interests include Process and Total Site Integration, Renewables in Industry, Industry Electrification, Sector Coupling, Process Simulation, Digital Twinning, Energy Saving, Heat Exchanger Networks, Heat Pumps, and Emissions Reduction.

Bohong Wang

Bohong Wang is an Associate Professor at Zhejiang Ocean University with expertise in energy systems engineering and pipeline transport engineering. He serves as an Associate Editor of the *Journal of Pipeline Systems Engineering and Practice* and a Guest Editor of the *Journal of Cleaner Production*, *Energy*, *Thermal Science and Engineering Progress*, and *Cleaner Energy Systems*. He was selected as a member of the "Youth Talent Training Program" of Zhejiang Province and a trainee of the Youth Talent Training Project of the China Institute of Navigation. He has led 10 national, provincial, and industrial projects related to the optimization of energy systems, oil–gas pipeline system optimization, reliability evaluation, and process integration. He has developed software in the field of heat integration, pipeline simulation, and hydraulic optimization. He has published more than 130 journal and conference papers, with an h-index of 31, and is ranked in the World's Top 2% of Scientists. He also serves as a reviewer for 60 international journals.

Timothy Gordon Walmsley

Tim Walmsley is a Senior Lecturer in Chemical and Process Engineering in the School of Engineering and Co-Director of the Ahuora Centre for Smart Energy Systems at the University of Waikato, Hamilton, New Zealand. His research fuses thermodynamics with digitalisation to create practical, high-impact concepts and roadmaps for industrial decarbonisation that lead to economic benefits and satisfy emissions targets. Within the Ahuora Centre, he leads research on thermal demand reduction and heat pump integration and directs the software development team, comprised of chemical, process and software engineering postdocs and students. The flagship output from this development work is the Ahuora digital twin platform—a modular, scalable, intuitive web application for multiscale process integration, modelling and optimisation.

energies

MDPI

Article

Improving the Economic Efficiency of Heat Pump Integration into Distillation Columns of Process Plants Applying Different Pressures of Evaporators and Condensers

Stanislav Boldyryev [1,*], Mariia Ilchenko [2] and Goran Krajačić [1]

[1] Faculty of Mechanical Engineering and Naval Architecture, The University of Zagreb, 10000 Zagreb, Croatia; goran.krajacic@fsb.unizg.hr

[2] Department of Integrated Technologies, Processes and Apparatuses, National Technical University "Kharkiv Polytechnic Institute", 61000 Kharkiv, Ukraine; mariia.ilchenko@khpi.edu.ua

* Correspondence: stanislav.boldyryev@fsb.unizg.hr

Abstract: The electrification of process industries is one of the main challenges when building a low-carbon society since they consume huge amounts of fossil fuels, generating different emissions. Heat pumps are some of the key players in the industrial sector of the carbon-neutral market. This study proposes an approach to improve the economic feasibility of heat pumps within process plants. Initial energy targeting with grand composite curves was used and supplemented with the detailed design of an evaporator and a compressor for different condensation and evaporation pressures. The trade-off between the capital cost of the heat pump and the electricity cost was investigated, and optimal configurations were selected. This case study investigates the gas fractioning unit of a polymer plant, where three heat pumps are integrated into distillation columns. The results demonstrate that the heat recovery is 174 MW and requires an additional 37.9 MW of electricity to reduce the hot utility by 212 MW. The selection of the evaporation and condensation pressures of heat pumps allows 21.5 M EUR/y to be saved for 7 years of plant operation. The emission-saving potential is estimated at 1.89 $ktCO_2/y$.

Keywords: process integration; heat pump; heat exchangers; industry electrification; energy saving; emission reduction; economic assessment

1. Introduction

The use of energy-saving technologies plays an important role in sustainable development and reducing environmental impact. One such technology that has become widely used is heat pumps.

A heat pump (HP) makes it possible to use low-potential heat sources and convert them into high-potential heat that can be used in production processes and the integration of renewable energies [1]. Heat pumps can be configured for different conditions and production requirements, making them the ideal choice for a variety of industries.

The use of heat pumps plays an important role in achieving energy efficiency, reducing the use of fossil energy resources, reducing negative emissions, and protecting the environment. This enables industrial companies to reduce their ecological footprint, improve market competitiveness, and contribute to sustainable development.

Heat pumps are used extensively in a variety of industries—primarily in the chemical, processing, and food industries. The design strategy for a heat pump-assisted distillation system was proposed many years ago [2], but its importance increased recently when environmental issues became a hot topic. Van de Bor et al. [3] investigated the recovery and upgrading of low-potential heat sources using HPs (to generate useful process heat) and low-temperature heat engines (to generate electricity) and considered the relevance and impact of wet compression on the performance of the HP. Yuan et al. [4] made a comparison of the specific energy consumption parameters of heat pump dryers, combustion-heated dryers, and electrically heated dryers. Tveit studied the application of a high-temperature HP and gave a comparative assessment of the reduction in CO_2 emissions from heat pumps and boilers operating on natural gas using the example of a dairy plant [5]. Similarly, Eurne et al. [6] made a comparison of a heat pump water-heater (ASHPWH) system with other options: a natural gas boiler, electricity, a liquefied petroleum gas (LPG) instant water heater, and a solar water heater with an electric or natural gas backup for a dairy farm.

A case study on the brewing process [7] considers an HP system with two parallel heat sources at different temperatures and times. The analysis of the proposed HP system shows that the CO_2 production from the consumption of electrical energy is reduced by 60%.

The use of HPs is also widespread in the agricultural sector. An analysis of the use of various types of HPs for heating greenhouses, heating water used for greenhouse heating, heating water used for watering, and hot technical water preparation included an assessment of the technical aspects and cost-effectiveness of implementation [8–10]. Heat pumps can also be used for district cooling in residential sectors during the summer through the use of existing district heat systems [11]. Ünal et al. [12] consider the energy, exergic, exergoeconomic and exergo-environmental analysis of an underfloor heating system integrated with the a geothermal HP. The authors investigate the distribution of losses in the system over the elements of the system separately, and the equivalent CO_2 emissions of heating a greenhouse with natural gas and an HP system was compared. Chiriboga et al. confirmed the possibility of designing and building a coupled geothermal HP [13]. Based on the results of the assessment of the energy potential of the solar and geothermal sources, the energy balance in the greenhouse was calculated to determine the parameters of the geothermal HP using the vapor compression cycle.

Using a coal-fired power plant as an example, Zhang et al. considered a cogeneration system based on an organic Rankine cycle (ORC) and absorption heat pump (AHP) to improve power output and heating capacity [14]. The efficiency analysis showed that this cogeneration system can increase the power output and heating capacity of the plant. Cao et al. compared the efficiency of different high-temperature HP systems to recover the heat from wastewater from an oil field and produce hot water [15]. The analysis of the data obtained on system energy consumption and efficiency provides recommendations for the selection of a suitable heat recovery system with high heat output for industrial applications.

Another example of the application of an open absorption heat pump (OAHP) system combined with flash evaporation for coal-fired flue gas is the work of Zhang et al. [16]. An exergy analysis of the proposed schemes showed an improvement in the exergy efficiency of the optimised systems.

Su et al. performed thermodynamic modelling and performance evaluation of a heat pump dryer by combining liquid desiccant dehumidification and mechanical vapour recompression [17]. The comparison of the working principles and the performance of the proposed system and the reference system showed that the proposed scheme improves the energy efficiency of the heat pump drying system.

The potential for using an HP with CO_2 as a working medium for the apple drying process was considered in [18]. An analysis of the system simulation results showed that the use of a closed-loop system is effective, but also leads to an increased drying time.

In the oil and chemical industry, it is common to use traditional distillation systems to separate mixtures of liquids. However, this process requires considerable use of fossil fuels as a heat source.

Waheed et al. looked at a de-ethanisation unit of a Nigerian refinery as an example and enhanced the vapour recompression heat pump (VRHP) models that were developed to reduce the heat loss and heat pump size [19]. These strategies are based on reducing the heat differential across the heat pump by utilizing the process stream within the system, the external process stream, and the utility streams.

For the production of n-butyl acetate and isopropyl alcohol, Liu et al. proposed a heat pump-assisted dividing wall column for a reactive distillation system and a heterogeneous azeotropic distillation system [20]. HP-assisted dividing wall columns are beneficial for cases where clean and highly effective electricity generation technologies are adopted and long-term profitability is considered.

Long et al. proposed an energy-efficient sequence for the natural gas liquid fractionation process. A hybrid heat pump-assisted system with a side reboiler and was proposed to maximise energy efficiency [21].

Long et al. considered different HP configurations to improve the energy efficiency of distillation columns for separating R-410A and R-22 [22]. Top vapour superheating was proposed for improving the performance of the HP configuration, as well as for protecting the compressor from liquid leakage. The possibility of replacing the throttle valve with a hydraulic turbine, which would reduce the operating costs, is being considered.

Zhu et al. analysed the separation process of a cyclohexane/sec-butyl alcohol/water azeotropic mixture by extractive distillation [23]. The distillation process is optimised based on a sequential and iterative optimization algorithm. For further energy saving, several energy-optimised processes are proposed: the thermally coupled extractive distillation process (TCED), the heat pump extractive distillation process (HPED), and the heat pump combined with thermal coupling extractive distillation process (HPCWTCED).

Šulgan et al. reviewed the production of ethyl acetate using an HP and presented a multi-objective evaluation based on energy requirements, economic analysis, and safety analysis [24]. As a result, the use of HPs is highly recommended in both the conventional process and in reactive columns with a separation unit. Since a higher level of process integration is achieved with an HP, economic aspects are improved, but at the same time, the safety aspects are worsened.

The application of HPs in the distillation process may be the solution for process electrification. The case study by Boldyryev et al. analysed natural gas liquid processing and assessed electrified thermal utility [25]. The targeting for appropriate HP placement resulted in increased heat recovery and a reduction in energy cost by up to 41%.

Florian Schlosser et al. reviewed HPs and identified concepts for their integration across industries and processes based on the grand composite curve (GCC) and demonstrated the saving potential [26]. Kim et al. proposed an optimal heat exchange network (HEN) with HPs in a wastewater heat recovery system in the textile industry [27]. The authors considered a two-step approach to design a heat exchange network and made an economic evaluation to minimise costs and maximise energy efficiency.

Case studies of a milk spray dryer by Walmsley et al. [28] and Gai et al. [29] focused on the modelling and optimization of an HP for convective dryers considering pinch design principles. Different schemes for the integration of the drying process were considered, as well as the optimization of the operating parameters for maximum efficiency. Lincoln et al. presented a fully electric milk evaporation system developed through an effective Process Integration and Electrification design method [30]. A sensitivity analysis of the final process design was conducted, which showed that it applied to a wide range of operating conditions. Klinac et al. proposed a pinch-based Total Site Heat Integration (TSHI) method, which is used for multi-level heat pump integration options at a meat processing site [31]. The results of the Total Site approach in coke-to-chemicals demonstrated the appropriate placement of HPs within inter-plant integration and showed a fast payback of 1.04 years [32].

Hegely and Lang conducted research and found that several levels of heat integration allow the energy consumption of a bioethanol plant to be reduced [33]. With the external energy demand and total annual cost of the different configurations, the authors concluded that the application of a heat pump is not recommended because of its high investment cost.

Cox et al. investigated the technical and economic performance of high-temperature heat pumps for use in the U.S. dairy industry [34]. A model was created to estimate the coefficient of performance (COP), internal rate of return (IRR), net present value (NPV), and payback period (PBP). Capital costs, operations and maintenance (O&M) costs, heat pump lifetime, electricity prices, natural gas prices, and the cost of carbon were varied to conduct a parametric study of the factors affecting the break-even price of high-temperature HPs.

Lu et al. investigated a high-temperature cascade HP system using low-potential heat from wastewater to produce steam for industrial processes, developing a mathematical model of the system to analyse thermodynamic performance and economic efficiency [35].

Martínez-Rodríguez et al. considered the thermo-economic optimization with a solar thermal-assisted heat pump and a storage system [36]. Here, two case studies were considered as examples: the dairy industry and a 2G bioethanol plant. A thermal and economic evaluation of the system was carried out to determine the supply of the heat load at the required process temperature under different conditions and operating temperatures in the evaporator. Based on the established thermodynamic states, the operating conditions of each HP component were determined.

Studies by Chen et al. [37], Zühlsdorf et al. [38], and Gómez-Hernández et al. [39] investigated the use of different types of refrigerants by comparing the performance of HP systems in terms of economics: the total cost rate, investment and operating costs, capital costs of equipment, cost of CO_2 penalty, energy and exergy efficiency, and NPV. Gudjonsdottir and Infante Ferreira considered a wet compression–resorption heat pump (CRHP) which operates NH_3-H_2O and NH_3-CO_2-H_2O systems [40]. The simple pay-back period for replacing an existing boiler with a CRHP system was calculated, depending on gas and electricity prices forecasted, the total investment costs, the installation costs, the annual fuel consumption costs, the operation and maintenance costs, and the capital return coefficient. Urbanucci et al. proposed and analysed the integration of a high-temperature HP into a trigeneration system [41]. An exergy analysis was carried out to compare the proposed energy system with the traditional one (separate production, cogeneration, and trigeneration). The economic evaluation was analysed using the methodology of the present value of electricity.

Studies by Wolf et al. [42] and Zuberi et al. [43] consider the thermodynamic and economic possibilities of using high-temperature and steam-generating HPs in the pulp and paper, textile, and automotive industries. The costs of consumption, investment, heat, and maintenance are compared, taking into account the energy price and interest rate. Wu et al. proposed a capacity-regulated HTHP system using a twin-screw compressor for waste heat recovery [44]. Here, an economic comparison is made between the HTHP system and steam heating: the thermal power of energy consumption, unit price, operating costs, capital costs, savings percentage, and payback period.

Gangar et al. [45] and Yang et al. [46] evaluated the technical and economic feasibility of producing energy, steam, and regenerative low-potential heat energy using mechanical vapour compression (MVC) heat pumps and absorption heat pumps, respectively. Lee et al. present an extensive economic analysis and environmental impact assessment of heat pump-assisted distillation under different conditions and scenarios: feed composition, plant capacity, and fuel price [47]. Capital and operating costs, the percentage of energy savings, the payback period, and the total annual costs were calculated and a sensitivity analysis were performed.

Hou et al. considered the optimization of tobacco-drying HPs in recovering waste heat from a monocrystal silicon furnace [48]. The issue was analysed from the energy and economic side, investigating the influence of the heat exchange area on the system performance. A thermo-economic and economic model of the system was developed and experimentally verified.

The widespread use of HPs is also due to their positive impact on the environment. For example, in cyclohexane/sec-butyl alcohol/water azeotropic separation systems, the use of the heat pump extractive distillation process and heat pump combined with thermal coupling extractive distillation process can reduce CO_2 emissions by 77.07% and 81.08%, respectively [23]. Hegely and Lang carried out an optimization of the higher pressure of the pressure-swing distillation of a maximum azeotropic mixture of water–ethylenediamine and came to the conclusion that using vapour recompression heat pumps is very favourable environmentally: CO_2 emissions and Eco-indicator 99 are reduced by 44 and 95%, respectively [49]. Liu et al. studied the production of n-butyl acetate and isopropyl alcohol and showed that CO_2 emissions, which are associated with primary energy demands, can be reduced by using heat pump-assisted dividing wall columns, especially when renewable resources such as hydropower are used to produce electricity [20]. Lee et al. assessed the environmental impact of using heat pump-assisted distillation in a gas fractionation unit. By switching to an alternative configuration using a heat pump, the greenhouse gas emissions can be reduced by 31% (splitter) and 32% (de-propaniser) compared to those of conventional distillation [47]. Janković et al. reviewed a process for improving the downstream processing of bioethanol obtained from syngas fermentation. The authors concluded that the implementation of the heat-integrated system and heat pump resulted in a 62.4% reduction in cooling water demand, and reduced total CO_2 emissions by 60.4% and 82.6% for grey and green electricity, respectively [50].

The analysis underlines the prospective application of HPs in industrial production belonging to different branches. Thus, it is shown that the majority of researchers are interested in the conjugate decision on questions regarding increasing the efficiency of thermal installations and the achievement of ecological stability. In terms of their performance, heat pumps markedly exceed almost all other available technologies, and the feasibility of their application is confirmed by technical and economic calculations by comparing many parameters.

The core contribution of this study lies In its targeted, in-depth exploration of the integration of heat pumps within the entire petrochemical process. The subject of heat pump integration has been extensively studied in industrial contexts, but this work tries to fill a significant gap in the existing literature by focusing on the economic trade-off and untapped potential for appropriately targeting heat pump integration in applying different pressure in evaporators and condensers. This work analyses the application of HPs within the industrial processes, updating the targeting procedure based on a GCC. Different condensation and evaporation pressures of HPs are used for the detailed design of HP heat exchangers. The trade-off between energy cost and capital cost for obtaining a detailed configuration of heat pumps is analysed and the best economic solution is selected. It is applied for the gas fractioning process of polymer plants, where three heat pumps are analysed, utilising the waste heat of distillation column condensers to column reboilers. The results of the case study provide the background for the discussion on the authors' hypothesis on improving the economic feasibility of the application of HPs.

The remaining parts of this paper give the framework and a description of the method (Section 2), followed by case studies on the gas fractioning process of polymer plants (Section 3), a discussion of the results in Section 4, and the conclusions.

2. Materials and Methods

The method proposed in the current study is based on the scientific hypothesis that there is a trade-off between energy and capital costs when applying heat pumps to industrial processes. The application of heat pumps in distillation columns is more complex due to the phase changes of the heat carriers on both sides of the heat exchanger. The reduced capital costs, energy costs, and total annual cost of heat pumps are analysed for different condensation and evaporation temperatures/pressures of the refrigerant.

The algorithm of the proposed methodology is as follows:

1. A definition of the process streams that should be heated and cooled by the heat pump with the use of the GCC (Figure 1);
2. The placement of the heat pump and the initial energy targets for both thermal energy and power, specifying the current (ΔT_{min}) between process and utility (refrigerant);
3. A simulation of the heat pump in the Aspen HYSYS 12 [51] environment under an acceptable pressure drop in the condenser and reboiler;

4. The calculation of the detailed configurations and capital costs of the condenser and reboiler based on the simulation results;
5. The calculation of the compressor's capital cost;
6. The calculation of the annualised capital cost of the HP equipment using the cost factors;
7. The calculation of the total annualised cost (TAC);
8. Changing the refrigerant pressure in the compressor inlet/outlet and a repetition of the previous steps;
9. The selection of the HP configuration with min TAC;
10. Performing a sensitivity analysis of the results by applying different electricity prices.

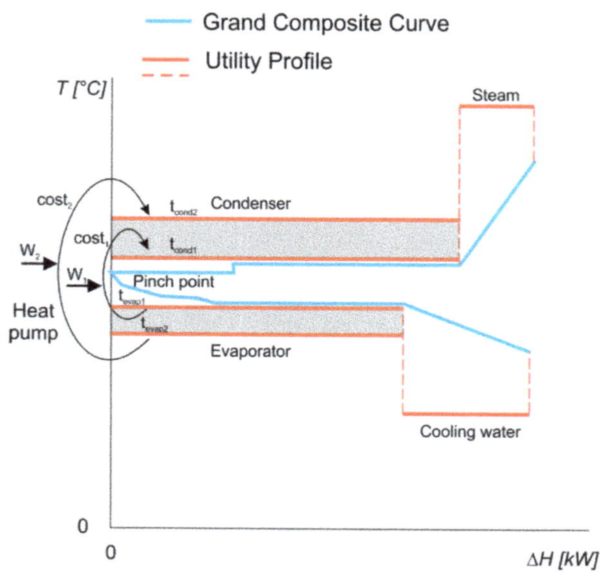

Figure 1. Different options for HP integration within industrial processes.

The minimum total annual cost was found for three heat pumps integrated into the natural gas liquid process plant. The costs were assessed by changing the evaporation and condenser pressure of the heat pumps. The cost correlations were investigated in the pressure range constrained by the process stream requirements from one side. For the evaporators, the minimum acceptable pressure is limited by a minimum acceptable pressure that is recommended to be no less than ambient pressure to avoid the injection of ambient air into the heat pump circuit. For condensers, the high pressure was selected at 150% of the initial pressure found from targeting with the grand composite curve.

The algorithm is based on the following mathematical formulations:

Mass and heat balances are calculated from Equations (1) and (2).

$$\sum_{j=1}^{J} M_j^{In} = \sum_{J=1}^{J} M_j^{Out} + M_{losses} \quad (1)$$

$$\sum_{j=1}^{J} \left(h_j^{In} M_j^{In} \right) + \sum_{i=1}^{I} W_i = \sum_{j=1}^{J} \left(h_j^{Out} M_j^{Out} \right) + Q_{losses} \quad (2)$$

The heat transfer area is defined for shell-and-tube heat exchangers from Equation (3):

$$Q = UA\Delta T_{LM} Ft \quad (3)$$

where Ft is equal to 1 for a full counter-current and ΔT_{LM} is defined in Equation (4):

$$\Delta T_{LM} = \frac{(T_{inH} - T_{outC}) - (T_{outH} - T_{inC})}{ln\frac{(T_{inH} - T_{outC})}{(T_{outH} - T_{inC})}} \tag{4}$$

The overall heat transfer coefficient is defined by the film of condensation outside a horizontal tube with Kern correlations [52]:

$$h_C = 0.725\left(\frac{k_L^3 \rho_L^2 \Delta H_{VAP} g}{d_O \mu_L \Delta T}\right)^{\frac{1}{4}} \tag{5}$$

and the evaporation of kettle and horizontal thermosyphon reboilers is due to [53]

$$h_{NB} = 0.182 P_C^{0.67} q^{0.7} \left(\frac{P}{P_C}\right)^{0.17} \tag{6}$$

The detailed designs of the condenser/reboiler and its price were found with the Aspen EDR Heat Exchanger software V12.

The COP of the HP was found from simulation results in Aspen HYSYS:

$$COP_{HP} = \frac{Q_{HP} + W}{W} \tag{7}$$

The capital cost of the compressor was estimated based on Chemical Engineering Indexes and Marshall and Swift using Equation (8):

$$CAPEX_{Comp} = 98,400\left(\frac{W}{250}\right)^{0.46} f(t) \cdot f(p) \cdot f(m) \tag{8}$$

where f(m) is the correction factor for the construction materials, f(p) is the correction factor for the design pressure, and f(t) is the correction factor for the design temperature. The selected correlation accounts for the material of the compressor and the pressure and temperature levels; moreover, the correction factor is used for the capital cost assessment to provide a conservative cost estimation.

The annualised capital cost of the HP was obtained from the condenser, evaporator, and compressor costs using the fractional interest rate (FIR), loan period (NY), and Lang factor (Lang). The Lang factor accounts for the cost of installation, piping, the control system, insulation, engineering fees, and other costs.

$$ACC = (CAPEX_{Comp} + CAPEX_{Cond} + CAPEX_{Evap}) \cdot Lang \cdot \left(\frac{\mathrm{FIR}(1+\mathrm{FIR})^{\mathrm{NY}}}{(1+\mathrm{FIR})^{\mathrm{NY}} - 1}\right) \tag{9}$$

The energy cost was found using the electricity target of the HP and the average electricity cost:

$$EC = W \cdot c_e \tag{10}$$

The TAC was calculated based on the capital cost and energy cost obtained from Equations (9) and (10).

$$TAC = ACC + EC \tag{11}$$

3. The Case Study

3.1. Process Description

This case study examines the streams of the gas fractioning unit of the polymer plant and integrates three heat pumps into the process to electrify the thermal utility for the further use of renewable energies. This study examines the operation of three heat pumps using Refrig-21 as a refrigerant, a dichlorofluoromethane with the formula CHCl2F, a molecular weight of 102.9, a normal boiling point of 8.9 °C, and an ideal liquid density of 1363 kg/m^3. It is a colourless and odourless gas.

HP-1: The refrigerant vapour flow with a temperature of 57.26 °C and a pressure of 480 kPa enters the compressor, where the refrigerant vapour is compressed to 900 kPa. From the compressor, the refrigerant with a temperature of 96.31 °C is directed to the condenser, where the refrigerant is cooled and condensed due to heat exchange with heat consumers (flow with a temperature of 79.97 °C and a pressure of 1011 kPa). The condensed liquid refrigerant with a temperature of 82.29 °C and a pressure of 895 kPa passes through the control valve, where the refrigerant is throttled. Next, the refrigerant with a temperature of 57.65 °C and a pressure of 485 kPa enters the evaporator. The refrigerant evaporates, giving off the heat of evaporation to cold consumers (flow with a temperature of 66.46 °C and a pressure of 907 kPa).

HP-2: The refrigerant vapour flow with a temperature of 52.37 °C and a pressure of 420 kPa enters the compressor, where the refrigerant vapour is compressed to 900 kPa. From the compressor, the refrigerant with a temperature of 99.30 °C is directed to the condenser, where the refrigerant is cooled and condensed due to heat exchange with heat consumers (flow with a temperature of 79.97 °C and a pressure of 1011 kPa). The condensed liquid refrigerant with a temperature of 82.29 °C and a pressure of 895 kPa passes through the control valve, where the refrigerant is throttled. Next, the refrigerant with a temperature of 52.80 °C and a pressure of 425 kPa enters the evaporator. The refrigerant evaporates, giving off the heat of evaporation to cold consumers (flow with a temperature of 57.35 °C and a pressure of 1911 kPa).

HP-3: The refrigerant vapour flow with a temperature of 45.94 °C and a pressure of 380 kPa enters the compressor, where the refrigerant vapour is compressed to 2000 kPa. From the compressor, the refrigerant with a temperature of 157.9 °C is directed to the condenser, where the refrigerant is cooled and condensed due to heat exchange with heat consumers (flow with a temperature of 99.06 °C and a pressure of 775 kPa). The condensed liquid refrigerant with a temperature of 120.8 °C and a pressure of 1995 kPa passes through the control valve, where the refrigerant is throttled. Next, the refrigerant with a temperature of 46.43 °C and a pressure of 355 kPa enters the evaporator. The refrigerant evaporates, giving off the heat of evaporation to cold consumers (flow with a temperature of 54.33 °C and a pressure of 1914 kPa).

The thermal and physical properties of all process streams were simulated in the Aspen environment with the Peng–Robinson fluid package. The simulation details of the heat pumps are shown in Figure 2. The properties of the process streams are shown in Table 1. Parameters of condensers, reboilers, and compressors of heat pumps are presented in Table 2. The material and energy balances of the heat pumps are presented in Table 3. The interconnection between all three heat pumps and the PFD of the investigated natural gas liquid plant is shown in Appendix A, Figure A1.

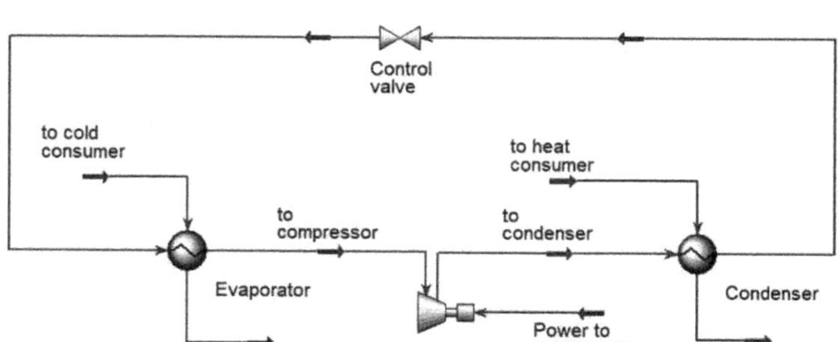

Figure 2. PFD of heat pumps simulated in Aspen Hysys 12.

Table 1. Composition of process streams.

Streams	Mass Flow, kg/h	Component Mass Fractions						
		Ethane	Propane	i-Butane	n-Butane	i-Pentane	n-Pentane	Refrig-21
HP-1								
To condenser	1,867,979	–	–	–	–	–	–	1.0000
To cold consumer	1,146,672	–	0.0364	0.5551	0.4085	–	–	–
To heat consumer	1,297,954	–	–	–	0.9987	0.0013	–	–
HP-2								
To condenser	389,690	–	–	–	–	–	–	1.0000
To cold consumer	254,492	0.0159	0.9092	0.0718	0.0032	–	–	–
To heat consumer	272,854	–	–	–	0.9990	0.0010	–	–
HP-3								
To condenser	1,493,798	–	–	–	–	–	–	1.0000
To cold consumer	761,491	0.0116	0.9863	0.0021	–	–	–	–
To heat consumer	1,131,018	–	–	–	–	0.4105	0.3355	0.2540

Table 2. Parameters of the heat pump equipment.

Parameters	Evaporator			Condenser		
	HP-1	HP-2	HP-3	HP-1	HP-2	HP-3
Duty, kW	95,889	19,703	62,185	106,692	22,430	82,863
Tube side feed mass flow, kg/h	1,867,979	389,690	1,493,798	1,867,980	389,690	1,493,798,
Shell side feed mass flow, kg/h	1,146,672	254,492	761,491	1,297,954	272,854	1,131,018,
Tube inlet temperature, °C	57.65	52.80	45.44	97.07	99.30	138.12
Tube outlet temperature, °C	58.00	52.37	44.95	82.29	82.29	105.04
Shell inlet temperature, °C	66.46	57.35	54.35	79.97	79.97	99.06
Shell outlet temperature, °C	62.97	54.00	47.83	79.80	79.81	102.30
Tube inlet pressure, kPa	485	425	345	900	900	1500
Tube outlet pressure, kPa	480	420	340	895	895	1495
Shell inlet pressure, kPa	907	1900	1914	1011	1011	775
Shell outlet pressure, kPa	902	1906	1909	1006	1006	770
	Compressor					
	HP-1	HP-2	HP-3			
Power, kW	10,804	2727	20,678			
COP	9.88	8.22	4.01			

Table 3. Mass and energy balance of heat pumps.

Inlet Streams			Outlet Streams		
Stream name	Mass flow, kg/h	Energy flow, kW	Stream name	Mass flow, kg/h	Energy flow, kW
		HP-1			
To cold consumer	1,146,672	−704,091	To cold consumer	1,146,672	−799,979
To heat consumer	1,297,954	−864,174	To heat consumer	1,297,954	−757,482
Power to compressor		10,804			
		HP-2			
To cold consumer	254,492	−166,615	To cold consumer	254,492	−186,317
To heat consumer	272,854	−181,666	To heat consumer	272,854	−159,236
Power to compressor		2727			
		HP-3			
To cold consumer	761,491	−500,131	To cold consumer	761,491	−562,317
To heat consumer	1,131,018	−772,202	To heat consumer	1,131,018	−689,339
Power to compressor		20,678			
Total flow	4,864,481	−3,154,670	Total flow	4,864,481	−3,154,670
			Imbalance	0.00%	-8.14×10^{-9}%

3.2. Scenarios

3.2.1. Scenario 1: Constant Pressure at the Evaporator and Varied Pressure at the Condenser

This scenario considers the constant pressure and temperature of the refrigerant at the evaporator outlet. The inlet pressure at the condenser is varied, which affects the energy utilisation, power consumption, and COP of the heat pump. The data for scenario 1 are presented in Table 4.

Table 4. Data for scenario 1.

	HP-1		HP-2		HP-3	
	Start point	Endpoint	Start point	Endpoint	Start point	Endpoint
Evaporator						
Inlet pressure, kPa	480		420		340	
Evaporation temperature, °C	57.26		52.37		44.95	
Condenser						
Inlet pressure, kPa	900	1300	900	1300	1500	1900
Condensation temperature, °C	82.29	99.08	82.29	99.08	106.00	118.11
Compressor						
Power consumption, kW	10,808	17,934	2.727	4.234	20,678	25,020
COP	9.87	5.95	8.22	5.30	4.01	3.31

3.2.2. Scenario 2: Constant Pressure at the Condenser and Varied Pressure at the Evaporator

This scenario considers the constant pressure and temperature of the refrigerant at the condenser outlet. The outlet pressure at the evaporator is varied, affecting the energy utilisation, power consumption, and COP of the heat pump. The data for scenario 2 are presented in Table 5.

Table 5. Data for scenario 2.

	HP-1		HP-2		HP-3	
	Start point	Endpoint	Start point	Endpoint	Start point	Endpoint
Evaporator						
Inlet pressure, kPa	480	101	420	120	340	101
Evaporation temperature, °C	57.26	9.55	52.37	13.13	44.95	8.53
Condenser						
Inlet pressure, kPa	900		900		1500	
Condensation temperature, °C	82.29		82.29		106.00	
Compressor						
Power consumption, kW	10,808	33,442	2727	6549	20,678	34,035
COP	9.87	3.19	8.22	3.43	4.01	2.44

3.2.3. Scenario 3: Both the Evaporator and Condenser Pressure of the Heat Pump Are Varied

Simultaneous pressure reduction in the evaporator and pressure increase in the condenser are applied. These changes affect the configuration of the condenser and the power consumption and COP of the heat pump.

The data for scenario 3 are presented in Table 6.

Table 6. Data for scenario 3.

	HP-1		HP-2		HP-3	
	Start point	Endpoint	Start point	Endpoint	Start point	Endpoint
Evaporator						
Inlet pressure, kPa	480	101	420	120	340	101
Evaporation temperature, °C	57.26	9.55	52.37	13.13	44.95	8.53
Condenser						
Inlet pressure, kPa	900	1300	900	1300	1500	1900
Condensation temperature, °C	82.29	99.08	82.29	99.08	106.00	118.11
Compressor						
Power consumption, kW	10,808	40,254	2727	8089	20,678	38,502
COP	9.87	2.65	8.22	2.77	4.01	2.15

3.3. Variables and Constraints

For the economic assessment of the HP operation parameters, the following variables were used:

- Y = 8670 h;
- Lang = 4.72;
- FIR = 0.1;
- NY = 7 years;
- $c_{e\ min}$ = 0.12 EUR/kWh (minimal EU price), $c_{e\ avg}$ = 0.21 EUR/kWh (average EU price), $c_{e\ max}$ = 0.39 EUR/kWh (maximal EU price) [54];
- Condenser pressure;
- Evaporator pressure;
- Compressor power;
- COP;
- Heat transfer area of the condenser;
- Heat transfer area of the evaporator.

The next constraints were used:

- Minimum pressure at the compressor inlet: 101.3 kPa;
- Maximum pressure at the compressor inlet of HP-1: 480 kPa;
- Maximum pressure at the compressor inlet of HP-2: 420 kPa;
- Maximum pressure at the compressor inlet of HP-3: 380 kPa;
- Minimum pressure at the compressor outlet of HP-1: 900 kPa;
- Minimum pressure at the compressor outlet of HP-2: 900 kPa;
- Minimum pressure at the compressor outlet of HP-3: 2000 kPa;
- The acceptable condenser pressure drop (tubes/shell) is 5 kPa;
- The acceptable evaporator pressure drop (tube/shell) is 5 kPa;
- Compressor efficiency (adiabatic) is 75%;
- Heat exchanger type is shell-and-tube for both evaporator and condenser;
- Tube type is plain;
- The tube material is carbon steel.

The design and cost of the condenser and evaporator were selected using the Aspen Exchanger Design & Rating application. The next Aspen databases were used for a capital cost assessment of the condenser and evaporator [55]:

- D_FXPRIV.PDA Private properties chemical databank properties;
- D_IDPRIV.PDA Private properties chemical databank index;
- D_VAPRIV.PDA Private properties chemical databank properties;
- N_MTLDEF.PDA Default materials for generic materials (ASME);
- N_MTLDIN.PDA Default materials for generic materials (DIN);
- N_MTLCDP.PDA Default materials for generic materials (AFNOR);
- N_PARTNO.PDA Part number assignment for the bill of materials;

- N_PRIVI.PDA Private properties materials databank index;
- N_PRIVP.PDA Private properties materials databank properties;
- N_STDLAB.PDA Fabrication standards, procedures, costs, etc.;
- N_STDMTL.PDA Fabrication standards as function of materials;
- N_STDOPR.PDA Fabrication operation efficiencies;
- N_STDWLD.PDA Fabrication welding standards;
- N_STDPRC.PDA Private materials prices.

4. Results and Discussion

4.1. Heat Pump 1

4.1.1. Average Electricity Price

The results of the TAC assessment of HP-1 for all three case studies are presented in Figure 3. The correlation of scenario 1 has a minimum TAC at a condenser pressure of 1050 kPa and the overall trend has extremum (Figure 3a). The heat transfer area of the condenser decreases due to the increase in the condensing pressure and temperature driving forces. The evaporator heat transfer area also decreases due to a reduction in the evaporator's heat duty. At the same time, the power consumption and compressor cost increase and, when the condenser pressure is above 1050 kPa, starts dominating in the capital cost share; the TAC also increases. These changes in capital cost distribution can be illustrated in Figure 4a. The correlation of scenario 2 shows that the TAC increases from the starting point of the condensing pressure 480 kPa to the endpoint of 101.3 kPa. The condenser cost dominates due to low driving forces and, as a result, the high heat transfer area of the condenser. The increase in driving force in the evaporator reduces its heat transfer area and capital cost when the pressure decreases from 480 kPa to 250 kPa. However, a further evaporator pressure decrease from 250 kPa to 101 kPa increases the heat transfer area due to it depending more on the stream velocity. The increases in the power consumption and compressor cost are not compensated for. The capital cost distribution of scenario 2 is represented in Figure 4b. The TAC in scenario 3 has also an increasing trend from the starting point to the end, when both the condensation and evaporation pressure are changed. The heat transfer area and capital cost of the condenser decrease as the condensation pressure increases, similar to scenario 1. The heat transfer area behaviour in the evaporator is similar to scenario 2. Changes in both the evaporator and condenser pressures do not compensate for the increase in electricity consumption and capital cost of the compressor, which is shown in Figures 4c and 5c. The minimum TAC of HP-1 of all three case studies occurred in scenario 1, and it is 33.26 M EUR .

The capital cost distribution between the compressor, evaporator, and condenser is shown in Figure 4. In scenario 1 (Figure 4a), the share of capital cost shifts between the condenser and compressor, while the evaporator cost remains almost the same from calculation point 3. The minimum TAC is observed for calculation point 4 (Figure 4a). The evaporator cost dominates in scenario 2 (Figure 4b) while decreasing the evaporation pressure. The share of the evaporator cost is between 62 and 72%. The compressor cost in scenario 3 (Figure 4c) is the highest in capital cost share and increases due to COP reduction.

The total cost share is shown in Figure 5, and the energy cost is about 50% for two calculation points; further, the decrease in COP energy cost is determinative. The minimum TAC of heat pump 1 is shown in Figure 5a, point 4, and the share of the energy cost is 72%.

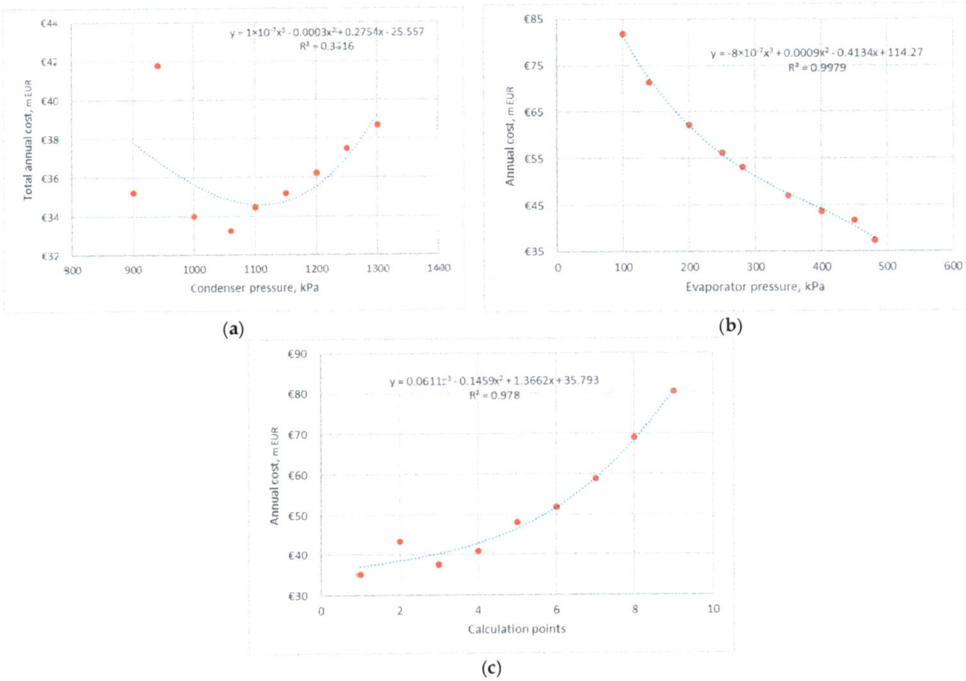

Figure 3. Total annual cost correlation of heat pump 1 for average electricity price: (**a**) scenario 1; (**b**) scenario 2; (**c**) scenario 3.

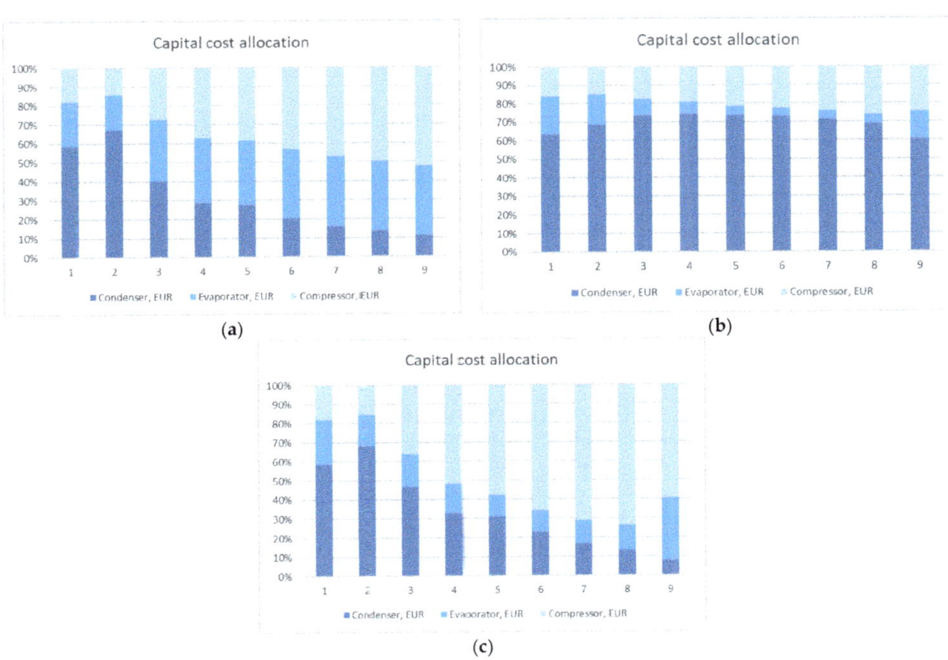

Figure 4. Capital cost allocation of heat pump 1: (**a**) scenario 1; (**b**) scenario 2; (**c**) scenario 3.

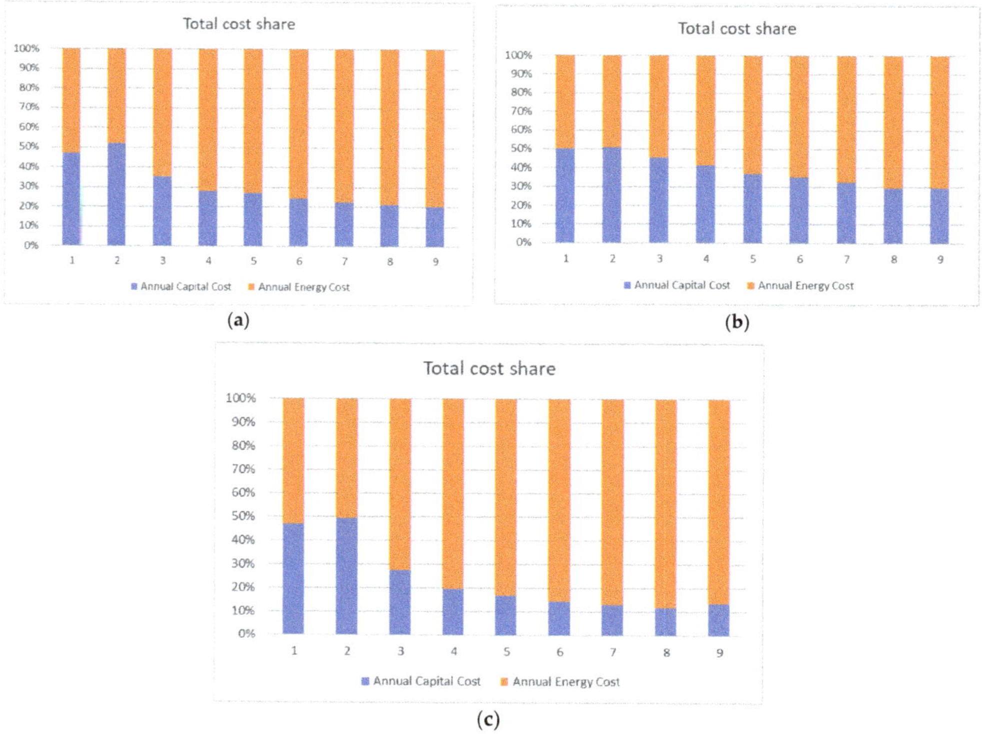

Figure 5. Total cost distribution of heat pump 1 for average electricity price: (**a**) scenario 1; (**b**) scenario 2; (**c**) scenario 3.

4.1.2. Sensitivity for Min and Max Electricity Prices

A basic analysis of all heat pumps was performed for the average electricity prices in the EU. Additional assessments were also carried out for the minimum and maximum EU electricity prices to see the sensitivity of the obtained results. Reducing the energy price moves the extremum of the TAC trendline of scenario 1 into the higher compressor pressure (Figure 6a), but the minimum of the calculation remains at the same point for a condensation pressure of 1050 kPa. The increase in energy price moves the min TAC point to a condensation pressure of 900 kPa, with an overall increasing trend (Figure 6b). The changes in the total cost share of case study 1 for the min and max energy prices are shown in Figure 7.

Changes in energy prices do not affect seriously the TAC cost trend of HP-1 in scenario 2, which is demonstrated in Figure 8; the minimum value remains for the condenser pressure of 480 kPa. The shares of energy and capital cost are shown in Figure 9.

The reduction in energy price affects the results of scenario 3 for HP-1, and an extremum appears for calculation point 3, with an evaporation pressure of 400 kPa and a condensation pressure of 1000 kPa (Figure 10a). This case can be seen in Figure 11a, and the share of the energy cost is 60%, which is less than that for the min TAC with average electricity prices. The increase in energy prices does not influence the nature of the trend line.

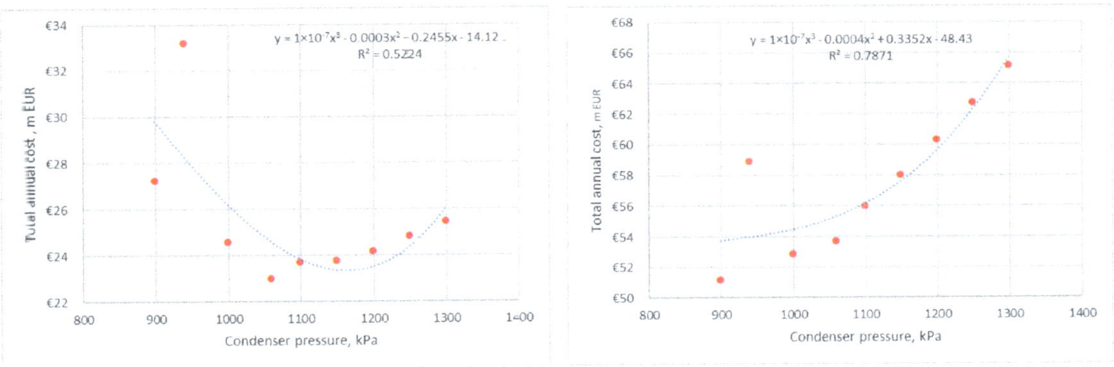

(a) (b)

Figure 6. Total annual cost correlation of heat pump 1 for scenario 1: (**a**) min electricity price; (**b**) max electricity price.

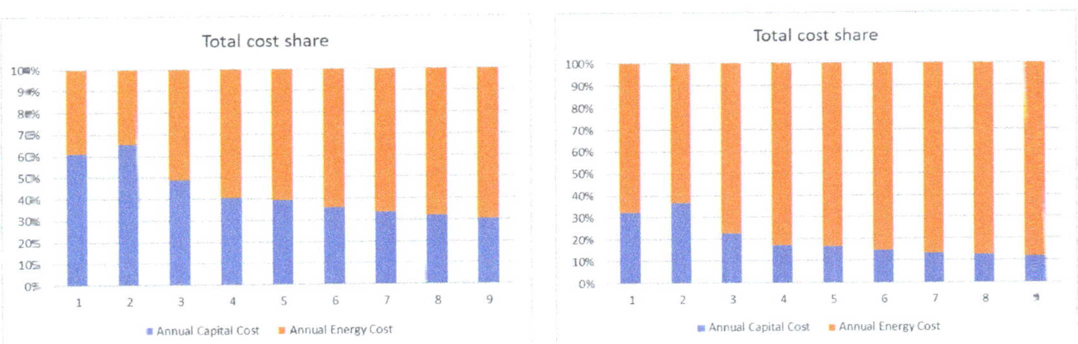

(a) (b)

Figure 7. Total cost distribution of heat pump 1 for scenario 1: (**a**) min electricity price; (**b**) max electricity price.

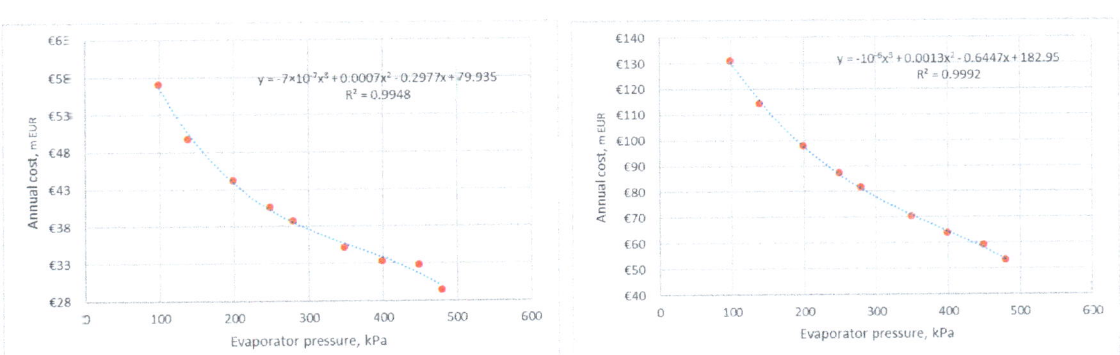

(a) (b)

Figure 8. Total annual cost correlation of heat pump 1 for scenario 2: (**a**) min electricity price; (**b**) max electricity price.

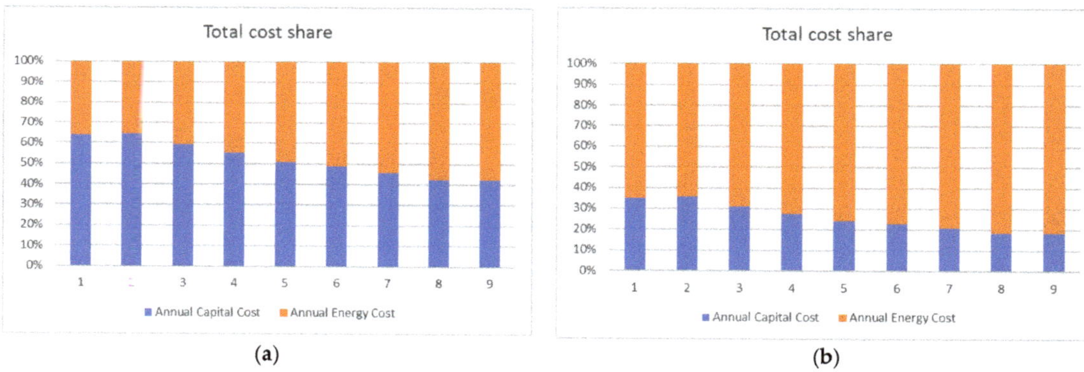

Figure 9. Total cost distribution of heat pump 1 for scenario 2: (**a**) min electricity price; (**b**) max electricity price.

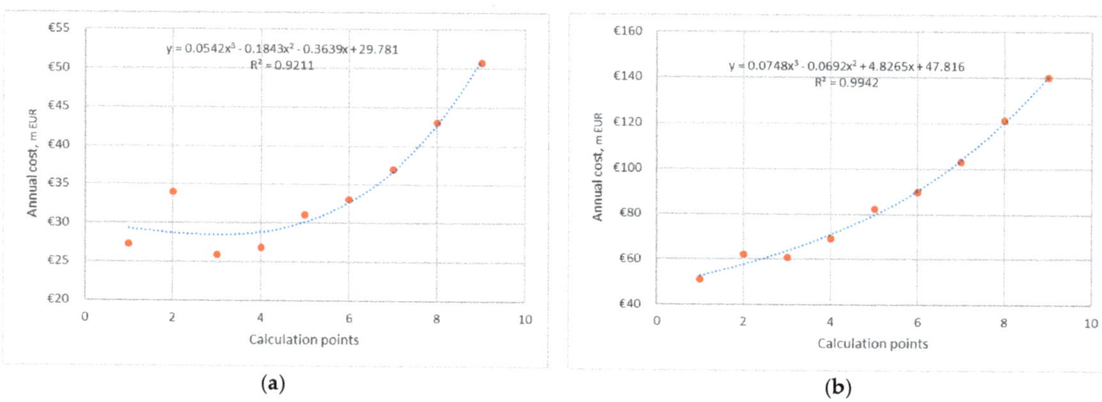

Figure 10. Total annual cost correlation of heat pump 1 for scenario 3: (**a**) min electricity price; (**b**) max electricity price.

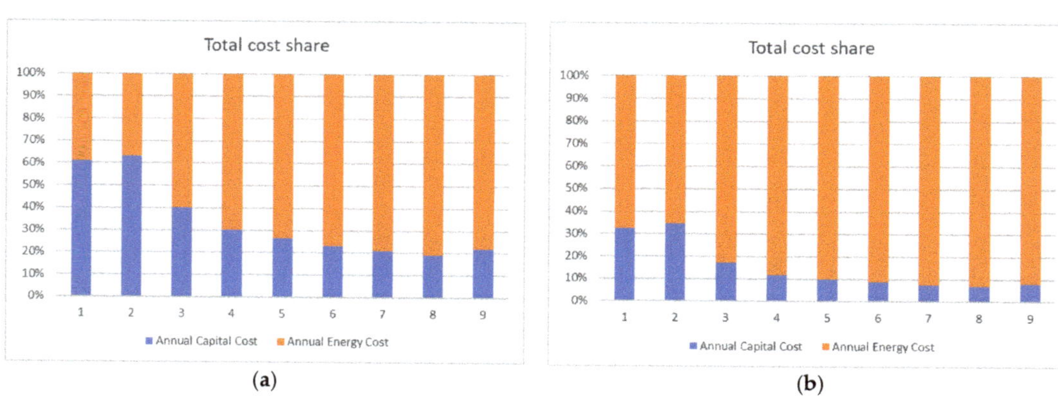

Figure 11. Total cost distribution of heat pump 1 for scenario 3: (**a**) min electricity price; (**b**) max electricity price.

4.2. Heat Pump 2

4.2.1. Average Electricity Price

The correlation of the TAC for HP-2 is more complicated than for HP-1. Figure 12 demonstrates the min TAC for scenario 1 (Figure 12a) and scenario 3 (Figure 12c), but scenario 3 has a more apparent extremum. The heat transfer area of the condenser decreases with the increase in the condensation pressure. The evaporator heat transfer area is not consistent with the heat duty reduction, and the compressor power increases with the condensation pressure increase. This is illustrated in Figure 13a. The trade-off can be found for a condenser pressure of 1050 kPa and an evaporator pressure of 420 kPa. In scenario 3, the extremum is more apparent due to the compressor cost dominating the capital cost distribution and energy cost in the total cost share. The heat transfer of the condenser and evaporator decreases; however, the evaporator heat transfer area slightly increases for pressures from 170 to 120 kPa due to a stream velocity reduction. This is illustrated in Figures 13c and 14c. The TAC in scenario 2 (Figure 12b) has an increasing trend but a decreasing evaporation pressure. The condenser cost is constant and dominates in this scenario and the TAC correlation has an increasing trend, and a small decrease in the evaporator heat transfer area is compensated for by the increase in the compressor cost and power consumption (Figures 13b and 14b). The minimum TAC is found for scenario 1, and it is 8.56 M EUR for a condensation pressure of 1050 kPa and an evaporation pressure of 420 kPa. The minimum TAC of scenario 3 is 8.77 M EUR , which is very close to the first one, and it is found for a condensation pressure of 950 kPa and an evaporation pressure of 390 kPa.

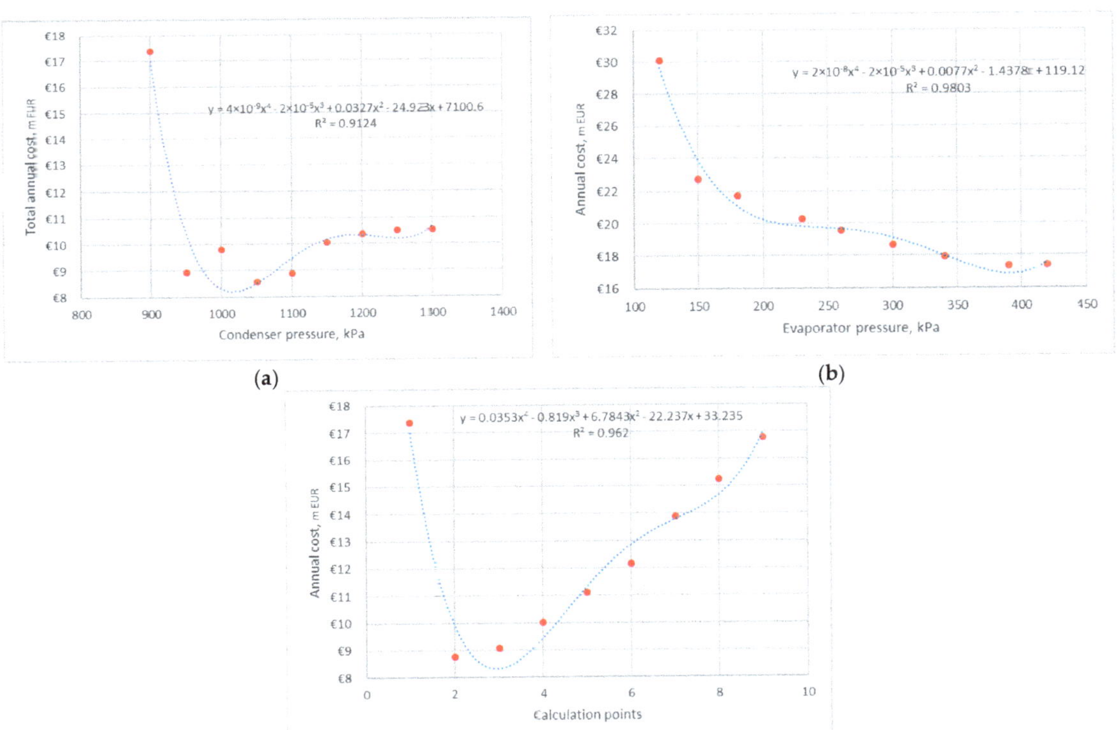

Figure 12. Total annual cost correlation of heat pump 2 for average electricity price: (**a**) scenario 1; (**b**) scenario 2; (**c**) scenario 3.

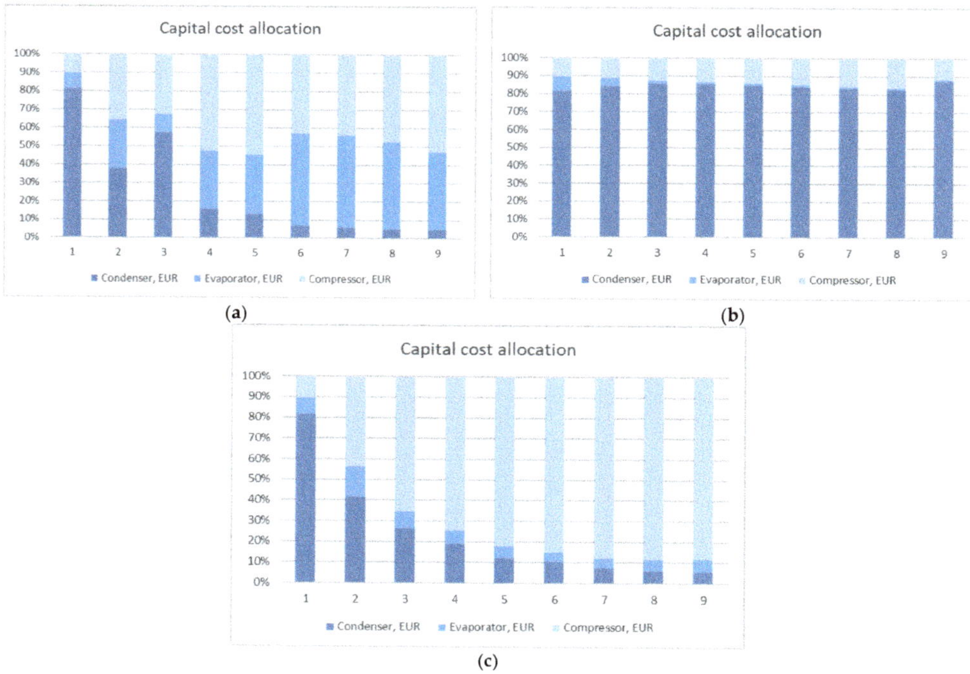

Figure 13. Capital cost allocation of heat pump 2: (**a**) scenario 1; (**b**) scenario 2; (**c**) scenario 3.

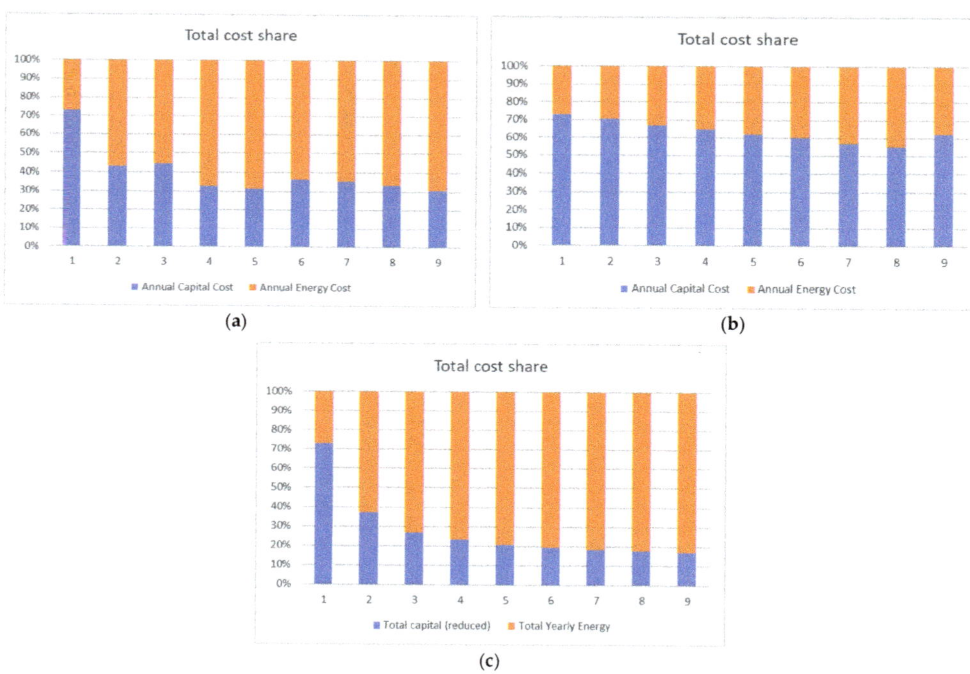

Figure 14. Total cost distribution of heat pump 2 for average electricity price: (**a**) scenario 1; (**b**) scenario 2; (**c**) scenario 3.

The distribution of capital cost shows the highest share of the condenser cost for scenario 2 (Figure 13b) and an increasing share of the compressor cost for scenario 3 (Figure 13c). The distribution of the equipment cost for scenario 1 is not linear (Figure 13a). The condenser cost decreases while the compressor and evaporator cost fluctuate.

The shares of energy and capital cost for scenario 2 are shown in Figure 14. It should be noted that the share of energy cost for the case studies with an extremum of TAC is 68% for scenario 1 (point 4) and 62% for scenario 3 (point 2).

4.2.2. Sensitivity for Min and Max Electricity Prices

The investigation of the results for scenario 1 obtained for the max EU electricity price demonstrates that the minimum TAC is observed for a condensation pressure of 950 kPa (Figure 15b). The min electricity price does not significantly affect the results of scenario 1 and the min TAC remains at 1050 kPa (condenser), confirmed by the correlation in Figure 15a. The share of the energy cost for the solutions with the min TAC of HP-2 is 54% for the min electricity prices and 71% for the max EU electricity prices (Figure 16).

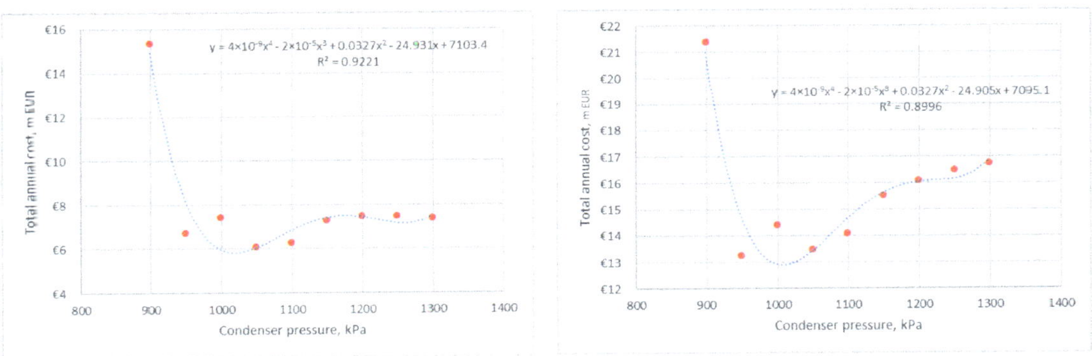

(a) (b)

Figure 15. Total annual cost correlation of heat pump 2 for scenario 1: (**a**) min electricity price; (**b**) max electricity price.

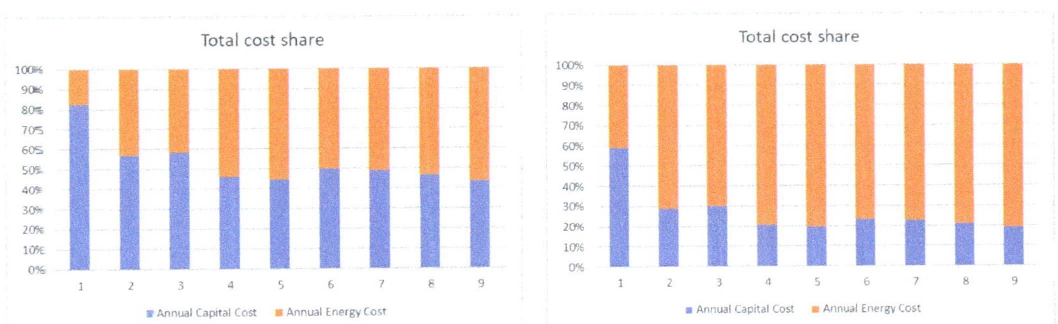

(a) (b)

Figure 16. Total cost distribution of heat pump 2 for scenario 1: (**a**) min electricity price; (**b**) max electricity price.

The results of scenario 2 for the max and min EU electricity prices are shown in Figures 17 and 18. The min energy prices move the min TAC point to an evaporation pressure of 380 kPa, but the extremum is not strongly marked. The max energy prices do not significantly affect the TAC trendline.

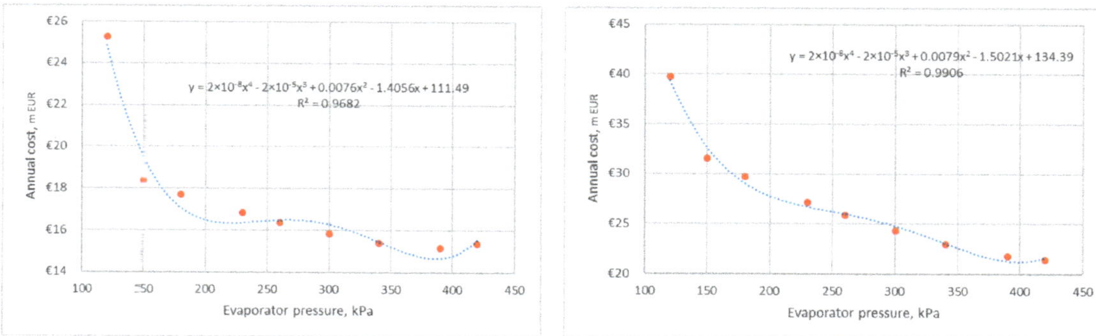

(a) (b)

Figure 17. Total annual cost correlation of heat pump 2 for scenario 2: (**a**) min electricity price; (**b**) max electricity price.

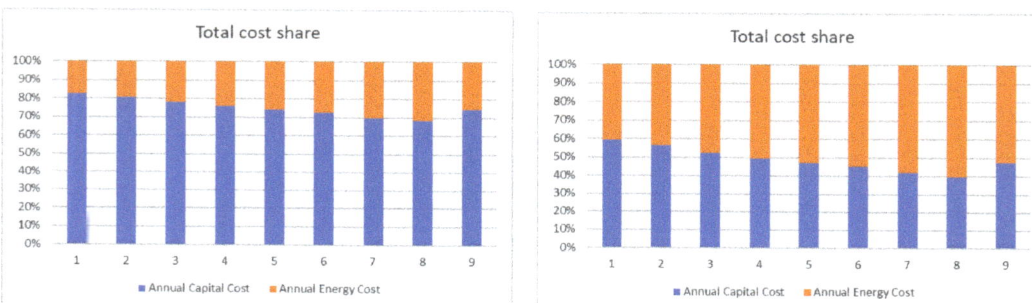

(a) (b)

Figure 18. Total cost distribution of heat pump 2 for scenario 2: (**a**) min electricity price; (**b**) max electricity price.

The results of scenario 3 for the min energy prices shift the min TAC to point 3 (condensation pressure: 1000 kPa; evaporation pressure: 340 kPa) (Figure 19a). The max energy prices remain at the min TAC in the same process condition (Figure 19b). The share of the energy cost is 61% for the min electricity price and 76% for the max electricity price (Figure 20).

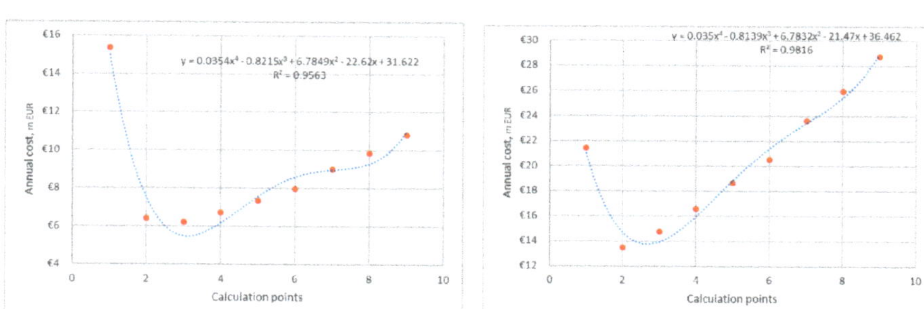

(a) (b)

Figure 19. Total annual cost correlation of heat pump 2 for scenario 3: (**a**) min electricity price; (**b**) max electricity price.

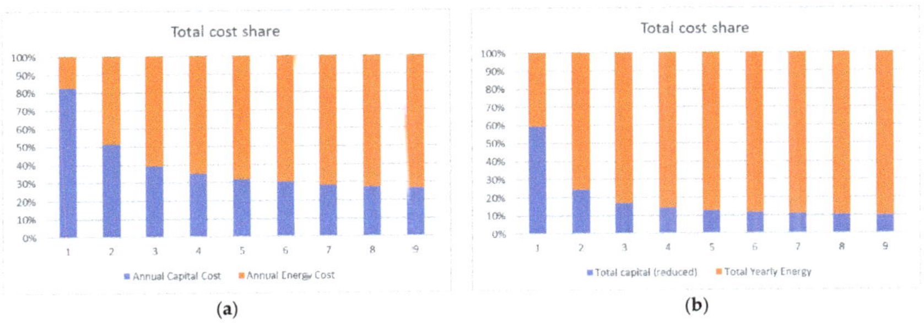

Figure 20. Total cost distribution of heat pump 2 for scenario 3: (**a**) min electricity price; (**b**) max electricity price.

4.3. *Heat Pump 3*

4.3.1. Average Electricity Price

The investigation of the TAC results for all scenarios of HP-3 shows that the correlation is monotonous and increases when the condensation pressure increases or the evaporation pressure decreases. This is well-illustrated in Figure 21. The heat transfer area of the condenser decreases with the increase in condensation pressure in the whole investigated range from 1500 kPa to 1900 kPa, due to an increase in the driving forces of heat transfer. The evaporator heat transfer area reduces the evaporation pressure from 340 to 180 kPa and increases it from 180 to 101 kPa. This can be justified by the effect of stream velocity and driving forces for different pressure ranges. However, the capital cost and TAC are mostly defined by compressor cost and energy cost. The situation is the same for the capital cost distribution, where the share of the compressor increases as the condenser pressure increases and the evaporator pressure reduces (Figure 22). There is a high share of energy cost in the TAC for all three scenarios, and this value is more than 80% (Figure 23). It can be justified by the low COP of HP-3 in comparison to the two previous ones.

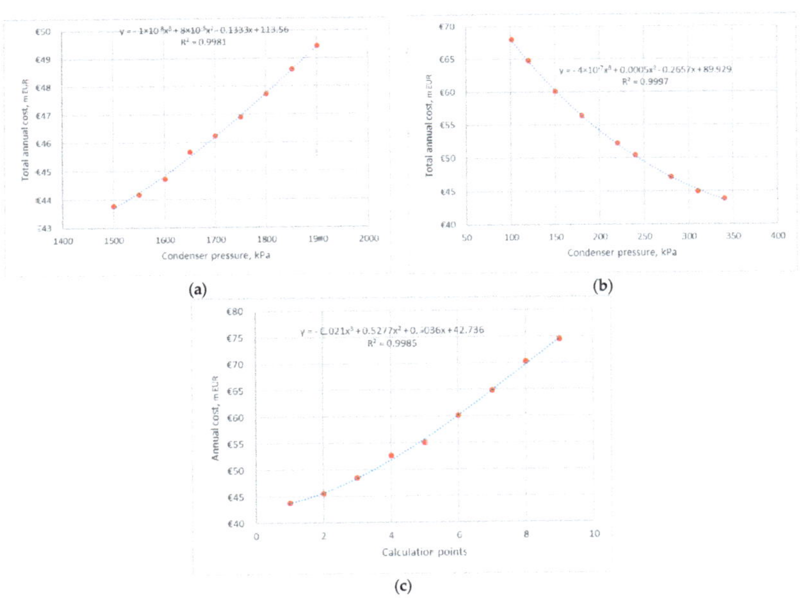

Figure 21. Total annual cost correlation of heat pump 3 for average electricity price: (**a**) scenario 1; (**b**) scenario 2; (**c**) scenario 3.

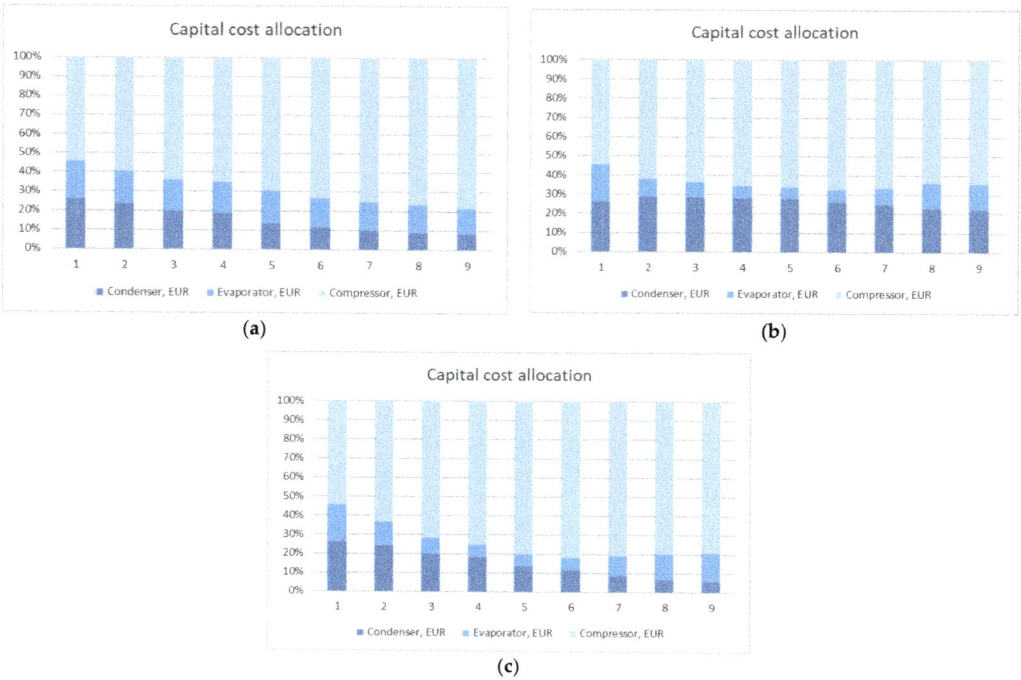

Figure 22. Capital cost allocation of heat pump 3: (**a**) scenario 1; (**b**) scenario 2; (**c**) scenario 3.

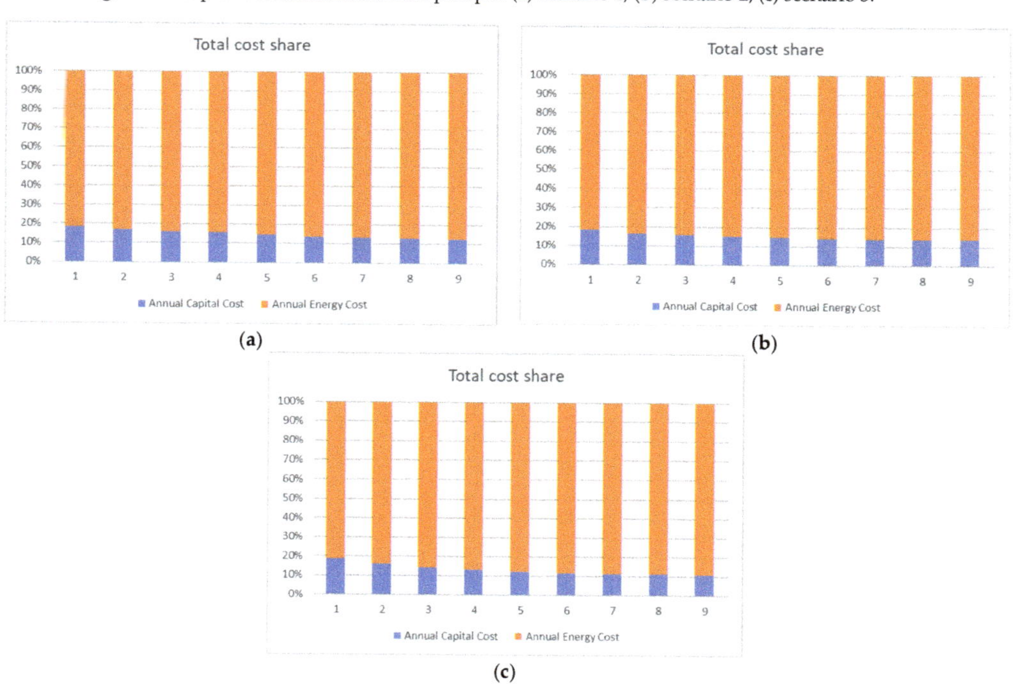

Figure 23. Total cost distribution of heat pump 3 for average electricity price: (**a**) scenario 1; (**b**) scenario 2; (**c**) scenario 3.

4.3.2. Sensitivity for Min and Max Electricity Prices

The investigation of the results for the min and max electricity prices does not change the overall picture and only changes the share of the electricity prices in the overall TAC distribution (Figures 24–29). The use of HP-3 is recommended for the operation conditions defined based on the GCC.

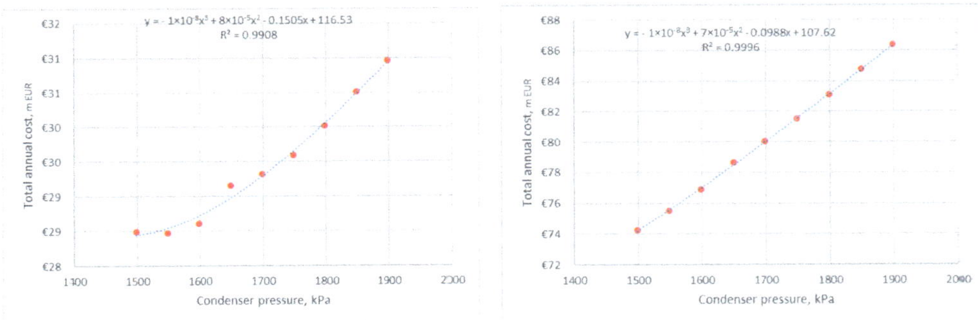

(**a**) (**b**)

Figure 24. Total annual cost correlation of heat pump 3 for scenario 1: (**a**) min electricity price; (**b**) max electricity price.

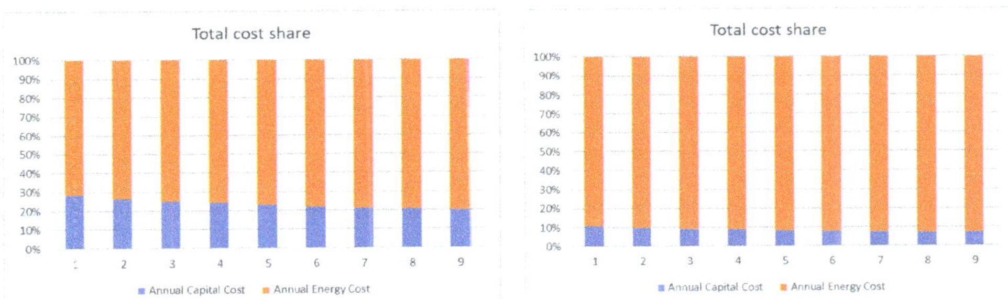

(**a**) (**b**)

Figure 25. Total cost distribution of heat pump 3 for scenario 1: (**a**) min electricity price; (**b**) max electricity price.

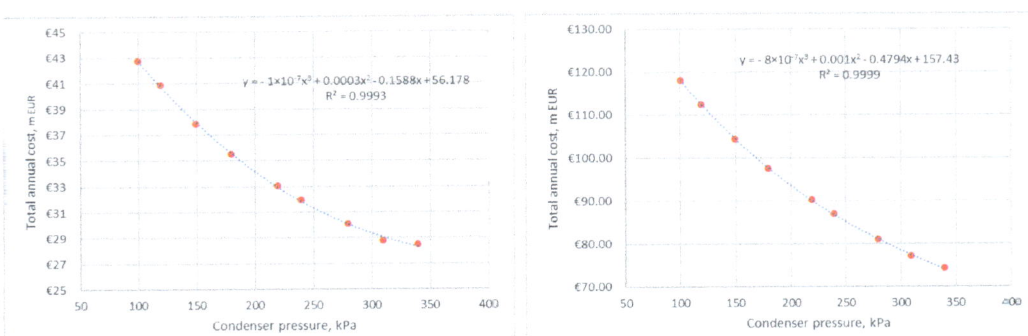

(**a**) (**b**)

Figure 26. Total annual cost correlation of heat pump 3 for scenario 2: (**a**) min electricity price; (**b**) max electricity price.

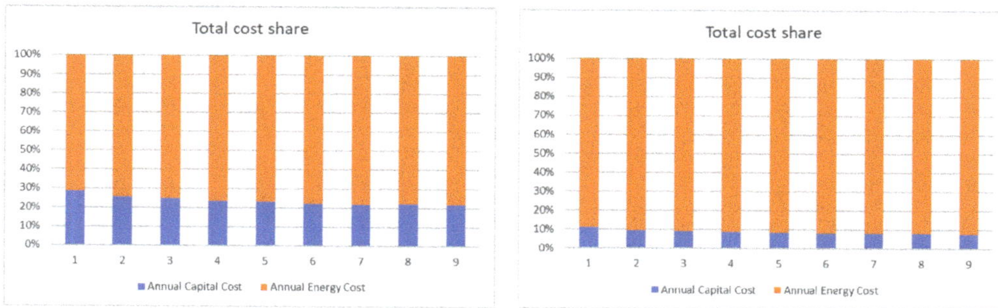

(a) (b)

Figure 27. Total cost distribution of heat pump 3 for scenario 2: (**a**) min electricity price; (**b**) max electricity price.

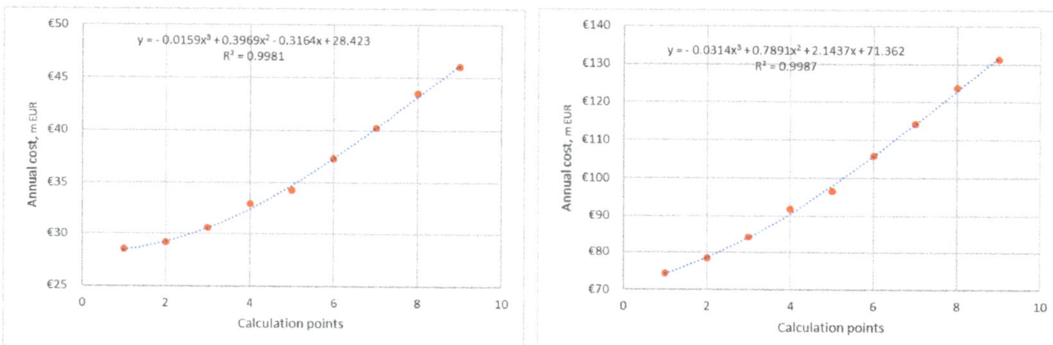

(a) (b)

Figure 28. Total annual cost correlation of heat pump 3 for scenario 3: (**a**) min electricity price; (**b**) max electricity price.

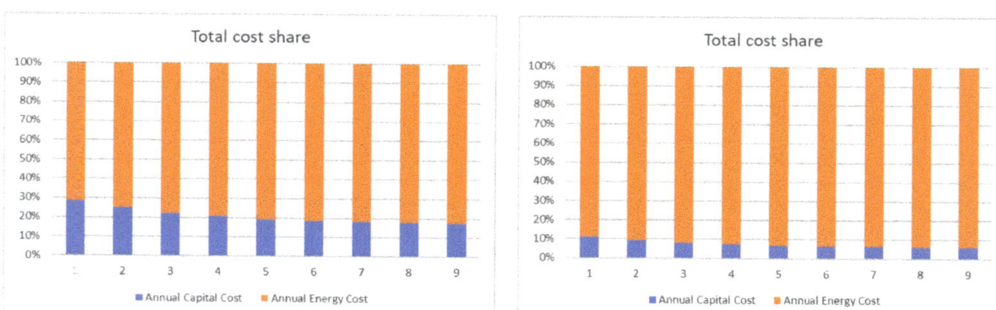

(a) (b)

Figure 29. Total cost distribution of heat pump 3 for scenario 3: (**a**) min electricity price; (**b**) max electricity price.

4.4. Impact of the Results

The obtained results of HP integration into the specific process of a polymer plant demonstrated that the energy targets obtained from the GCC were not always the best solution, especially in economic terms. This is very important when applying heat pumps of high capacity on an industrial scale; these results should be impressive for potential investors and plant owners. Table 7 demonstrates the results of the HP calculation for a gas fractioning unit of a polymer plant proposed by a modified targeting approach (optimised). These results are compared with the ones obtained by traditional targeting with the use of the GCC. HP-3 can be applied with its initial operation mode and the targets defined by the GCC. HP-1 and HP-2 can be adjusted by selection of the evaporator and condenser temperature to reduce the overall cost. In both cases, the condensation pressure should be changed, and the TAC reduction is 6% for HP-1 and 50% for HP-2.

Table 7. Comparison of the results of heat pump application.

	HP-1		**HP-2**		**HP-3**	
	Targeting (Base case)	Optimised	Targeting (Base case)	Optimised	Targeting (Base case)	Optimised
Evaporator						
Inlet temperature, °C	57.65	57.65	52.80	52.80	45.44	45.44
Outlet temperature, °C	58.00	58.00	52.37	52.37	44.95	44.95
Heat duty, kW	95,888	92,825	19,703	19,092	62,185	62,185
LMTD	7.32	7.32	2.66	3.09	5.12	5.12
Heat transfer area, m^2	24,029	22,921	6199	5672	9998	9998
Condenser						
Inlet temperature, °C	97.07	107.75	99.30	109.33	138.12	138.12
Outlet temperature, °C	82.29	89.57	82.29	89.14	106.04	106.04
Heat duty, kW	106,692	106,697	22,430	22,430	82,863	82,863
LMTD	7.46	17.17	8.08	17.41	16.45	16.45
Heat transfer area, m^2	67,957	17,229	72,576	2648	13,782	13,782
Compressor						
Inlet pressure, kPa	480	480	420	420	340	340
Outlet pressure, kPa	900	1060	900	1050	1500	1500
Power, kW	10,804	13,877	2727	3338	20,678	20,678
CCP	9.87	7.69	8.22	6.72	4.01	4.01
Economic indicators						
Annual capital cost, k EUR	16,619	9353	12,673	2811	8150	8150
Energy cost, k EUR/y [1]	18,611	23,901	4696	5747	35,608	35,608
Total annual cost, k EUR/y	35,231	33,255	17,369	8559	43,758	43,758

[1] Defined for average energy prices.

However, the energy cost and electricity consumption are higher than defined for the base case. This issue is important when electrifying the energy system to the grid and applying renewable energies. On the one hand, we have a more economically viable solution for heat pump applications, and, on another hand, more electricity is required from the grid. This issue should be additionally investigated in terms of the economic aspects of grid extension driven by renewable energies. It can be also used for the appropriate planning of energy needs in future energy systems.

In this work, the authors used the current average EU electricity for non-household users. The application of heat pumps presumes increased electricity exports by process plants, and the use of real market prices is more viable. For the sensitivity analysis of the results, namely, the min total annual cost, calculations were also made for the min and max EU electricity price. In the literature, the utility price assumption for heat pumping or vapour recompression is 18.72 USD/GJ [56]. But this value covers the generation, including thermodynamic efficiency without taxes, market prices, and other correction factors. This parameter should be specific for different regions and countries.

The present work addresses the changes in heat pump configuration, maintaining the same operation mode of distillation columns and finding the trade-off between capital and energy when integrating heat pumps in a specific industrial process. However, it should be noted that the reboilers of the de-propaniser and de-isobutaniser are driven by the heat pumps, as illustrated in Figure A1 (Appendix A). The reboiler of the debutaniser is more complicated; the existing reboiler works in parallel with one driven by the heat pump. The operation of these two units should be synchronised to keep the operation parameters of the distillation column. The regulation of the whole flowsheet should be additionally studied and optimised to provide process parameters within acceptable limits.

The selection of the compressor remains a big challenge when providing a concept design of chemical processes. Apart from power, the compressor cost depends on different factors, e.g., gas composition, pressure, temperature, etc. According to Luyben (2018), the estimation of compressor cost made by different researchers may differ by five times [57]. Furthermore, the impact of different process parameters is crucial, e.g., suction pressure, especially for multistage compression.

In this work, a sensitivity analysis of the economic results was performed by applying different electricity prices. It was applied for all calculated configurations of heat pumps. Due to the different electricity prices in European countries, it is important to understand the result variation and applicability. All heat pumps were analysed for min, max, and average electricity prices, as they define the economic efficiency of heat pump application. The influence of process conditions was not investigated and could be considered in potential future work.

5. Conclusions

The main findings of the current research involve a detailed exploration of heat pump integration within the natural gas liquid processing of a petrochemical plant. The features of a detailed simulation of the heat pump equipment were extensively studied in industrial contexts, and the economic trade-off was investigated for the appropriate targeting of heat pump integration, applying different pressures in the evaporator and condenser. The extremum of the TAC correlation was found for two of the three considered heat pumps.

The results of this work may be considered in both the theoretical and practical context of heat pump integration into industrial sites. First, the generalised results can be used for an update of the pinch-based targeting procedure to achieve better economic efficiency in heat pump integration. The simplified workflow can be as follows: (1) the construction of a GCC; (2) a TAC assessment of heat pump integration; (3) energy targets; (4) concept process design; and (5) economic efficiency. This may provide new insights for heat pump targeting within industrial clusters.

Second, practical arrangements can be provided based on the obtained results. The total annual costs of heat pump 1 and heat pump 2 were reduced by 6% and 50%, respectively, in comparison to the configuration of heat pumps targeted by grand composite curves. But it requires increases in electricity consumption of 22% and 18% for heat pump 1 and heat pump 2, as well as additional thermal utility for cooling increased by 3% for both heat pumps. The total annual cost for the application of both heat pumps is reduced by 21.5 M EUR , which may be impressive for investors when integrating heat pumps into a specific polymer plant. Heat pump 3 can be applied with the same operating conditions as defined by the grand composite curve. A detailed layout of condensers and evaporators was obtained that may be used for practical applications.

Third, an assessment of the investigated process electrification was carried out and the decarbonization potential was estimated based on the simulation results in the Aspen environment. The simulation details are summarised in Table 8. By connecting the observed natural gas liquid plant to a renewable electricity grid by applying heat pumps, the potential emission saving is estimated at 1.89 $ktCO_2/y$.

Despite these findings, the limitations should be identified and discussed. A common problem of heat pumps' efficiency in industry-related process economics and their application in the industrial environment is characterised based on COP criteria; this issue should be solved by the proper placement and targeting of heat pumps. The application of heat pumps usually leads to complications in the process flowsheet and supplemented control strategy that may lead to incorrect operation modes and start-up and shut down problems. Proper attention must be paid to the dynamic simulation of the control loops and proper control strategy development. High-temperature processes may also constrain heat pump application due to the physical properties of refrigerants.

Table 8. CO_2 saving due to application of heat pumps in natural gas liquid plant (simulated in Aspen environment with Peng–Robinson fluid package).

Item	Value
Energy saving (MW)	212.00
Natural gas heat value (kJ/kg mole)	802,518.00
Air/gas ratio (Std. Gas Vol. Flow)	10.90
Natural gas composition	Methane 100%
Natural gas burned (STD_m^3/h)	96,731.00
Steam generated (t/h)	291.20
Steam generation efficiency	0.95
CO_2 emission (t/h)	45.64
Specific CO_2 emission (tCO_2/MW)	0.215
Annual CO_2 saving (tCO_2/y)	1885.88

Apart from the investigated process, the application of heat pumps in other industrial processes that use distillation columns should be considered in future research, and correlations might be identified and generalised for wide application in different cases. Another important issue is related to the synthesis of heat exchanger networks when applying heat pumps in industrial processes. It generates a task for new research focusing on optimisation problems, simplification, and cost assessments. As mentioned above, the operability of the industrial process is crucial and may provide a stable final product. This aspect is worthy of additional investigations into the overall process sensitivity and system control parameters.

Since this work is targeted towards industry decarbonization, the integration of renewable energies remains challenging. Ideally, it would be aimed at grid extension, the variability of supply and demands, the coupling of industry and energy systems, and economic aspects of such integration.

Author Contributions: Conceptualization, S.B.; methodology, S.B.; software, M.I. and S.B.; validation, M.I. and G.K.; formal analysis, M.I.; investigation, S.B. and M.I.; resources, M.I. and G.K.; data curation, M.I.; writing—original draft preparation, M.I. and S.B.; writing—review and editing, S.B.; visualization, M.I. and S.B.; supervision, S.B.; project administration, S.B. and G.K.; funding acquisition, G.K. and S.B. All authors have read and agreed to the published version of the manuscript.

Funding: This research was funded by the European Commission under Project No. 101119793 LIFE22-CET-SET_HEAT.

Data Availability Statement: Data are available upon request.

Conflicts of Interest: The authors declare no conflicts of interest.

Nomenclature

A	Heat transfer area (m^2)
ACC	Annual capital cost (EUR/y)
c_e	Electricity prices (EUR/kWh)
$CAPEX_{Evap}$	Evaporator capital cost (EUR)
$CAPEX_{Comp}$	Compressor capital cost (EUR)
$CAPEX_{Cond}$	Condenser capital cost (EUR)
COP_{HP}	Coefficient of performance (n/d)
d_o	Outside diameter of tube (m)
EC	Annual energy cost (EUR/y)
$f(t)$	Correction factor for design temperature (n/d)
$f(p)$	Correction factor for design pressure (n/d)
$f(m)$	Correction factor for construction materials (n/d)

FIR	Fractional interest rate per year (%)
Ft	Correction factor of heat transfer (n/d)
g	Gravitational constant (9.81 m·s^{-2})
h_C	Condensing film coefficient (W·m^{-2}·K^{-1})
h_j^{In}	Enthalpy of the inlet stream (kJ kg^{-1})
h_j^{Out}	Enthalpy of the outlet stream (kJ kg^{-1})
h_{NB}	Nucleate boiling coefficient (W·m^{-2}·K^{-1})
J	Number of process streams, (n/d)
k_L	Thermal conductivity of the liquid (W·m^{-1}·K^{-1})
$Lang$	Lang factor (n/d)
M_j^{In}	Mass flow of the inlet stream (kg h^{-1})
M_j^{Out}	Mass flow of the outlet stream (kg h^{-1})
M_{losses}	Losses of mass flows (kg h^{-1})
NY	Bank loan period (y)
P	Operating pressure (kPa)
P_C	Liquid critical pressure (kPa)
Q	Heat duty (W)
q	Heat flux (W·m^{-2})
Q_{HP}	Heat absorbed by heat pump at low temperature (W)
Q_{losses}	Energy losses (W)
TAC	Total annual cost (EUR/y)
T_{inC}	Inlet temperature of the cold stream (°C)
T_{inH}	Inlet temperature of the hot stream (°C)
T_{outC}	Outlet temperature of the cold stream (°C)
T_{outH}	Outlet temperature of the hot stream (°C)
U	Heat transfer coefficient (W·m^{-2}·K^{-1})
W_i	Power (W)
ΔH_{VAP}	Latent heat (J·kg−1)
ΔT_{LM}	Logarithmic temperature difference (°C)
ΔT	Temperature difference across the condensate film (°C)
μ_L	Viscosity of the liquid (kg·m^{-1}·s^{-1})
ρ_L	Density of the liquid (kg·m^{-3})

Appendix A

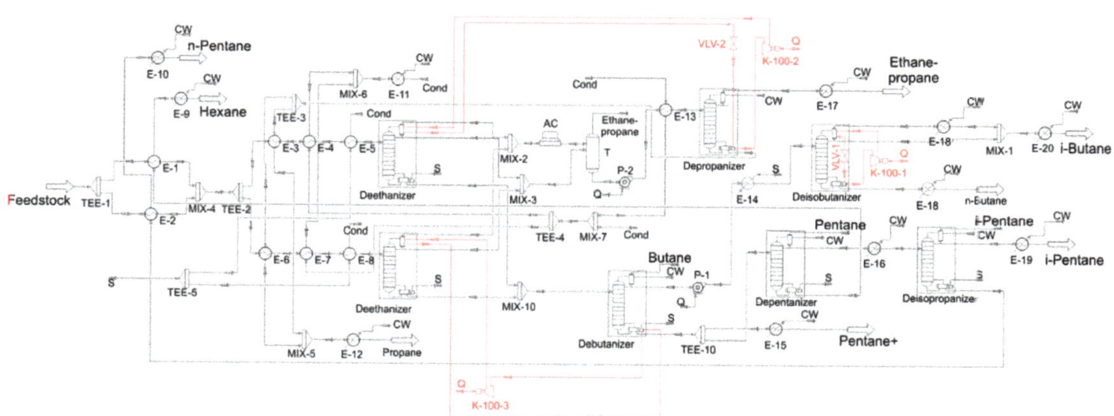

Figure A1. PFD of the natural gas liquid fractioning with heat pump integration. AC—air cooler; Cond—condensate; CW—cooling water; E—heat exchanger; K—compressor; MIX—mixer; TEE—splitter; P—pump; Q—electricity; S—steam; T—tank; VLV—expansion valve.

References

1. Barkaoui, A.-E.; Boldyryev, S.; Duic, N.; Krajacic, G.; Guzović, Z. Appropriate Integration of Geothermal Energy Sources by Pinch Approach: Case Study of Croatia. *Appl. Energy* **2016**, *184*, 1343–1349. [CrossRef]
2. Mészáros, I.; Fonyó, Z. Design Strategy for Heat Pump Assisted Distillation System. *J. Heat Recovery Syst.* **1986**, *6*, 469–476. [CrossRef]
3. van de Bor, D.M.; Infante Ferreira, C.A.; Kiss, A.A. Low Grade Waste Heat Recovery Using Heat Pumps and Power Cycles. *Energy* **2015**, *89*, 864–873. [CrossRef]
4. Yuan, G.; Chu, K.H. Heat Pump Drying of Industrial Wastewater Sludge. *Water Pract. Technol.* **2020**, *15*, 404–415. [CrossRef]
5. Tveit, T.-M. Application of an Industrial Heat Pump for Steam Generation Using District Heating as a Heat Source. 12the IEA Heat Pump Conference, Rotterdam, O.3.8.3. 2017. Available online: https://heatpumpingtechnologies.org/publications/o-3-8-3-application-of-an-industrial-heat-pump-for-steam-generation-using-district-heating-as-a-heat-source/ (accessed on 30 January 2024).
6. Byrne, P.S.; Carton, J.G.; Corcoran, B. Investigating the Suitability of a Heat Pump Water-Heater as a Method to Reduce Agricultural Emissions in Dairy Farms. *Sustainability* **2021**, *13*, 5736. [CrossRef]
7. Hermanucz, P.; Geczi, G.; Barotfi, I. Energy Efficient Solution in the Brewing Process Using a Dual-Source Heat Pump. *Therm. Sci.* **2022**, *26*, 2311–2319. [CrossRef]
8. Anifantis, A.; Colantoni, A.; Pascuzzi, S.; Santoro, F. Photovoltaic and Hydrogen Plant Integrated with a Gas Heat Pump for Greenhouse Heating: A Mathematical Study. *Sustainability* **2018**, *10*, 378. [CrossRef]
9. Nemś, A.; Nemś, M.; Świder, K. Analysis of the Possibilities of Using a Heat Pump for Greenhouse Heating in Polish Climatic Conditions—A Case Study. *Sustainability* **2018**, *10*, 3483. [CrossRef]
10. Słyś, D.; Pochwat, K.; Czarniecki, D. An Analysis of Waste Heat Recovery from Wastewater on Livestock and Agriculture Farms. *Resources* **2020**, *9*, 3. [CrossRef]
11. Babak, T.; Duić, N.; Khavin, G.; Boldyryev, S.; Krajačić, G. Possibility of Heat Pump Use in Hot Water Supply Systems. *J. Sustain. Dev. Energy Water Environ. Syst.* **2016**, *4*, 203–215. [CrossRef]
12. Ünal, F.; Akan, A.E.; DemİR, B.; Yaman, K. 4E Analysis of an Underfloor Heating System Integrated to the Geothermal Heat Pump for Greenhouse Heating. *Turk. J. Agric. For.* **2022**, *46*, 762–780. [CrossRef]
13. Chiriboga, G.; Capelo, S.; Bunces, P.; Guzmán, C.; Cepeda, J.; Gordillo, G.; Montesdeoca, D.E.; Carvajal, C.G. Harnessing of Geothermal Energy for a Greenhouse in Ecuador Employing a Heat Pump: Design, Construction, and Feasibility Assessment. *Heliyon* **2021**, *7*, e08608. [CrossRef]
14. Zhang, H.; Liu, Y.; Liu, X.; Duan, C. Energy and Exergy Analysis of a New Cogeneration System Based on an Organic Rankine Cycle and Absorption Heat Pump in the Coal-Fired Power Plant. *Energy Convers. Manag.* **2020**, *223*, 113293. [CrossRef]
15. Cao, X.-Q.; Yang, W.-W.; Zhou, F.; He, Y.-L. Performance Analysis of Different High-Temperature Heat Pump Systems for Low-Grade Waste Heat Recovery. *Appl. Therm. Eng.* **2014**, *71*, 291–300. [CrossRef]
16. Zhang, H.; Dong, Y.; Lai, Y.; Zhang, H.; Zhang, X. Waste Heat Recovery from Coal-Fired Boiler Flue Gas: Performance Optimization of a New Open Absorption Heat Pump. *Appl. Therm. Eng.* **2021**, *183*, 116111. [CrossRef]
17. Su, W.; Ma, D.; Lu, Z.; Jiang, W.; Wang, F.; Xiaosong, Z. A Novel Absorption-Based Enclosed Heat Pump Dryer with Combining Liquid Desiccant Dehumidification and Mechanical Vapor Recompression: Case Study and Performance Evaluation. *Case Stud. Therm. Eng.* **2022**, *35*, 102091. [CrossRef]
18. Jokiel, M.; Bantle, M.; Kopp, C.; Halvorsen Verpe, E. Modelica-Based Modelling of Heat Pump-Assisted Apple Drying for Varied Drying Temperatures and Bypass Ratios. *Therm. Sci. Eng. Prog.* **2020**, *19*, 100575. [CrossRef]
19. Waheed, M.A.; Oni, A.O.; Adejuyigbe, S.B.; Adewumi, B.A.; Fadare, D.A. Performance Enhancement of Vapor Recompression Heat Pump. *Appl. Energy* **2014**, *114*, 69–79. [CrossRef]
20. Liu, Y.; Zhai, J.; Li, L.; Sun, L.; Zhai, C. Heat Pump Assisted Reactive and Azeotropic Distillations in Dividing Wall Columns. *Chem. Eng. Process. Process Intensif.* **2015**, *95*, 289–301. [CrossRef]
21. Long, N.V.D.; Minh, L.Q.; Pham, T.N.; Bahadori, A.; Lee, M. Novel Retrofit Designs Using a Modified Coordinate Descent Methodology for Improving Energy Efficiency of Natural Gas Liquid Fractionation Process. *J. Nat. Gas Sci. Eng.* **2016**, *33*, 458–468. [CrossRef]
22. Long, N.V.D.; Han, T.H.; Lee, D.Y.; Park, S.Y.; Hwang, B.B.; Lee, M. Enhancement of a R-410A Reclamation Process Using Various Heat-Pump-Assisted Distillation Configurations. *Energies* **2019**, *12*, 3776. [CrossRef]
23. Zhu, Z.; Qi, H.; Shen, Y.; Qiu, X.; Zhang, H.; Qi, J.; Yang, J.; Wang, L.; Wang, Y.; Ma, Y.; et al. Energy-Saving Investigation of Organic Material Recovery from Wastewater via Thermal Coupling Extractive Distillation Combined with Heat Pump Based on Thermoeconomic and Environmental Analysis. *Process Saf. Environ. Prot.* **2021**, *146*, 441–450. [CrossRef]
24. Šulgan, B.; Labovský, J.; Variny, M.; Labovská, Z. Multi-Objective Assessment of Heat Pump-Assisted Ethyl Acetate Production. *Processes* **2021**, *9*, 1380. [CrossRef]
25. Boldyryev, S.; Kuznetsov, M.; Ryabova, I.; Krajačić, G.; Kaldybaeva, B. Assessment of Renewable Energy Use in Natural Gas Liquid Processing by Improved Process Integration with Heat Pumps. *e-Prime—Adv. Electr. Eng. Electron. Energy* **2023**, *5*, 100246. [CrossRef]
26. Florian, S.; Cordin, A. Walmsley Timothy Gordon Heat Pump Integration by Pinch Analysis for Industrial Applications: A Review. *Chem. Eng. Trans.* **2019**, *76*, 7–12. [CrossRef]

27. Kim, Y.; Lim, J.; Shim, J.Y.; Hong, S.; Lee, H.; Cho, H. Optimization of Heat Exchanger Network via Pinch Analysis in Heat Pump-Assisted Textile Industry Wastewater Heat Recovery System. *Energies* **2022**, *15*, 3090. [CrossRef]
28. Walmsley, T.G.; Klemeš, J.J.; Walmsley, M.R.; Atkins, M.J.; Varbanov, P.S. Varbanov Innovative Hybrid Heat Pump for Dryer Process Integration. *Chem. Eng. Trans.* **2017**, *57*, 1039–1044. [CrossRef]
29. Limei, G.; Sabev, V.P.; Gordon, W.T. Klemes Jiri Jaromir Process Integration Using a Joule Cycle Heat Pump. *Chem. Eng. Trans.* **2019**, *76*, 415–420. [CrossRef]
30. Lincoln, B.J.; Kong, L.; Pineda, A.M.; Walmsley, T.G. Process Integration and Electrification for Efficient Milk Evaporation Systems. *Energy* **2022**, *258*, 124885. [CrossRef]
31. Klinac, E.; Carson, J.K.; Hoang, D.; Chen, Q.; Cleland, D.J.; Walmsley, T.G. Multi-Level Process Integration of Heat Pumps in Meat Processing. *Energies* **2023**, *16*, 3424. [CrossRef]
32. Ulyev, L.; Kapustenko, P.; Vasilyev, M.; Boldyryev, S. Total Site Integration for Coke Oven Plant. *Chem. Eng. Trans.* **2013**, *35*, 235–240. [CrossRef]
33. Hegely, L.; Lang, P. Reduction of the Energy Demand of a Second-Generation Bioethanol Plant by Heat Integration and Vapour Recompression between Different Columns. *Energy* **2020**, *208*, 118443. [CrossRef]
34. Cox, J.; Belding, S.; Lowder, T. Application of a Novel Heat Pump Model for Estimating Economic Viability and Barriers of Heat Pumps in Dairy Applications in the United States. *Applied Energy* **2022**, *310*, 118499. [CrossRef]
35. Lu, Z.; Yao, Y.; Liu, G.; Ma, W.; Gong, Y. Thermodynamic and Economic Analysis of a High Temperature Cascade Heat Pump System for Steam Generation. *Processes* **2022**, *10*, 1862. [CrossRef]
36. Martínez-Rodríguez, G.; Díaz-de-León, C.; Fuentes-Silva, A.L.; Baltazar, J.-C.; García-Gutiérrez, R. Detailed Thermo-Economic Assessment of a Heat Pump for Industrial Applications. *Energies* **2023**, *16*, 2784. [CrossRef]
37. Chen, T.; Kyung Kwon, O. Experimental Analyses of Moderately High-Temperature Heat Pump Systems with R245fa and R1233zd(E). *Energy Eng.* **2022**, *119*, 2231–2242. [CrossRef]
38. Zühlsdorf, B.; Bühler, F.; Mancini, R.; Cignitti, S. High Temperature Heat Pump Integration Using Zeotropic Working Fluids for Spray Drying Facilities. 12the IEA Heat Pump Conference, Rotterdam, O.3.9.3. 2017. Available online: https://backend.orbit.dtu.dk/ws/portalfiles/portal/132571251/O.3.9.3_High_Temperature_Heat_Pump_Integration_using_Zeotropic_Working_F....pdf (accessed on 30 January 2024).
39. Gómez-Hernández, J.; Grimes, R.; Briongos, J.V.; Marugán-Cruz, C.; Santana, D. Carbon Dioxide and Acetone Mixtures as Refrigerants for Industry Heat Pumps to Supply Temperature in the Range 150–220 °C. *Energy* **2023**, *269*, 126821. [CrossRef]
40. Gudjonsdottir, V.; Infante Ferreira, C.A. Technical and Economic Analysis of Wet Compression–Resorption Heat Pumps. *Int. J. Refrig.* **2020**, *117*, 140–149. [CrossRef]
41. Urbanucci, L.; Bruno, J.C.; Testi, D. Thermodynamic and Economic Analysis of the Integration of High-Temperature Heat Pumps in trigeneration systems. *Applied Energy.* **2019**, *238*, 516–533. [CrossRef]
42. Wolf, M.; Detzlhofer, T.; Proll, T. A Comparative Study of Industrial Heat Supply Based on Second-Law Analysis and Operating Costs. *Therm. Sci.* **2018**, *22*, 2203–2213. [CrossRef]
43. Zuberi, M.J.S.; Hasanbeigi, A.; Morrow, W. Techno-Economic Evaluation of Industrial Heat Pump Applications in US Pulp and Paper, Textile, and Automotive Industries. *Energy Effic.* **2023**, *16*, 19. [CrossRef]
44. Wu, X.; Xing, Z.; He, Z.; Wang, X.; Chen, W. Performance Evaluation of a Capacity-Regulated High Temperature Heat Pump for Waste Heat Recovery in Dyeing Industry. *Appl. Therm. Eng.* **2016**, *93*, 1193–1201. [CrossRef]
45. Gangar, N.; Macchietto, S.; Markides, C.N. Recovery and Utilization of Low-Grade Waste Heat in the Oil-Refining Industry Using Heat Engines and Heat Pumps: An International Technoeconomic Comparison. *Energies* **2020**, *13*, 2560. [CrossRef]
46. Yang, L.; Ren, Y.; Wang, Z.; Hang, Z.; Luo, Y. Simulation and Economic Research of Circulating Cooling Water Waste Heat and Water Resource Recovery System. *Energies* **2021**, *14*, 2496. [CrossRef]
47. Lee, J.; Son, Y.; Lee, K.; Won, W. Economic Analysis and Environmental Impact Assessment of Heat Pump-Assisted Distillation in a Gas Fractionation Unit. *Energies* **2019**, *12*, 852. [CrossRef]
48. Hou, J.; Mao, C.; Xu, Y. Thermoeconomic Multiobjective Optimization of Tobacco Drying Heat Pump Recovering Waste Heat from Monocrystal Silicon Furnace Based on SVR ANN Model in Southwest China. *Energy Sci. Eng.* **2023**, *11*, 2351–2365. [CrossRef]
49. Hegely, L.; Lang, P. Optimisation of the Higher Pressure of Pressure-Swing Distillation of a Maximum Azeotropic Mixture. *Energy* **2023**, *271*, 126939. [CrossRef]
50. Janković, T.; Straathof, A.J.J.; Kiss, A.A. Advanced Downstream Processing of Bioethanol from Syngas Fermentation. *Sep. Purif. Technol.* **2023**, *322*, 124320. [CrossRef]
51. Aspen HYSYS | Process Simulation Software | AspenTech. Available online: https://www.aspentech.com/en/products/engineering/aspen-hysys (accessed on 31 January 2024).
52. Kern's Process Heat Transfer, 2nd Edition | Wiley. Available online: https://www.wiley.com/en-us/Kern's+Process+Heat+Transfer,+2nd+Edition-p-9781119364825 (accessed on 29 January 2024).
53. Hewitt, G.F. *Handbook of Heat Exchanger Design*; Begell House: New York, NY, USA, 1992; ISBN 978-1-56700-000-9.
54. Electricity Price Statistics—Statistics Explained. Available online: https://ec.europa.eu/eurostat/statistics-explained/index.php?title=Electricity_price_statistics (accessed on 31 January 2024).
55. Aspen Exchanger Design and Rating (EDR) | AspenTech. Available online: https://www.aspentech.com/en/products/engineering/aspen-exchanger-design-and-rating (accessed on 31 January 2024).

56. Turton, R.; Shaeiwitz, J.A.; Bhattacharyya, D.; Whiting, W.B. *Analysis, Synthesis, and Design of Chemical Processes*, 5th ed.; Pearson Education: London, UK, 2021. Available online: https://www.pearson.com/en-us/subject-catalog/p/analysis-synthesis-and-design-of-chemical-processes/P200000000551/9780137459483 (accessed on 14 February 2024).
57. Luyben, W.L. Capital Cost of Compressors for Conceptual Design. *Chem. Eng. Process.—Process Intensif.* **2018**, *126*, 206–209. [CrossRef]

energies

MDPI

Article

Simultaneous Optimization of Work and Heat Exchange Networks

Nidret Ibrić [1,2,*], Chao Fu [3] and Truls Gundersen [1]

1 Department of Energy and Process Engineering, Norwegian University of Science and Technology (NTNU), Kolbjørn Hejes v 1B, NO-7034 Trondheim, Norway; truls.gundersen@ntnu.no
2 Faculty of Technology, University of Tuzla, Urfeta Vejzagića 8, 75000 Tuzla, Bosnia and Herzegovina
3 SINTEF Energy Research, Sem Sælands vei 11, NO-7034 Trondheim, Norway; chao.fu@sintef.no
* Correspondence: nidret.ibric@ntnu.no or nidret.ibric@untz.ba

Abstract: This paper introduces a simultaneous optimization approach to synthesizing work and heat exchange networks (WHENs). The proposed work and heat integration (WHI) superstructure enables different thermodynamic paths of pressure and temperature-changing streams. The superstructure is connected to a heat exchanger network (HEN) superstructure, enabling the heat integration of hot and cold streams identified within the WHI superstructure. A two-step solution strategy is proposed, consisting of initialization and design steps. In the first step, a thermodynamic path model based on the WHI superstructure is combined with a model for simultaneous optimization and heat integration. This nonlinear programming (NLP) model aims to minimize operating expenditures and provide an initial solution for the second optimization step. In addition, hot and cold streams are identified, enabling additional model reduction. In the second step of the proposed solution approach, a thermodynamic path model is combined with the modified HEN model to minimize the network's total annualized cost (TAC). The proposed mixed integer nonlinear programming (MINLP) model is validated by several examples, exploring the impact of the equipment costing and annualization factor on the optimal network design. The results from these case studies clearly indicate that the new synthesis approach proposed in this paper produces solutions that are consistently similar to or better than the designs presented in the literature using other methodologies.

Keywords: heat exchanger network; work and heat integration; mathematical programming; superstructure optimization; thermodynamic path

1. Introduction

A substantial amount of energy in the form of heat and work is used in the chemical and petrochemical, metals and metallurgical, pulp and paper, food processing and other industries. Therefore, particular attention has been given to energy savings in the past half century. This has become more significant as we shift to using renewable energy sources in these industries. Heat integration (HI) has been a major development in resource conservation, providing a systematic conceptual approach to finding the best energy targets that can be obtained in any given process. The combination of mathematical programming (MP) with HI has made the methodology more easily applicable to large-scale problems of HI. Heat exchanger network (HEN) synthesis has been a popular research area for more than forty-five years [1].

In the synthesis of classical HENs, assuming constant pressure, heat is recovered from hot process streams to cold process streams, whose identity is known before optimization. When the pressure of the process streams changes, the stream's identity can also change due to the increasing temperature when compressing the stream or decreasing temperature when expanding the stream. For example, a process stream identified as a cold stream that needs heating can be compressed, resulting in an increased temperature; thus, the same stream can then become hot. A similar conclusion can be drawn for the hot stream that needs expansion. As the hot stream expands, its temperature decreases due to cooling; thus, the stream can become cold after expansion. Because work is consumed in changing the pressure of the streams, this relatively new research concept is referred to as the work and heat exchange network (WHEN). In WHENs, pressure and temperature are two equally important and related parameters. This complex relationship is described in detail by Fu et al. [2]. The authors identified the compressor/expander inlet temperature as a crucial parameter affecting the work consumed/produced and, consequently, the hot and cold utility consumption. Two main approaches are used to solve WHEN problems: pinch analysis (PA) and MP. PA is a graphical approach that uses fundamental thermodynamic insights, and MP is based on the mathematical optimization of superstructures composed of different work and heat integration opportunities. In addition, the two approaches can also be combined to solve the problem.

Aspelund et al. [3] introduced an extended traditional PA with exergy analysis called Extended Pinch Analysis and Design (ExPAnD), which relies on a set of heuristics and engineering insights. A set of heuristic rules guiding the design is proposed for different categories of streams (pressure-changing, temperature-changing or phase-changing streams). The authors use the exergy efficiency to measure the quality of the optimal design. In a series of papers, Fu and Gundersen proposed and proved a set of fundamental theorems for the integration of compressors [4] and expanders [5] in the above ambient processes that result in designs with minimum exergy consumption. Fu et al. [6] highlighted the importance of thermodynamic insights in understanding the problem and fundamental concepts, but they also presented limitations, especially when addressing the network cost. For this particular reason, the MP approach has attracted more research interest.

Wechsung et al. [7] proposed a state-space superstructure consisting of pressure-changing stages separately for hot and cold streams and a pinch operator for heat integration, capable of handling variable temperatures. However, compression and expansion are achieved at the pinch temperature according to the proposed ExPanD methodology [3], which might yield sub-optimal solutions. The objective function minimizes exergy consumption. Huang and Karimi [8] proposed a multistage superstructure with separate HEN and work exchange network (WEN) blocks within each stage. The superstructure is distinct for low-pressure streams that are compressed and high-pressure streams that are expanded. Stream splitting enables utility compression/expansion, valves and single-shaft turbine compressors (SSTCs) with the same optimized inlet temperature. The HEN superstructure is a stage-wise superstructure that includes non-isothermal mixing [9]. The objective function minimizes the total annualized cost (TAC) of the network. Onishi et al. [10] proposed a multistage work integration superstructure with heat integration between the WEN stages. The HEN model is based on the well-known stage-wise superstructure by Yee and Grossmann [11]. Streams are classified as high-pressure and low-pressure streams, for which the WEN superstructure differs. The WEN stage consists of parallel branches of turbines and compressors on a common SSTC. Turbines are associated with stages for high-pressure streams, enabling work recovery from high-pressure to low-pressure streams, with an additional utility turbine and an expansion valve. For low-pressure streams, there is an equal number of compressors corresponding to the number of turbines and utility compressors on a single shaft. Compared to the work of Huang and Karimi [8], parallel SSTC units are possible, including utility compressors and turbines. Helper motors and turbine generators exist within SSTC units to deal with an excess or shortage of energy. The streams are sent to the HEN between the WEN stages to improve the pressure recovery. The objective function minimizes the TAC of a combined HEN and WEN. However, in the work of Onishi et al. [12], the classical stage-wise superstructure for HEN synthesis [11] is intertwined with the predefined pressure change network for hot and cold streams. In this approach, the hot and cold streams follow specific pressure change routes with a maximum of three stages of expansion and compression. Unlike most MP deterministic approaches, modeled using algebraic languages such as the General Algebraic Modeling

System (GAMS), Pavão et al. [13] proposed a new meta-heuristic approach, a matrix-based implementation of a WHEN superstructure, inspired by previous works [7,14] coupled with an enhanced stage-wise HEN superstructure [15]. This work was later extended [16], including additional structural possibilities for work and heat integration with practical constraints (upper/lower bounds on temperature) for pressure change units, as well as a number of coupled units (compressors and turbines) per shaft. A framework for simultaneous WHI based on a mixed integer nonlinear programming (MINLP) model was proposed by Nair et al. [17]. In the proposed framework, the streams are unclassified, and the pressure change is also allowed for constant pressure streams. In addition, the authors considered the phase changes of streams with phase-based property calculations. In the proposed superstructure, each stream starts at a specific pressure and temperature and passes through a series of WHEN stages to reach its final pressure. A WHEN stage comprises a typical stage-wise HEN as in Huang et al. [9] and pressure change units consisting of SSTCs and valves. A similar superstructure to those proposed by Huang and Karimi [8] and Nair et al. [17] has been developed by Onishi et al. [18], with an approach combining MP with a pinch location method [19,20] to obtain an optimal network design with the minimum TAC. The authors proposed a generalized disjunctive programming (GDP) model for the selection of pressure change equipment in the work integration stages and the identification of unclassified streams. The model is also suitable for the handling of unknown inlet and outlet temperatures. Although the authors claim global optimality, Fu et al. [6] obtained better solutions with their graphical approach that minimizes the exergy consumption rather than the TAC of the network.

Li et al. [21] proposed an alternative approach to presenting all potential flowsheet configurations for WHENs with building blocks. These blocks consist of block interiors representing the mixing, splitting, utility heating/cooling and boundaries between adjacent blocks that allow for work and heat integration. Contrary to other proposed superstructures, this approach does not require work and heat integration stages and includes equipment such as two-stream and multi-stream heat exchangers, compressors/expanders and SSTCs. The model is formulated as an MINLP with the objective function of minimizing the TAC. Pavão et al. [22] used a pinch-based approach to define pressure-changing routes based on capital and operating cost targets before detailed WHEN synthesis and provided an efficient starting point for WHEN synthesis. The solution approach is a hybrid meta-heuristic method based on stochastic methods to define modifications to the pressure manipulation route. The obtained pressure-changing routes are fixed, and then the HI problem is solved with a specified heat recovery approach temperature (HRAT) using the approach of "spaghetti" design to reduce the heat transfer area [23]. Previous steps are used as an initial design for the simultaneous WHEN problem. This paper highlights the importance of targeting before the design step, while Yu et al. [24] presented alternative formulations for the simultaneous optimization and heat integration of WHENs. A two-step sequential design procedure was proposed by Lin et al. [25]. In the first step of the proposed approach, a stage-wise superstructure, in which the unclassified streams subsequently go through the HEN and WEN stages, is used to perform a targeting procedure with an extended PA method under the assumption of vertical heat transfer. When the optimal thermodynamic path is identified with the proposed GDP model, the HEN is synthesized using a stage-wise HEN superstructure and an MINLP model [11]. The solution approach was later modified [26] with an efficient optimization strategy by modifying the targeting step to identify the optimal thermodynamic paths of process streams using genetic algorithms and a golden section search. The HEN design is determined using a global search algorithm in the design phase. Lin et al. [27] presented a solution procedure to obtain globally optimal WHEN designs based on a minimal WHEN system using an enumeration procedure on predefined thermodynamic paths. In addition, the properties of the streams are calculated using cubic equations of state; however, because the authors used a reduced minimal WHEN representation to obtain simple designs, the presented solutions are not global solutions to the problem, as later shown in this paper.

Most of the previous research has been focused on minimizing the TAC of the network. Yang et al. [28] presented a multi-objective MINLP model for the synthesis of WHENs, including constant-pressure and pressure-changing streams with multi-stage compression. However, their proposed stage-wise superstructure includes only the compression of predefined cold streams. The Pareto front of the solutions has opposing objectives: minimizing the TAC and the exergy consumption. A similar superstructure was created to address the expansion of hot, high-pressure streams [29], later modified to include a simple model for the steam Rankine cycle for a utility system in order to account for multiple utilities [30]. Braccia et al. [31] combined the WHEN superstructure proposed by Huang and Karimi [8] with the HEN superstructure of Yee and Grossmann [11], enabling a change in stream identities with explicit modeling, avoiding additional binary variables and nonlinearities. This allows the pressure-changing streams to act as a low-pressure hot stream and a high-pressure cold stream at different stages in the proposed network. For a more detailed and in-depth analysis of thermodynamic-based and optimization-based methods for WHEN synthesis, the reader is referred to a recent review paper by Yu et al. [32]. The authors present a review of application studies and the equipment used in work and heat integration, as well as a critical review of the studies regarding the methods, pressure change equipment, property models and objective function.

There are several limitations in previous research on work and heat integration. The fundamental thermodynamic approach of Fu and Gundersen [4,5] only considered energy performance (exergy), neglecting the investment cost or TAC. Some optimization-based approaches [17,18] use superstructures that do not include certain topological combinations related to HEN configurations. Other methodologies have used a predefined thermodynamic path for pressure-changing streams [7,27], while some studies only consider the compression of cold streams [28] and/or the expansion of hot streams [29].

This work presents a novel WHI superstructure based on feasible thermodynamic paths involving the multi-choice mixing and splitting of streams. In the presented superstructure, the streams are unclassified but identified as hot and cold within different superstructure elements. A two-step solution strategy consisting of a targeting/initialization step and a design step is proposed. Contrary to much previous research, where the targeting is used to fix the thermodynamic paths, this work uses the targeting model for stream identification and the initialization of the MINLP model for the simultaneous synthesis of the WHEN, performed in the second step. In addition, most studies use the stage-wise superstructure [11] and its modifications to design the HEN within the WHEN. In this work, we use a modified superstructure from Floudas et al. [33] that gives additional heat integration opportunities when integrated with a WHI superstructure. Furthermore, the HRAT temperature required for the targeting step can be changed to provide different initializations and generate a set of local solutions to the MINLP model.

2. Problem Formulation

Figure 1 shows the conceptual formulation of the work and heat exchange network (WHEN) synthesis problem. Given is a set of process streams $s \in S$ with specified supply and target parameters (pressure and temperature). The process streams are connected to the heat exchanger network (HEN), consisting of heat exchangers, heaters (H) and coolers (C), enabling heat integration, and the work exchange network (WEN), consisting of compressors, expanders and expansion valves, enabling pressure changes in process streams. The HEN and WEN are interconnected, enabling the simultaneous optimization of work and heat integration. The system requires hot and cold utilities for additional heating and cooling within the HEN and electricity required/produced by running compressors and expanders.

The objective is to synthesize the optimal network design with the minimum TAC. This implies finding the optimal placement of the pressure change equipment within the HEN. The following assumptions are imposed to simplify the synthesis procedure.

- Steady-state operation is considered.

- All process streams are ideal gaseous streams.
- Compression and expansion are reversible and adiabatic (i.e., isentropic).
- Expansion through the valve is isenthalpic with a constant Joule–Thompson coefficient.
- One hot and one cold utility is assumed.
- The compressor and expander isentropic efficiency is considered constant.
- The heat capacity flow rates are constant.
- Heat and pressure losses are neglected.
- The costs of mixers and expansion valves are negligible.

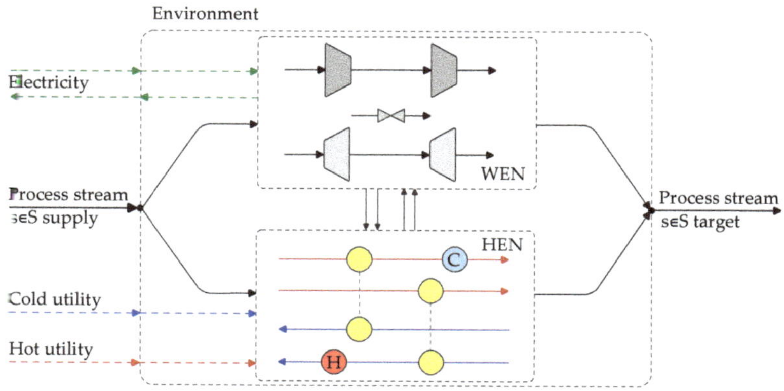

Figure 1. Conceptual problem formulation.

3. Methodology

3.1. Possible Thermodynamic Paths

To create a superstructure, we start with the feasible thermodynamic paths (trajectories) shown in Figure 2, previously studied in the literature [34] for the change in state for process streams. Let us observe a state change ($1 \rightarrow 2$) where the stream requires compression and heating (low-pressure cold stream), assuming that the isentropic efficiency is 1 (ideal isentropic compression). The stream can be compressed from its initial state and cooled afterwards, following the path ($1 \rightarrow a \rightarrow 2$). This trajectory requires compression at higher temperatures, and more work is consumed for the compression. Still, it produces higher outlet temperatures in the stream (high-quality heat) that can be used for heat integration. An additional increase in the compressor inlet temperature creates better-quality heat (path $1 \rightarrow e \rightarrow f \rightarrow 2$). However, it may not be practical to compress the stream at higher temperatures due to limitations in the operating conditions of the compressors. To reduce the compressor inlet temperature, it is possible to cool down the compressor inlet stream (at constant pressure), followed by the compression of the stream (path $1 \rightarrow b \rightarrow 2$). An additional decrease in the compressor inlet temperature requires additional heating after the compression of the stream (path $1 \rightarrow c \rightarrow d \rightarrow 2$). Reducing the compressor inlet temperature consumes less compression work, but the compressor outlet temperature decreases, and the quality of the heat is lower.

If we now consider that the stream is compressed to a pressure higher than required, the possible number of thermodynamic paths increases significantly. The stream can be cooled and compressed to pressure $p > p_2$ ($1 \rightarrow c \rightarrow h$) and, after this, expanded in a pressure reduction valve and cooled to the required temperature ($h \rightarrow a \rightarrow 2$). Expansion can also be done in an expander, producing the work ($h \rightarrow i \rightarrow 2$).

Similarly, the diagram shown in Figure 2 can also be used to analyze the thermodynamic paths of a high-pressure hot stream (change of state $2 \rightarrow 1$). The same trajectories can be followed but in the reverse direction of the pressure change. However, the expansion of the process stream has the opposite effect compared to compression. Increasing the inlet temperature of the expander increases the work production in the expander. On the other hand, high-quality heat is not available for heat integration because it is used to produce work. The thermodynamic path (2 → a → 1) shows the expansion process ending at a specified target temperature T_1. An additional increase in the expander inlet temperature produces more work but requires the additional cooling of the stream leaving the expander (path 2 → f → e → 1). Thermodynamic path 2 → b → 1 starts with expansion at the supply state and requires additional heating to reach the target state, and thermodynamic path 2 → d → c → 1 includes cooling before expanding the stream in the expander. Expansion to the required pressure can be done by using an expansion valve (isenthalpic process); however, work is not being produced in this case, and there is a significant loss in the exergy of the stream. After the expansion, additional cooling is required (path 2 → g → 1). As can be seen, many possible thermodynamic paths exist for the change of state 1 → 2 or 2 → 1. In theory, if we assume that change of state 1 → 2 can be achieved by combining compression, expansion, cooling and heating at any feasible pressure and temperature, as long as we start at state (1) and end up at state (2), there can be an infinite number of trajectories. However, most of them would be impractical.

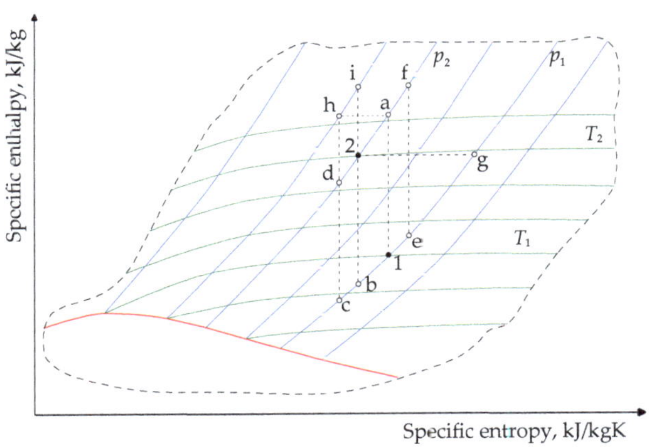

Figure 2. Hypothetical thermodynamic paths for compression and expansion of streams.

It should be noticed that the objective of the feasible thermodynamic paths stage of the method proposed by Yu et al. [34] is to fix the stream data before performing traditional HEN synthesis. In contrast, the feasible thermodynamic paths stage of the present method only serves to identify process streams and initialize the WHEN synthesis, where the process stream variables (pressure, temperature and branch flow rates) are allowed to change. This paper addresses the problem of finding the optimal trajectory for pressure-changing streams and simultaneously optimizing the HEN by minimizing the TAC of the network. We begin by synthesizing the superstructure based on the trajectories shown in Figure 2.

3.2. Work and Heat Integration (WHI) Superstructure

A superstructure shown in Figure 3 has been designed based on practical and feasible thermodynamic paths. This superstructure enables the exploration of various trajectories for stream s from the supply to the target state. In the proposed superstructure, the multi-choice splitting of stream s enables the heating or cooling of a stream before compression or expansion. We use the term multi-choice splitting because hot and cold stream branches can exist simultaneously (not a single-choice selection). It is important to highlight that the stream can remain isothermal when there is no temperature change in the heat exchangers. Multi-stage compression with inter-stage cooling enables the pressure increase of stream s, allowing for additional savings in the work required for compression (not shown in Figure 2). Conversely, the pressure reduction is achieved through multi-stage expansion with inter-stage heating, maximizing the work output, or with an expansion valve. The bypass option is incorporated to accommodate non-pressure-changing streams. The heating and cooling of process stream s leaving the pressure-changing stage $k = |K|$ is enabled by the multi-choice mixing of stream s, where hot and cold branches can exist simultaneously with multi-choice selection.

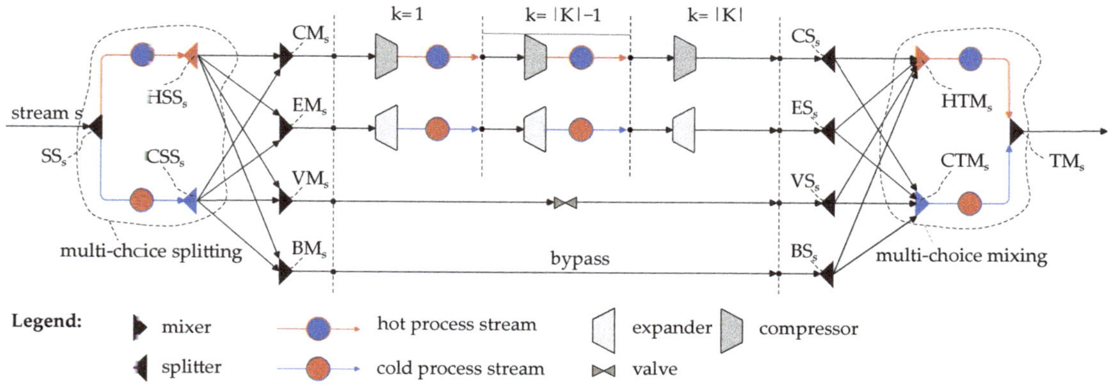

Figure 3. Work and heat integration (WHI) superstructure.

The proposed superstructure enables the heating and cooling of streams at the supply pressure, pressure change or bypass for non-pressure-changing streams, and the heating and cooling of streams at the target pressure. The proposed superstructure captures heat integration opportunities within and outside the pressure-changing stages. Additionally, it enables the search for the optimal placement of the pressure-changing equipment to minimize the network's TAC. The number of pressure-changing stages is set arbitrarily by the user.

The symbols used for the mixing and splitting units in the superstructure of Figure 3 have the following meanings:

SS_s Supply splitter for stream s;
HSS_s Supply splitter for a potential hot stream s;
CSS_s Supply splitter for a potential cold stream s;
CM_s Mixer before compressor for stream s;
EM_s Mixer before expander for stream s;
VM_s Mixer before valve for stream s;
BM_s Mixer before bypass for stream s.

The splitters CS_s, ES_s, VS_s and BS_s have similar meanings for stream s after the corresponding operation (compression, expansion, valve and bypass):

HTM_s Target mixer for a potential hot stream s;
CTM_s Target mixer for a potential cold stream s;
TM_s Target mixer for stream s.

3.3. *Heat Exchanger Network (HEN) Superstructure*

The HEN superstructure is based on the superstructure proposed by Floudas et al. [33], with small modifications regarding the introduction of a bypass connection for isothermal streams (Figure 4). Figure 4a shows the superstructure design for a hot stream i matched with cold streams j in exchanger $i - j$. The initial splitter S_i allows the parallel arrangement of the heat exchangers and mixers/splitters (EM_i and ES_i) in front of/behind each exchanger $i - j$, enabling the configuration of heat exchangers in series. The additional cooling of a hot stream i is possible in cooler $i - C$. The superstructure design for cold stream j matched with hot streams i in heat exchanger $i - j$ is shown in Figure 4b. However, the HEN superstructure in Floudas et al. [33] is derived based on the previous solution of two consecutive targeting models to minimize the utility consumption and number of matches. Thus, the model is easier to solve when knowing the supply and target temperatures of all hot and cold streams within the network and the corresponding minimum number of matches (with given duties) used to design the superstructure. In contrast, when designing the WHEN superstructure, the identity and temperature of the streams can be changed due to compression and expansion, and thus the inlet–outlet temperatures are optimization variables. This makes the synthesis problem much more difficult to solve. In addition, when the HEN superstructure (Figure 4) is coupled with the WHI superstructure (Figure 3), it enables more heat integration opportunities than the standalone HEN superstructure.

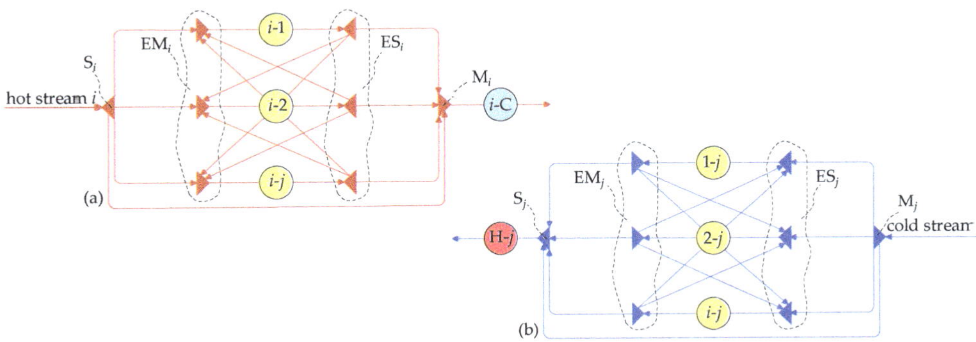

Figure 4. The HEN superstructure for (**a**) hot stream i (**b**) cold stream j.

3.4. Model

The overall model consists of three models: the thermodynamic path model (M_1), the heat integration (HI) model M_2 of Duran and Grossmann [19] and the HEN model M_3. To address the WHEN synthesis problem, the models are connected using appropriate equations to identify the streams and connect the models, as presented in Appendix A.1. A detailed description of the proposed models follows.

3.4.1. The Thermodynamic Path Model

The thermodynamic path model (M_1) is based on the WHI superstructure (Figure 3) that enables the identification of streams and the optimum placement for the pressure change equipment.

Multi-choice splitting. Stream s is unclassified, which means that it can be either hot or cold. Thus, according to Equation (1), the stream is divided in splitter SS$_s$ into fractions of the heat capacity flow rate serving as both hot and cold streams.

$$m_s = m_s^{(\mathrm{hs})} + m_s^{(\mathrm{cs})}, \quad \forall s \in S \tag{1}$$

These split fractions of hot and cold stream s are directed to the mixers before pressure change stage $k = 1$. However, the stream cannot be of low and high pressure simultaneously. Thus, auxiliary binary parameters (CINT, EINT, BINT) are introduced with assigned values based on the supply ($p_s^{(\mathrm{in})}$) and target pressure ($p_s^{(\mathrm{out})}$). For example, if the supply pressure is lower than the target pressure ($p_s^{(\mathrm{in})} < p_s^{(\mathrm{out})}$), the stream requires compression, and thus CINT = 1, and EINT = 0 and BINT = 0 for a specified stream s. Otherwise, if $p_s^{(\mathrm{in})} > p_s^{(\mathrm{out})}$ EINT = 1, CINT = 0 and BINT = 0. In the case of of $p_s^{(\mathrm{in})} = p_s^{(\mathrm{out})}$, then BINT = 1.

Equations (2) and (3) represent the mass balance equations for the splitters of hot (HSS_s) and cold (CSS_s) supply streams.

$$m_s^{(hs)} = \underset{CINT=1}{m_s^{(hs\to comp)}} + \underset{EINT=1}{m_s^{(hs\to exp)}} + \underset{EINT=1}{m_s^{(hs\to valve)}} + \underset{BINT=1}{m_s^{(hs\to bypass)}}, \quad \forall s \in S \tag{2}$$

$$m_s^{(cs)} = \underset{CINT=1}{m_s^{(cs\to comp)}} + \underset{EINT=1}{m_s^{(cs\to exp)}} + \underset{EINT=1}{m_s^{(cs\to valve)}} + \underset{BINT=1}{m_s^{(cs\to bypass)}}, \quad \forall s \in S \tag{3}$$

Mixers in front of pressure change stages. Streams from splitters (HSS_s and CSS_s) are directed to the pressure change stages or act as bypass streams if $p_s^{(in)} = p_s^{(out)}$. Several mixers for the compressor (CM_s), expander (EM_s), valve (VM_s) and bypass stream (BM_s) can be identified in the superstructure shown in Figure 3, for which the mass and heat balances are given by Equations (4)–(11).

$$m_s^{(comp)} = m_s^{(hs\to comp)} + m_s^{(cs\to comp)}, \quad \forall s \in S, CINT = 1 \tag{4}$$

$$m_s^{(comp)} T_s^{(comp,in)} = m_s^{(hs\to comp)} T_s^{(hs,out)} + m_s^{(cs\to comp)} T_s^{(cs,out)}, \quad \forall s \in S, CINT = 1 \tag{5}$$

$$m_s^{(exp)} = m_s^{(hs\to exp)} + m_s^{(cs\to exp)}, \quad \forall s \in S, EINT = 1 \tag{6}$$

$$m_s^{(exp)} T_s^{(exp,in)} = m_s^{(hs\to exp)} T_s^{(hs,out)} + m_s^{(cs\to exp)} T_s^{(cs,out)}, \quad \forall s \in S, EINT = 1 \tag{7}$$

$$m_s^{(valve)} = m_s^{(hs\to valve)} + m_s^{(cs\to valve)}, \quad \forall s \in S, EINT = 1 \tag{8}$$

$$m_s^{(valve)} T_s^{(valve,in)} = m_s^{(hs\to valve)} T_s^{(hs,out)} + m_s^{(cs\to valve)} T_s^{(cs,out)}, \quad \forall s \in S, EINT = 1 \tag{9}$$

$$m_s^{(bypass)} = m_s^{(hs\to bypass)} + m_s^{(cs\to bypass)}, \quad \forall s \in S, BINT = 1 \tag{10}$$

$$m_s^{(bypass)} T_s^{(bypass)} = m_s^{(hs\to bypass)} T_s^{(hs,out)} + m_s^{(cs\to bypass)} T_s^{(cs,out)}, \quad \forall s \in S, BINT = 1 \tag{11}$$

Compression stages. Multi-stage compression with inter-stage cooling enables a pressure increase in the compression stages. The compression work in stage k is given by Equation (12).

$$W_{s,k}^{(comp)} = m_s^{(comp)} \left(T_{s,k}^{(comp,out)} - T_{s,k}^{(comp,in)} \right), \quad \forall s \in S, \forall k \in K, CINT = 1 \tag{12}$$

The ideal temperature increase following compression is directly linked to the increase in pressure as given by Equation (13) under the assumption of isentropic compression, ideal gas, and constant specific heat capacity. The actual temperature increase accounting for the compressor's isentropic efficiency is given by Equation (14).

$$T_{id,s,k}^{(comp,out)} = T_{s,k}^{(comp,in)} \left(\frac{p_{s,k}^{(comp,out)}}{p_{s,k}^{(comp,in)}} \right)^{\frac{\kappa-1}{\kappa}}, \quad \forall s \in S, \forall k \in K, CINT = 1 \tag{13}$$

$$T_{s,k}^{(comp,out)} = T_{s,k}^{(comp,in)} + \left(\frac{T_{id,s,k}^{(comp,out)} - T_{s,k}^{(comp,in)}}{\eta} \right), \quad \forall s \in S, \forall k \in K, CINT = 1 \tag{14}$$

Equations (15) and (16) set the pressure and temperature constraints for each compression stage $k \in K$.

$$T_{s,k}^{(comp,out)} \geq T_{s,k}^{(comp,in)}, \quad \forall s \in S, \forall k \in K, CINT = 1 \tag{15}$$

$$p_{s,k}^{(comp,out)} \geq p_{s,k}^{(comp,in)}, \quad \forall s \in S, \forall k \in K, CINT = 1 \tag{16}$$

Temperature and pressure equality constraints apply for the compression stages $k \in K$ as given by Equations (17)–(23).

$$T_{s,k=1}^{(comp,in)} = T_s^{(comp,in)}, \quad \forall s \in S, CINT = 1 \tag{17}$$

$$T_{s,k}^{(\text{comp,in})} = T_{s,k-1}^{(\text{IC,out})}, \quad \forall s \in S, k \neq 1, \text{CINT} = 1 \tag{18}$$

$$T_{s,k}^{(\text{comp,out})} = T_{s,k}^{(\text{IC,in})}, \quad \forall s \in S, k \neq |K|, \text{CINT} = 1 \tag{19}$$

$$T_{s,k=|K|}^{(\text{comp,out})} = T_{s}^{(\text{comp,out})}, \quad \forall s \in S, \text{CINT} = 1 \tag{20}$$

$$p_{s,k=1}^{(\text{comp,in})} = p_{s}^{(\text{in})}, \quad \forall s \in S, \text{CINT} = 1 \tag{21}$$

$$p_{s,k}^{(\text{comp,in})} = p_{s,k-1}^{(\text{comp,out})}, \quad \forall s \in S, k \neq 1, \text{CINT} = 1 \tag{22}$$

$$p_{s,k=|K|}^{(\text{comp,out})} = p_{s}^{(\text{out})}, \quad \forall s \in S, \text{CINT} = 1 \tag{23}$$

Expansion stages. Multi-stage expansion with inter-stage heating enables a pressure decrease in the stages of expansion. The expansion work in stage k is given by Equation (24).

$$W_{s,k}^{(\text{exp})} = m_{s}^{(\text{exp})} \left(T_{s,k}^{(\text{exp,in})} - T_{s,k}^{(\text{exp,out})} \right), \quad \forall s \in S, \forall k \in K, \text{EINT} = 1 \tag{24}$$

The temperature decrease following the expansion is directly linked to the decrease in pressure as given by Equation (25) under the assumption of ideal expansion. The actual temperature decrease accounting for the expander's isentropic efficiency is given by Equation (26).

$$T_{\text{id},s,k}^{(\text{exp,out})} = T_{s,k}^{(\text{exp,in})} \left(\frac{p_{s,k}^{(\text{exp,out})}}{p_{s,k}^{(\text{exp,in})}} \right)^{\frac{\kappa-1}{\kappa}}, \quad \forall s \in S, \forall k \in K, \text{EINT} = 1 \tag{25}$$

$$T_{s,k}^{(\text{exp,out})} = T_{s,k}^{(\text{exp,in})} - \eta \left(T_{s,k}^{(\text{exp,in})} - T_{\text{id},s,k}^{(\text{exp,out})} \right), \quad \forall s \in S, \forall k \in K, \text{EINT} = 1 \tag{26}$$

Equations (27) and (28) set the temperature and pressure constraints for each expansion stage $k \in K$.

$$T_{s,k}^{(\text{exp,in})} \geq T_{s,k}^{(\text{exp,out})}, \quad \forall s \in S, \forall k \in K, \text{EINT} = 1 \tag{27}$$

$$p_{s,k}^{(\text{exp,in})} \geq p_{s,k}^{(\text{exp,out})}, \quad \forall s \in S, \forall k \in K, \text{EINT} = 1 \tag{28}$$

Temperature and pressure equality constraints apply for the expansion stages $k \in K$ as given by Equations (29)–(35).

$$T_{s,k=1}^{(\text{exp,in})} = T_{s}^{(\text{exp,in})}, \quad \forall s \in S, \text{EINT} = 1 \tag{29}$$

$$T_{s,k}^{(\text{exp,in})} = T_{s,k-1}^{(\text{IH,out})}, \quad \forall s \in S, k \neq 1, \text{EINT} = 1 \tag{30}$$

$$T_{s,k}^{(\text{exp,out})} = T_{s,k}^{(\text{IH,in})}, \quad \forall s \in S, k \neq |K|, \text{EINT} = 1 \tag{31}$$

$$T_{s,k=|K|}^{(\text{exp,out})} = T_{s}^{(\text{exp,out})}, \quad \forall s \in S, \text{EINT} = 1 \tag{32}$$

$$p_{s,k=1}^{(\text{exp,in})} = p_{s}^{(\text{in})}, \quad \forall s \in S, \text{EINT} = 1 \tag{33}$$

$$p_{s,k}^{(\text{exp,in})} = p_{s,k-1}^{(\text{exp,out})}, \quad \forall s \in S, k \neq 1, \text{EINT} = 1 \tag{34}$$

$$p_{s,k=|K|}^{(\text{exp,out})} = p_{s}^{(\text{out})}, \quad \forall s \in S, \text{EINT} = 1 \tag{35}$$

Expansion valves. Pressure reduction in expansion valves is an isenthalpic process. The temperature change due to change in pressure is determined by the Joule–Thompson expansion coefficient (μ), as given by Equation (36). The sign of the Joule–Thomson coefficient determines whether the process is characterized as cooling ($\mu < 0$) or heating ($\mu > 0$).

$$T_s^{(\text{valve,out})} = T_s^{(\text{valve,in})} + \mu\left(p_s^{(\text{valve,in})} - p_s^{(\text{valve,out})}\right), \quad \forall s \in S, \text{EINT} = 1 \tag{36}$$

Equation (37) gives an inequality constraint linked to pressure changes during expansion, while Equations (38) and (39) represent equality constraints for the supply and target pressures.

$$p_s^{(\text{valve,in})} \geq p_s^{(\text{valve,out})}, \quad \forall s \in S, \text{EINT} = 1 \tag{37}$$

$$p_s^{(\text{valve,in})} = p_s^{(in)}, \quad \forall s \in S, \text{EINT} = 1 \tag{38}$$

$$p_s^{(\text{valve,out})} = p_s^{(\text{out})}, \quad \forall s \in S, \text{EINT} = 1 \tag{39}$$

Splitters after pressure change stages. Fractions of process stream s exiting a pressure change or bypass stages are directed to a multi-choice mixer, enabling heating and cooling. The mass balance equations for the compressor (CS_s), expander (ES_s), expansion valve (VS_s) and bypass (BS_s) splitters are provided by Equations (40)–(43).

$$m_s^{(\text{comp})} = m_s^{(\text{comp}\to\text{ht})} + m_s^{(\text{comp}\to\text{ct})}, \quad \forall s \in S, \text{CINT} = 1 \tag{40}$$

$$m_s^{(\text{exp})} = m_s^{(\text{exp}\to\text{ht})} + m_s^{(\text{exp}\to\text{ct})}, \quad \forall s \in S, \text{EINT} = 1 \tag{41}$$

$$m_s^{(\text{valve})} = m_s^{(\text{valve}\to\text{ht})} + m_s^{(\text{valve}\to\text{ct})}, \quad \forall s \in S, \text{EINT} = 1 \tag{42}$$

$$m_s^{(\text{bypass})} = m_s^{(\text{bypass}\to\text{ht})} + m_s^{(\text{bypass}\to\text{ct})}, \quad \forall s \in S, \text{BINT} = 1 \tag{43}$$

Multi-choice mixers on the target side. The cooling and/or heating of stream s exiting the pressure change or bypass stage is enabled within multi-choice mixers for the streams acting as hot (mixer HTM_s) and cold (mixer CTM_s). The mass and heat balance equations of mixer HTM_s are given by Equations (44) and (45).

$$m_s^{(\text{ht})} = \underset{\text{CINT}=1}{m_s^{(\text{comp}\to\text{ht})}} + \underset{\text{EINT}=1}{m_s^{(\text{exp}\to\text{ht})}} + \underset{\text{EINT}=1}{m_s^{(\text{valve}\to\text{ht})}} + \underset{\text{BINT}=1}{m_s^{(\text{bypass}\to\text{ht})}}, \quad \forall s \in S \tag{44}$$

$$m_s^{(\text{ht})} T_s^{(\text{ht,in})} = \underset{\text{CINT}=1}{m_s^{(\text{comp}\to\text{ht})} T_s^{(\text{comp,out})}} + \underset{\text{EINT}=1}{m_s^{(\text{exp}\to\text{ht})} T_s^{(\text{exp,out})}} + \underset{\text{EINT}=1}{m_s^{(\text{valve}\to\text{ht})} T_s^{(\text{valve,out})}} + \underset{\text{BINT}=1}{m_s^{(\text{bypass}\to\text{ht})} T_s^{(\text{bypass})}}, \quad \forall s \in S \tag{45}$$

CTM_s exists within the proposed superstructure for the stream acting as a cold stream, leaving a pressure change or bypass stage mixer. The mass and heat balance equations of mixer CTM_s are given by Equations (46) and (47).

$$m_s^{(\text{ct})} = \underset{\text{CINT}=1}{m_s^{(\text{comp}\to\text{ct})}} + \underset{\text{EINT}=1}{m_s^{(\text{exp}\to\text{ct})}} + \underset{\text{EINT}=1}{m_s^{(\text{valve}\to\text{ct})}} + \underset{\text{BINT}=1}{m_s^{(\text{bypass}\to\text{ct})}}, \quad \forall s \in S \tag{46}$$

$$m_s^{(\text{ct})} T_s^{(\text{ct,in})} = \underset{\text{CINT}=1}{m_s^{(\text{comp}\to\text{ct})} T_s^{(\text{comp,out})}} + \underset{\text{EINT}=1}{m_s^{(\text{exp}\to\text{ct})} T_s^{(\text{exp,out})}} + \underset{\text{EINT}=1}{m_s^{(\text{valve}\to\text{ct})} T_s^{(\text{valve,out})}} + \underset{\text{BINT}=1}{m_s^{(\text{bypass}\to\text{ct})} T_s^{(\text{bypass})}}, \quad \forall s \in S \tag{47}$$

Target mixers (TM_s). In the final TM_s mixer, the hot and cold streams are mixed isothermally as the temperature of the hot and cold streams leaving the heat exchanger is set to the stream target temperature $T_s^{(\text{out})}$. Equation (48) gives the mass balance of mixer TM_s.

$$m_s = m_s^{(\text{ht})} + m_s^{(\text{ct})}, \quad \forall s \in S \tag{48}$$

3.4.2. The HI Model

The HI model M_2 is the simultaneous optimization and heat integration model from Duran and Grossmann [19] based on the pinch location method, where the inequalities given by Equations (49) and (50) constrain the hot utility consumption.

$$q_{\mathrm{HU}} \geq \sum_j f_j\left[\max\left(0, T_j^{(\mathrm{out})} - \left(T_{i'}^{(in)} - \mathrm{HRAT}\right)\right) - \max\left(0, T_j^{(\mathrm{in})} - \left(T_{i'}^{(\mathrm{in})} - \mathrm{HRAT}\right)\right)\right] - \sum_i f_i\left[\max\left(0, T_i^{(\mathrm{in})} - T_{i'}^{(\mathrm{in})}\right) - \max\left(0, T_i^{(\mathrm{out})} - T_{i'}^{(\mathrm{in})}\right)\right], \quad \forall i' \tag{49}$$

$$q_{\mathrm{HU}} \geq \sum_j f_j\left[\max\left(0, T_j^{(\mathrm{out})} - T_{j'}^{(\mathrm{in})}\right) - \max\left(0, T_j^{(\mathrm{in})} - T_{i'}^{(\mathrm{in})}\right)\right] - \sum_i f_i\left[\max\left(0, T_i^{(\mathrm{in})} - \left(T_{j'}^{(\mathrm{in})} + \mathrm{HRAT}\right)\right) - \max\left(0, T_i^{(\mathrm{out})} - \left(T_{j'}^{(\mathrm{in})} + \mathrm{HRAT}\right)\right)\right], \quad \forall j' \tag{50}$$

The cold utility consumption is determined by the global heat balance given by Equation (51).

$$q_{\mathrm{HU}} + \sum_i ec_i = q_{\mathrm{CU}} + \sum_j ec_j \tag{51}$$

where the heat content of the hot (ec_i) and cold (ec_j) streams is given by Equations (52) and (53).

$$ec_i = f_i\left(T_i^{(\mathrm{in})} - T_i^{(\mathrm{out})}\right), \quad \forall i \tag{52}$$

$$ec_j = f_j\left(T_j^{(\mathrm{out})} - T_j^{(\mathrm{in})}\right), \quad \forall j \tag{53}$$

A smooth approximation of the max operators [35] was used to handle discontinuous derivatives in Equations (49) and (50).

3.4.3. The HEN Model

The HEN model (M_3) is based on the superstructure shown in Figure 4, which originated from the HEN model in Floudas et al. [33]. The initial hot stream splitter (S_i) enables stream splitting for a parallel configuration of heat exchangers. The mass balance of splitter S_i is given by Equation (54).

$$f_i = \sum_j f_{i,j}^{(\mathrm{in}\to\mathrm{ehs})} + f_i^{(\mathrm{bypass})}, \quad \forall i \tag{54}$$

The mixers (EM_i) and splitters (ES_i) for the heat exchangers enable the connection between the heat exchangers, facilitating different heat exchange configurations (serial and parallel). The mass and heat balances of mixer EM_i are given by Equations (55) and (56). Equation (57) gives the mass balance of splitter ES_i.

$$f_{i,j}^{(\mathrm{ehs})} = f_{i,j}^{(\mathrm{in}\to\mathrm{ehs})} + \sum_{\substack{j' \\ j'\neq j}} f_{j,i,j'}^{(\mathrm{ehs})}, \quad \forall i, \forall j \tag{55}$$

$$f_{i,j}^{(\mathrm{ehs})} T_{i,j}^{(\mathrm{ehs,in})} = f_{i,j}^{(\mathrm{in}\to\mathrm{ehs})} T_i^{(\mathrm{in})} + \sum_{\substack{j' \\ j'\neq j}} f_{j,i,j'}^{(\mathrm{ehs})} T_{i,j'}^{(\mathrm{ehs,out})}, \quad \forall i, \forall j \tag{56}$$

$$f_{i,j}^{(\mathrm{ehs})} = f_{i,j}^{(\mathrm{ehs}\to\mathrm{out})} + \sum_{\substack{j' \\ j'\neq j}} f_{j',i,j}^{(\mathrm{ehs})}, \quad \forall i, \forall j \tag{57}$$

The fractions of the streams leaving heat exchangers $i - j$ are directed to the final mixer M_i, enabling additional cooling in cooler i–C. The mass and heat balances of mixer M_i are given by Equations (58) and (59).

$$f_i = \sum_j f_{i,j}^{(\text{ehs}\to\text{out})} + f_i^{(\text{bypass})}, \quad \forall i \tag{58}$$

$$f_i T_i^{(\text{m,out})} = \sum_j f_{i,j}^{(\text{ehs}\to\text{out})} T_{i,j}^{(\text{ehs,out})} + f_i^{(\text{bypass})} T_i^{(\text{in})}, \quad \forall i \tag{59}$$

Analogous equations apply for the cold stream superstructure shown in Figure 4b. The mass balance of splitter S_j is given by Equation (60).

$$f_j = \sum_i f_{i,j}^{(\text{in}\to\text{ecs})} + f_j^{(\text{bypass})}, \quad \forall j \tag{60}$$

The mass and heat balances of mixer EM_j are given by Equations (61) and (62). Equation (63) gives the mass balance of splitter ES_j.

$$f_{i,j}^{(\text{ecs})} = f_{i,j}^{(\text{in}\to\text{ecs})} + \sum_{\substack{i' \\ i'\neq i}} f_{i,i',j}^{(\text{ecs})}, \quad \forall i, \forall j \tag{61}$$

$$f_{i,j}^{(\text{ecs})} T_{i,j}^{(\text{ecs,in})} = f_{i,j}^{(\text{in}\to\text{ecs})} T_i^{(\text{in})} + \sum_{\substack{i' \\ i'\neq i}} f_{i,i',j}^{(\text{ecs})} T_{i',j}^{(\text{ecs,out})}, \quad \forall i, \forall j \tag{62}$$

$$f_{i,j}^{(\text{ecs})} = f_{i,j}^{(\text{ecs}\to\text{out})} + \sum_{\substack{i' \\ i'\neq i}} f_{i',i,j}^{(\text{ecs})}, \quad \forall i, \forall j \tag{63}$$

The fractions of the streams leaving heat exchangers $i - j$ are directed to the final mixer M_j, enabling additional heating in heater H–j. The mass and heat balances of mixer M_j are given by Equations (64) and (65).

$$f_j = \sum_i f_{i,j}^{(\text{ecs}\to\text{out})} + f_j^{(\text{bypass})}, \quad \forall j \tag{64}$$

$$f_j T_j^{(\text{m,out})} = \sum_j f_{i,j}^{(\text{ecs}\to\text{out})} T_{i,j}^{(\text{ecs,out})} + f_j^{(\text{bypass})} T_j^{(\text{in})}, \quad \forall j \tag{65}$$

For the individual heat exchangers $i - j$, Equations (66) and (67) define the heat balance equations for the hot and cold streams.

$$q_{i,j} = f_{i,j}^{(\text{ehs})} \left(T_{i,j}^{(\text{ehs,in})} - T_{i,j}^{(\text{ehs,out})} \right), \quad \forall i, \forall j \tag{66}$$

$$q_{i,j} = f_{i,j}^{(\text{ecs})} \left(T_{i,j}^{(\text{ecs,out})} - T_{i,j}^{(\text{ecs,in})} \right), \quad \forall i, \forall j \tag{67}$$

The heat balance for cooler $i -$ C is given by Equation (68), and Equation (69) gives the heat balance for heater H–j.

$$q_i^{(\text{C})} = f_i \left(T_i^{(\text{m,out})} - T_i^{(\text{out})} \right), \quad \forall i \tag{68}$$

$$q_j^{(\text{H})} = f_j \left(T_j^{(\text{out})} - T_j^{(\text{m,out})} \right), \quad \forall i \tag{69}$$

The global heat balances for hot stream i and cold stream j are given by Equations (70) and (71).

$$f_i \left(T_i^{(\text{in})} - T_i^{(\text{out})} \right) = \sum_j q_{i,j} + q_i^{(\text{C})}, \quad \forall i \tag{70}$$

$$f_j \left(T_j^{(\text{out})} - T_j^{(\text{in})} \right) = \sum_i q_{i,j} + q_j^{(\text{H})}, \quad \forall j \tag{71}$$

The temperature constraints for the exchangers are given by constraints (72)–(77).

$$\Delta Th_{i,j} \leq T_{i,j}^{(\text{ehs,in})} - T_{i,j}^{(\text{ecs,out})} + \Gamma(1 - z_{i,j}), \quad \forall i, \forall j \tag{72}$$

$$\Delta Tc_{i,j} \leq T_{i,j}^{(\text{ehs,out})} - T_{i,j}^{(\text{ecs,in})} + \Gamma(1 - z_{i,j}), \quad \forall i, \forall j \tag{73}$$

$$\Delta Th_i^{(\text{C})} \leq T_i^{(\text{m,out})} - T_{\text{CU}}^{(\text{out})} + \Gamma(1 - z_i), \quad \forall i \tag{74}$$

$$\Delta Tc_i^{(\text{C})} \leq T_i^{(\text{out})} - T_{\text{CU}}^{(\text{in})} + \Gamma(1 - z_i), \quad \forall i \tag{75}$$

$$\Delta Th_j^{(\text{H})} \leq T_{\text{HU}}^{(\text{in})} - T_j^{(\text{out})} + \Gamma(1 - z_j), \quad \forall j \tag{76}$$

$$\Delta Tc_j^{(\text{H})} \leq T_{\text{HU}}^{(\text{out})} - T_j^{(\text{m,out})} + \Gamma(1 - z_j), \quad \forall j \tag{77}$$

Constraints (78)–(82) apply to the existence of equipment in the optimal design, and constraints (83) and (84) are related to the existence of hot and cold stream branches connected to the heat exchangers.

$$q_{i,j} \leq q_{i,j}^{(\text{ub})} z_{i,j}, \quad \forall i, \forall j \tag{78}$$

$$q_i^{(\text{C})} \leq q_i^{(\text{C,ub})} z_i, \quad \forall i \tag{79}$$

$$q_j^{(\text{H})} \leq q_j^{(\text{H,ub})} z_j, \quad \forall j \tag{80}$$

$$W_{s,k}^{(\text{comp})} \leq W_{s,k}^{(\text{comp,ub})} z_{s,k}^{(\text{comp})}, \quad \forall s, \forall k \tag{81}$$

$$W_{s,k}^{(\text{exp})} \leq W_{s,k}^{(\text{exp,ub})} z_{s,k}^{(\text{exp})}, \quad \forall s, \forall k \tag{82}$$

$$f_{i,j}^{(\text{ehs})} \leq f_{i,j}^{(\text{ehs,ub})} z_{i,j}, \quad \forall i, \forall j \tag{83}$$

$$f_{i,j}^{(\text{ecs})} \leq f_{i,j}^{(\text{ecs,ub})} z_{i,j}, \quad \forall i, \forall j \tag{84}$$

In addition, inequality constraints (85)–(88) are related to the inlet and outlet temperatures of the hot and cold streams in the heat exchangers, coolers and heaters, respectively.

$$T_{i,j}^{(\text{ehs,in})} \geq T_{i,j}^{(\text{ehs,out})}, \quad \forall i, \forall j \tag{85}$$

$$T_{i,j}^{(\text{ecs,in})} \leq T_{i,j}^{(\text{ecs,out})}, \quad \forall i, \forall j \tag{86}$$

$$T_i^{(\text{m,out})} \geq T_i^{(\text{out})}, \quad \forall i \tag{87}$$

$$T_j^{(\text{m,out})} \leq T_j^{(\text{out})}, \quad \forall j \tag{88}$$

3.5. Objective Function

The objective function is given by Equation (89) for the combined NLP model ($M_1 + M_2$), which includes the operational cost for energy (electricity and hot and cold utilities).

$$\min Z_1 = \sum_{\substack{s,k \\ \text{CINT}=1}} W_{s,k}^{(\text{comp})} C_{\text{el}}^{(\text{cons})} - \sum_{\substack{s,k \\ \text{EINT}=1}} W_{s,k}^{(\text{exp})} C_{\text{el}}^{(\text{prod})} + q_{\text{HU}} C_{\text{HU}} + q_{\text{CU}} C_{\text{CU}} \tag{89}$$

The objective function of the MINLP model ($M_1 + M_3$) for a simultaneous WHEN is given by Equation (90), which includes the operational cost for energy (electricity and hot and cold utilities) and the investment cost for the equipment (compressors, turbines and heat exchangers). The investment cost for the valves is considered negligible in this study.

$$\min Z_2 = \sum_{\substack{s,k\\ \text{CINT}=1}} W_{s,k}^{(\text{comp})} C_{\text{el}}^{(\text{cons})} - \sum_{\substack{s,k\\ \text{EINT}=1}} W_{s,k}^{(\text{exp})} C_{\text{el}}^{(\text{prod})} + \sum_i q_i^{(\text{C})} C_{\text{CU}} + \sum_j q_j^{(\text{H})} C_{\text{HU}}$$
$$+ af\left(\text{FBM}_{\text{comp}} \sum_{\substack{s,k\\ \text{CINT}=1}} \left(az_{\text{comp}} + b\left(W_{s,k}^{(\text{comp})}\right)^n\right) + \text{FBM}_{\text{exp}} \sum_{\substack{s,k\\ \text{EINT}=1}} \left(az_{\text{exp}} + b\left(W_{s,k}^{(\text{exp})}\right)^n\right)\right.$$
$$\left. + \text{FBM}_{\text{exc}} \sum_{i,j} \left(az_{i,j} + bA_{i,j}^n\right) + \text{FBM}_{\text{exc}} \sum_i \left(az_i + bA_i^n\right) + \text{FBM}_{\text{exc}} \sum_j \left(az_j + bA_j^n\right)\right) \quad (90)$$

where the heat transfer areas for the heat exchangers, coolers and heaters are given as follows, using Chen's approximation of the logarithmic mean temperature difference [36,37]:

$$A_{i,j} = \frac{q_{i,j}\left(\frac{1}{h_i} + \frac{1}{h_j}\right)}{\left(\Delta Th_{i,j} \Delta Tc_{i,j} \frac{\Delta Th_{i,j} + \Delta Tc_{i,j}}{2}\right)^{1/3}}$$

$$A_i = \frac{q_i^{(\text{C})}\left(\frac{1}{h_i} + \frac{1}{h_{\text{CU}}}\right)}{\left(\Delta Th_i^{(\text{C})} \Delta Tc_i^{(\text{C})} \frac{\Delta Th_i^{(\text{C})} + \Delta Tc_i^{(\text{C})}}{2}\right)^{1/3}}$$

$$A_j = \frac{q_j^{(H)}\left(\frac{1}{h_j} + \frac{1}{h_{\text{HU}}}\right)}{\left(\Delta Th_j^{(\text{H})} \Delta Tc_j^{(\text{H})} \frac{\Delta Th_j^{(\text{H})} + \Delta Tc_j^{(\text{H})}}{2}\right)^{1/3}}$$

The investment annualization factor [38], for a given fractional interest rate (ir) and equipment depreciation period (dp), is obtained from

$$af = \frac{ir(1+ir)^{dp}}{(1+ir)^{dp} - 1}$$

3.6. Model Limitations and Possible Improvements

The following limitations in the proposed model should be addressed in the future development of the model.

- The model only includes pressure manipulation equipment for gaseous streams; to include liquid streams, additional pressure change branches should be included to enable the pumping of fluids and liquid expansion.
- Only one hot and one cold utility is considered, suggesting that the model's extension should integrate the WHEN network with the utility network, enabling the extraction of utilities at different temperature levels.
- Only utility expanders and compressors are considered; there is a need to include SSTCs to enable additional work integration opportunities.
- Realistic efficiencies of compressors and expanders should be considered.
- The phase change (evaporation/condensation) should be considered.

3.7. Solution Approach

The synthesis of WHENs is a non-convex optimization problem that is very difficult to solve as the scale of the problem increases. In this work, two relatively complex superstructures, a WHI superstructure and a HEN superstructure, are entangled. Models related to these superstructures are nonconvex and nonlinear, with a nonlinear objective function including capital investment for the equipment. Thus, finding global or even good local solutions can be quite challenging. To solve the optimization problem, a two-step solution strategy is employed, shown in Figure 5. In the first step, the combined NLP model ($M_1 + M_2$) is solved with the objective function given by Equation (89). The solution from the first step provides efficient initialization for the second MINLP model. In addition, the NLP model identifies hot and cold streams, reducing the number of heat exchanger matches (see Appendix A.2) in the second optimization step. In the second step of the proposed solution strategy, a combined MINLP model ($M_1 + M_3$) is solved with the objective function given by Equation (90). Models $M_1 + M_2$ and $M_1 + M_3$ are connected using equations to identify streams within the WHI superstructure presented in Appendix A.1. The model is developed in the GAMS [39] and solved with Conopt4/Ipopt/Snopt as the NLP solvers for the first NLP model and SBB as the MINLP solver with Conopt4 as a subsolver. The model can be solved in iterations (optional) or multiple runs by assigning new values for HRAT in each run, providing different initial solutions for the MINLP model and generating a set of local solutions from which the best one can be selected.

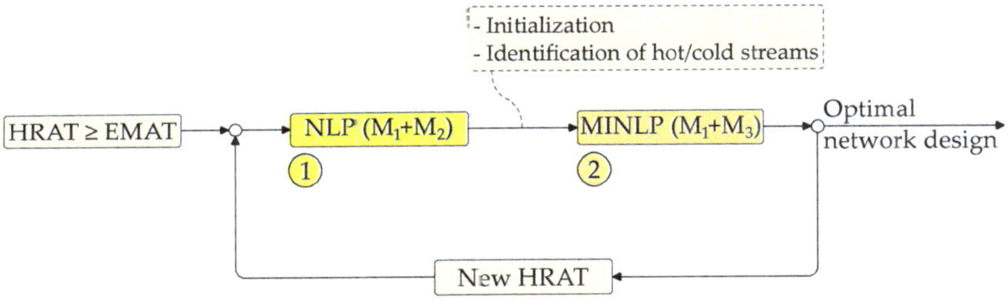

Figure 5. Strategy for solving the optimization problem.

4. Examples

This section provides four examples to demonstrate the methodology and analyze the impacts of the electricity cost, equipment cost and annualized investment cost on the optimal network design. Table 1 gives the parameters and utility cost data for the studied examples, and Table 2 shows the equipment cost data. The cost data from the literature are normalized to the same year (2022) using the Chemical Engineering Plant Cost Index (CEPCI), and the data are fitted to the same model equation $a + bS^n$, where a, b and n are coefficients, and S is the equipment attribute. The nonlinear regression software Sigma Plot v 14.0 [40] is used for data fitting.

Table 1. Parameters and cost data for the studied examples.

Parameter	Unit	Examples 1–3	Example 4
Heating utility cost	k$/kWy	0.377	0.337
Cooling utility cost	k$/kWy	0.1	0.1
Electricity cost (consumed)	k$/KWy	0.45505	0.45505
Electricity cost (produced)	k$/KWy	0.45505	0.36403
Joule–Thompson coefficient of expansion	K/MPa	1.961	-
Isentropic efficiency	-	1	1

Table 2. Equipment cost data in k$ adjusted with CEPCI = 819 for 2022.

Reference	*S* Range	Equipment	*a*	*b*	*n*	R^2
Seider et al. [41]	149–22,371	COMP	-	4.0422	0.8	1
	15–3728	EXP	-	1.1097	0.81	1
	4–1115	STHE	23.0784	0.1165	1	0.9989
			20.7139	0.208	0.9166	1
Peters et al. [42]	75–3000	COMP	-	1.8337	0.9435	1
	100–3000	EXP	-	6.9313	0.5889	1
	10–1000	STHE	12.342	0.17	1	0.9991
			9.0307	0.2889	0.9239	0.9997
Couper et al. [43]	149–22,371	COMP	-	19.235	0.62	1
	15–3728	EXP	-	0.9731	0.81	1
	14–1115	STHE	7.0232	0.2479	1	0.9996
			9.5624	0.1785	1.0469	0.9999
Woods [44]	2–4000	COMP	-	2.198	0.9	1
	746–22,371	EXP	-	1.7929	0.8	1
	20–1150	STHE	-	2.8628	0.65	1
Towler and Sinnott [45]	75–30,000	COMP	888.122	30.625	0.6	1
	1000–15,000	EXP *	1026.8	0.1968	1	1
	10–1000	STHE	49	0.1072	1.2	1
Turton [46]	450–3000	COMP	-	4.2380	0.71	0.9993
	100–3000	EXP	−2140.823	1641.81	0.0712	0.9983
	20–1000	STHE	34.1951	0.0876	1.1532	0.9999

* Shah et al. [47]. For compressors and expanders, the equipment attribute is in kW; for heat exchangers, it is in m^2.

4.1. Example 1

This example considers the integration of compressors into a HEN, optimizing the compressor inlet temperature. Data for this example are taken from the literature [4] and presented in Table 3. Only the S2 stream is pressure-changing, while S1 and S3 are non-pressure-changing streams. All streams are considered an ideal gas with polytropic exponent $k = 1.4$. The EMAT is set to be 20 K. In addition, the annualization factor for the investment is calculated for an interest rate of 8% over a period of 10 years. The example is solved by considering the different investment cost equations given in Table 2. In addition, a basic sensitivity analysis is performed, comparing the network designs for different electricity costs.

Table 3. Stream data for Example 1.

Stream	$T_s^{(in)}$ (K)	$T_s^{(out)}$ (K)	mc_p (kW/K)	$p_s^{(in)}$ (MPa)	$p_s^{(out)}$ (MPa)	h (kW/m^{2}K)
S1	637	333	2	0.1	0.1	0.1
S2	288	523	1	0.1	0.3	0.1
S3	472	543	4	0.1	0.1	0.1
HU	673	673	-	-	-	1
CU	288	288	-	-	-	1

Table 4 shows the optimization results for Example 1 using different equipment cost equations from Table 2. Note that the key performance indicators (utility consumption and OPEX) are the same in all cases, resulting in the same network design. However, the CAPEX is shown to be significantly different for the four cases; thus, the TAC can be overestimated or underestimated, showing the importance of having quality cost data.

Table 4. Network performance indicators for Example 1 with different cost equations.

Indicator	Unit	Seider et al. [41]	Couper et al. [43]	Peters et al. [42]	Towler and Sinnott [45]
No. of HEs	-	5	5	5	5
No. of COMP	-	1	1	1	1
HU demand	kW	235.587	235.587	235.587	235.587
CU demand	kW	135	135	135	135
Compression work	kW	174.41	174.41	174.41	174.41
Total CAPEX	k$	407.928	598.092	363.073	1908.525
CAPEX	k$/y	60.685	89.133	54.109	284.427
OPEX	k$/y	181.683	181.683	181.683	181.683
TAC	k$/y	242.368	270.816	235.792	466.110

References point to the cost equations in Table 2.

Figure 6 shows the optimal network design for Example 1. The S1 stream is a hot stream cooled at a constant pressure in two heat exchangers (with heat loads 360 kW and 185 kW) and one cooler (135 kW) to obtain the stream target temperature. The S2 stream is not compressed at the lowest temperature but at the pinch temperature. Thus, the S2 stream changes its identity; stream S2 is cold before and hot after the compression. The compression heat is used to preheat the cold S stream. The S3 stream is a cold stream integrated within two parallel heat exchangers (with heating loads of 360 kW and 124.41 kW); additionally, a heater is required with a heat load of 235.59 kW to meet the stream's target temperature.

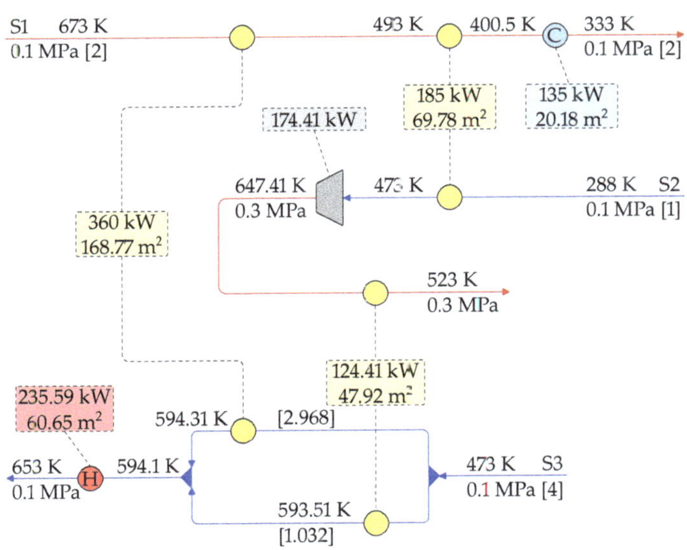

Figure 6. Optimal network design for Example 1.

Table 5 compares the results of two case studies presented in the literature by Fu and Gundersen [4] with our approach. Case A is a network design where compression starts at ambient temperature, and Case B is compression at the pinch, as suggested by the proposed pinch methodology in their work, minimizing the exergy consumption of the system. The network design in this work is the same as in Case B, with very small, negligible differences in the network design parameters, possibly related to rounding errors.

Table 5. Network performance indicators for Example 1.

Indicator	Unit	Fu and Gundersen [4] Case A	Fu and Gundersen [4] Case B	This Work
HU demand	kW	410	235.5	235.59
CU demand	kW	241.3	135.0	135.0
Compression work	kW	106.3	174.5	174.41
Exergy consumption	kW	340.8	309.2	309.2
Total CAPEX	k$	1742.0	1908.73	1908.52
CAPEX	k$/y	259.610	284.457	284.427
OPEX	k$/y	227.073	181.691	181.683
TAC	k$/y	486.682	466.148	466.110

In addition, a basic sensitivity analysis is performed for a value that is ±50% of the base value for electricity cost (0.45505 k$/kWy). The optimization results are presented in Table 6. For an electricity cost that is −50% of the base value, the optimal network design is the same as for the base electricity cost. However, for an electricity cost of +50% of the base electricity cost, the optimal network design (Figure 7a) exhibits reduced work consumption (106.2 kW compared to 174.41 kW for the base cost design) due to the compression of stream S2 at ambient temperature. This reduces the heat recovery as the stream leaving the compressor now acts as a cold stream requiring heating; thus, increased utility consumption is needed. Compared to the solution obtained by Fu and Gundersen [4] for the case of ambient compression (see Figure 7b), the optimum network design exhibits the same exergy consumption (340.8 kW) but a slightly lower TAC (501.723 k$/y compared to 506.375 k$/y) due to the reduced number of heat exchangers and thus the HEN investment cost.

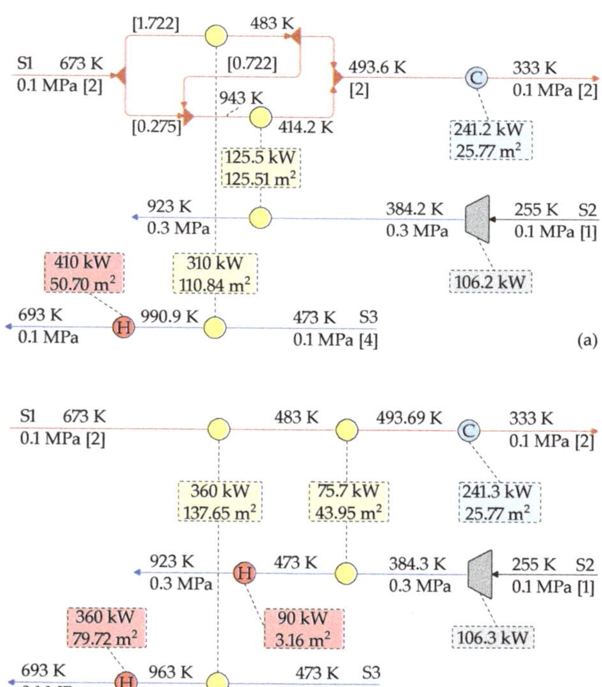

Figure 7. Optimal network design for Example 1 considering electricity cost of 0.6826 $/kWy. (**a**) Optimum design in this paper, (**b**) optimum design from Fu and Gundersen [4] Case A.

Table 6. Sensitivity analysis for Example 1 regarding electricity cost.

Indicator	Unit	Electricity Cost (k$/kWy)		
		0.2275	0.45505	0.6826
No. of HEs	-	5	5	4
HU demand	kW	235.59	235.59	410
CU demand	kW	135.0	135.0	241.2
Compression work	kW	174.41	174.41	106.2
Exergy consumption	kW	309.2	309.2	340.74
Total CAPEX	k$	1908.52	1908.52	1681.17
CAPEX	k$/y	284.427	284.427	250.543
OPEX	k$/y	141.995	181.683	251.18
TAC	k$/y	426.422	466.110	501.723

4.2. Example 2

This example considers the integration of expanders into the HEN, optimizing the expander inlet temperature. Data for this example are taken from the literature [5] and presented in Table 7. Only the S1 stream is pressure-changing, while S2, S3 and S4 are non-pressure-changing hot and cold streams. All streams are considered ideal gases with polytropic exponent $k = 1.4$. The EMAT is set to be 20 K. In addition, the investment annualization factor is varied from 0.08 to 1 to analyze its impact on the optimal network design. Equipment cost data are based on the equation from Couper et al. [43] presented in Table 2.

Table 7. Stream data for Example 2.

Stream	$T_s^{(in)}$ (K)	$T_s^{(out)}$ (K)	mc_p (kW/K)	$p_s^{(in)}$ (MPa)	$p_s^{(out)}$ (MPa)	h (kW/m^{2}K)
S1	673	333	3	0.3	0.1	0.1
S2	603	353	9	0.1	0.1	0.1
S3	288	493	6	0.1	0.1	0.1
S4	413	653	8	0.1	0.1	0.1
HU	673	673	-	-	-	1
CU	288	288	-	-	-	1

Table 8 shows the optimization results for the optimal designs with different annualization factor values. Lower values of the annualization factor (e.g., 0.08) give designs with higher total investment costs for the expanders and compressors. However, the annualized investment cost is lower because the investment is distributed over a longer depreciation time. In addition, the hot and cold utility consumption is minimal compared to other solutions. The annualization factor is a trade-off parameter between the operational and investment costs; increasing the annualization factor places more weight on the investment, and, thus, the total investment decreases while the annualized investment increases. This reduces the work produced in the expander and increases the hot and cold utility consumption. Thus, for a value of $af = 1$, we have the lowest total investment cost, corresponding to the minimum work produced and the highest utility consumption. However, because the total investment is the lowest, we have a simpler network design with fewer heat exchangers.

Figure 8 shows the optimal network design with the annualization factor $af = 0.1$. The S1 stream is a hot stream cooled to almost 498 K in a series of two heat exchangers with heat loads of 210 kW and 313.59 kW, respectively, and then expanded from 0.3 MPa to 0.1 MPa, producing 402.86 kW of work. An additional heat exchanger with a load of 26.41 kW and a cooler with a load of 67.14 kW are placed after the expander to reach the target temperature of 333 K. The S2 stream is cooled in a series of four heat exchangers (with a total heat load of 2250 kW). The S3 stream is completely integrated within a series of four heat exchangers (with a total heat load of 1230 kW), without external heating or cooling, while the S4 cold stream requires a series of three heat exchangers (with a total heat load of 1570 kW) and an additional heater with a heat load of 350 kW. The optimal network design consists of a total of seven heat exchangers (heat recovery 2800 kW), one heater (350 kW) and one cooler (67.14 kW). Figure 9 shows the optimal network design for a case with an annualization factor of 0.2. The optimal network design now has fewer heat exchangers (7 vs. 9), less work produced and more utilities consumed, for the reasons previously explained. Figure 10 shows an even simpler optimal network design for an annualization factor $af = 1$, with the additional expense in the OPEX and significantly increased utility consumption, where the income from expansion no longer covers the hot and cold utility costs, as indicated by the positive OPEX value in the last column of Table 8.

Table 8. Optimal network performance indicators for Example 2. A comparison regarding different annualization factors.

Indicator	Unit	Annualization Factor					
		0.08	0.1	0.2	0.3	0.5	1
No. of HE		9	9	7	7	6	6
HU demand	kW	350	350	350	366.55	401.46	491.70
CU demand	kW	67.07	67.14	81.44	98.99	153.09	243.33
Expansion work	kW	402.93	402.86	388.56	387.55	368.37	368.37
Total CAPEX	k$	644.88	643.857	598.897	567.351	493.432	435.125
CAPEX	k$/y	51.59	64.386	119.779	170.205	246.716	435.125
OPEX	k$/y	−44.696	−44.658	−36.721	−28.268	−0.965	42.079
TAC	k$/y	6.894	19.727	83.058	141.937	245.751	477.204

Table 9 compares the results with those from the literature [5]. The previous authors presented an exergy-pinch-based methodology to find the optimum network design with the minimum exergy consumed by the system. Different cases, A, B and C, correspond to different network designs based on the proposed set of rules for the expansion of streams at different temperatures. The cost for the network designs has been recalculated based on their optimal design. Network design C (Example 5, Case C in their work [5]) has the lowest exergy consumption (but the highest exergy production) but, interestingly, also the lowest TAC of the three designs. However, because the authors did not minimize the TAC of the network, our design exhibits a lower TAC, with fewer heat exchangers and only one expander, but with increased exergy consumption (decreased exergy production).

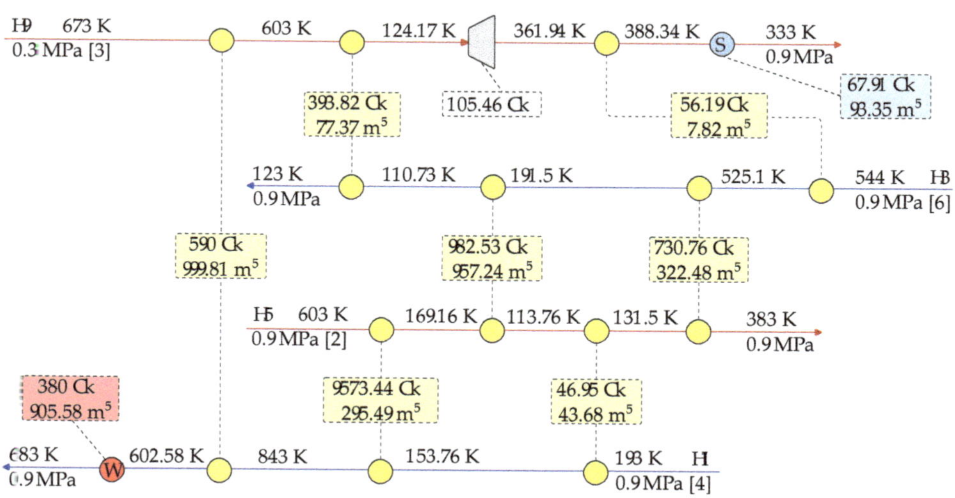

Figure 8. Optimal network design for Example 2 ($af = 0.1$).

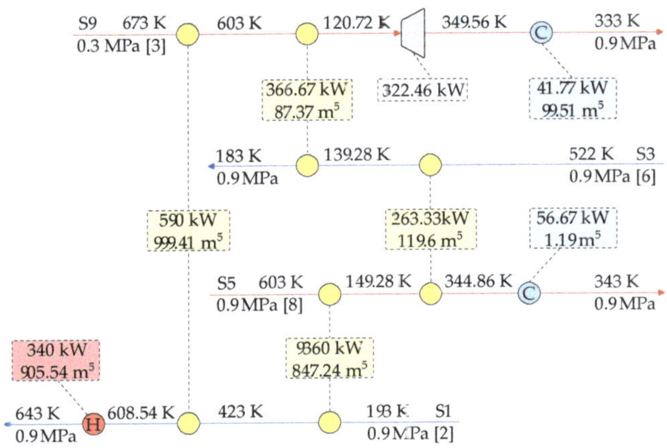

Figure 9. Optimal network design for Example 2 ($af = 0.2$).

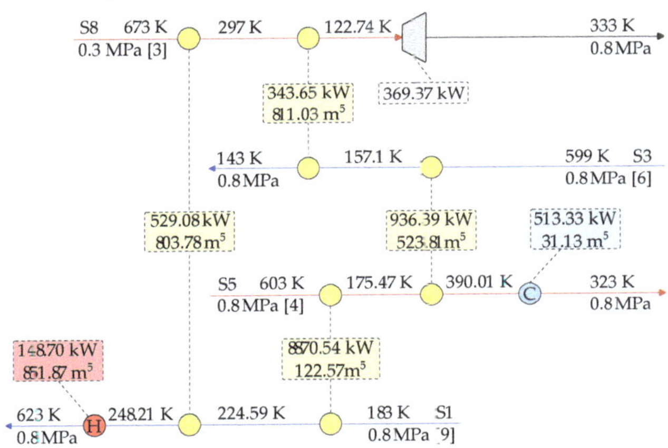

Figure 10. Optimal network design for Example 2 ($af = 1$).

Table 9. Network performance indicators and results comparison for Example 2 ($af = 0.1$).

Indicator	Unit	Fu and Gundersen [5] Case A	Fu and Gundersen [5] Case B	Fu and Gundersen [5] Case C	This Work
No. of HEs	-	7	8	9	7
No. of EXP	-	1	1	2	1
HU demand	kW	694	637.5	350	350
CU demand	kW	270	270	63.5	67.14
Expansion work	kW	543.9	487.5	406.6	402.86
Exergy consumption	kW	−147.0	−122.9	−206.4	−188.34
Total CAPEX	k$	651.771	599.452	772.952	643.857
CAPEX	k$/y	65.177	59.945	77.295	64.386
OPEX	k$/y	41.131	45.496	−46.727	−44.658
TAC	k$/y	106.308	105.441	30.568	19.727

4.3. Example 3

In this example, compression and expansion are considered simultaneously. The stream data for Example 3 are given in Table 10, provided from the literature [48], with the equipment cost data given in Table 2, using the cost equations from Couper et al. [43] and Towler and Sinnott [45]. In addition, we consider the direct and indirect installed costs for the equipment accounted for by bare module factors $FBM_{exc} = 3.29$, $FBM_{comp} = 2.8$ and

$FBM_{exp} = 3.5$ from the literature [46]. The annualization factor for the investment is 0.18. The isentropic efficiency of the compressors and expanders is 1, and The EMAT is 20 K.

Table 10. Stream data for Example 3.

Stream	$T_s^{(in)}$ (K)	$T_s^{(out)}$ (K)	mc_p (kW/K)	$p_s^{(in)}$ (MPa)	$p_s^{(out)}$ (MPa)	h (kW/m²K)
S1	673	308	2	0.2	0.1	0.1
S2	593	433	4	0.1	0.1	0.1
S3	383	308	3	0.1	0.1	0.1
S4	288	653	3	0.1	0.2	0.1
S5	463	523	10	0.1	0.1	0.1
HU	673	673	-	-	-	1
CU	288	288	-	-	-	1

Figure 11 shows the optimal network design with the minimum TAC for the cost equation from Couper et al. [43]. The optimal design exhibits a rather complicated network including five heat exchangers, one heater and two coolers, with hot and cold utility consumption of 350.61 kW and 304.92 kW, respectively. The expansion of stream S1 starts at 375.46 K (the work produced is 134.91 kW) with the outlet stream at the required target temperature. The compression of stream S4 starts at ambient temperature, 288 K, with the work consumed being 189.23 kW. Interestingly, stream S3 is not integrated because it is the hot stream at the lowest temperature and requires cooling of 225 kW. The OPEX of the network is 187.387 k\$/y with an annualized investment of 409.508 k\$/y. The TAC of the network is 596.895 k\$/y.

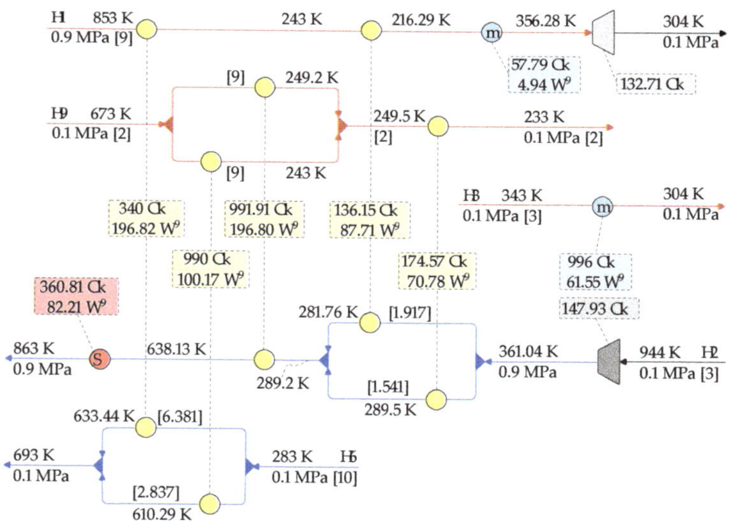

Figure 11. Optimal network design for Example 3 (cost equation from Couper et al. [43] with $af = 0.18$).

Example 3 is also solved using a different cost equation from Towler and Sinnott [45] in Table 2, showing that the selection of the cost equation does affect the optimal design, in contrast to the solutions obtained for Example 1, where all cost equations considered gave the same optimal network design. Figure 12 shows the optimal network design in this case. This design is much simpler, including three heat exchangers, one heater and two coolers, with hot and cold utility consumption of 485.77 kW and 401.44 kW, respectively. The expansion of stream S1 now starts at a higher temperature of 483 K (the work produced is 173.55 kW vs. 134.9 kW), with the outlet stream requiring additional cooling (176.44 kW) to achieve the target temperature. The compression of stream S4 also starts at ambient temperature, 288 K, for a total amount of work consumed of 189.23 kW (the same as in the design shown in Figure 11). Stream S3 is not integrated. Table 11 compares the results with the literature, where the authors [48] presented three cases of optimal designs using a systematic approach based on thermodynamic insights to minimize the exergy consumption of the system. The cost for their designs was recalculated using the cost data provided for this example.

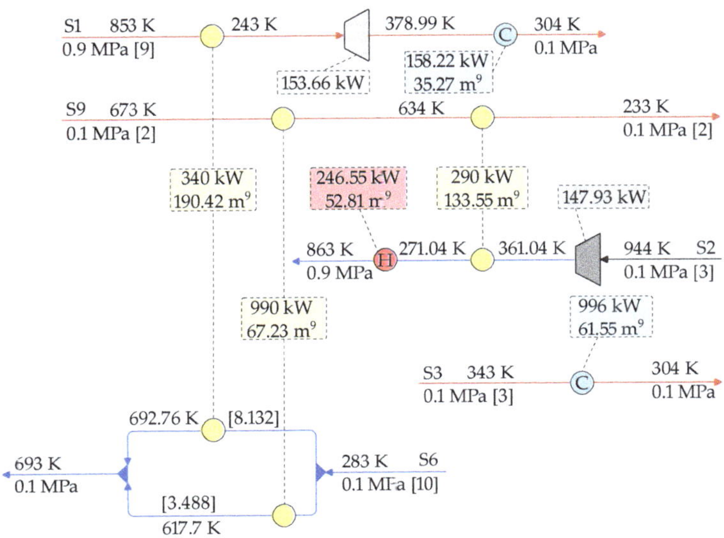

Figure 12. Optimal network design for Example 3 (cost equation from Towler and Sinnott [45] with $af = 0.18$).

Table 11. Network performance indicators for Example 3 (cost equation from Towler and Sinnott [45]).

Indicator	Unit	Fu and Gundersen [48] Case I	Fu and Gundersen [48] Case II	Fu and Gundersen [48] Case III	This Work
No. of HEs	-	8	8	11	6
No. of EXP	-	1	1	2	1
No. of COMP	-	1	1	2	1
Hot utility demand	kW	591.8	119.4	37.6	485.71
Cold utility demand	kW	439.4	150	91.7	225
Compression work	kW	189.3	304.2	312.4	189.23
Expansion work	kW	241.8	173.6	158.3	173.55
Exergy consumption	kW	286.0	198.9	175.6	293.6
Total CAPEX	k$	10,037.51	10,824.71	18,246.25	9574.04
CAPEX	k$/y	1806.75	1948.45	3284.32	1723.33
OPEX	k$/y	243.120	119.445	93.45	230.392
TAC	k$/y	2049.87	2067.89	3377.794	1953.742

In addition, we use the cost equation from Towler and Sinnott [45], with the annualization factor now having a value $af = 0.08$. This places more weight on the operational expenditures, and, as a result, we have an entirely different network design with a reduced OPEX compared to the solution with $af = 0.18$. The optimal network design shown in Figure 13 exhibits cold utility consumption of only 96.02 kW, without using hot utilities. This results in a network design requiring much more compression power (351.96 kW vs. 189.23 kW). Expansion starts at a lower temperature; thus, the work produced is reduced (155.94 kW vs. 177.55 kW). Notice that the S3 stream is fully integrated and does not require cold utilities. The TAC of this optimal design is 973.29 k$/y, compared to the best design from Fu and Gundersen [48] (Case II recalculated for $af = 0.08$), with a

TAC 985.42 k\$/y. The difference in cost is about 1.2%. The reason for the much smaller difference in cost is that the designs from Fu and Gundersen [48] favor the cases in which less weight is placed on the investment cost. In these cases, their methodology gives very good results (cost-wise), even if the objective is to minimize the exergy consumption.

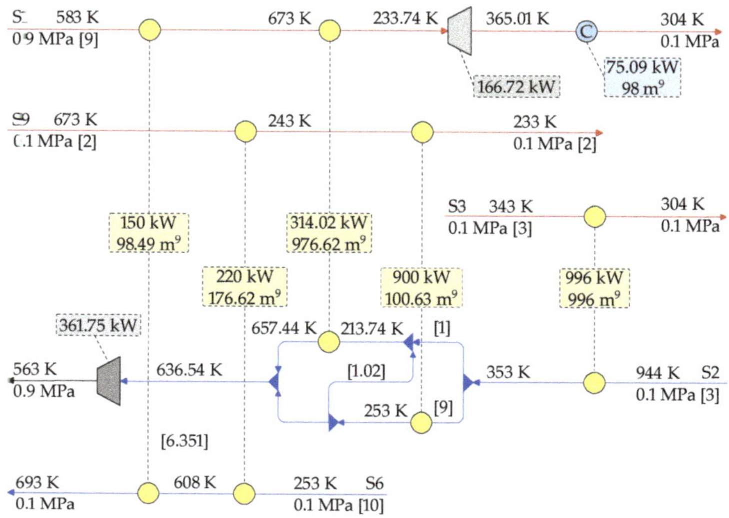

Figure 13. Optimal network design for Example 3 (cost equation from Towler and Sinnott [45] with $af = 0.08$).

4.4. Example 4

The previous example considered both compression and expansion simultaneously. However, only one stream is expanded, and one stream is compressed. Thus, Yu et al. [34] modified Example 3 to create a problem wherein all streams are subject to a pressure change. In addition, the supply and target temperatures are the same for all streams, so it is unclear which streams should be hot or cold before optimization. The proposed superstructure in this work enables the handling of unclassified streams; each can be hot or cold before and after the pressure change stages. The stream data for this example are given in Table 12. In addition, to compare the results, the cost data given in Table 13 are provided from the literature [27]. The original cost equation for the heat exchangers in the reference is $93,500.12 + 602.96A + 0.149A^2$ and is converted to the standardized model $a + bA^n$ used in this paper. The R^2 of the fit is 0.9999, making the error negligible compared to the original equation. The annualization factor for the capital investment is 0.18. The isentropic efficiency of the compressors and expanders is 1. The exchanger's minimum approach temperature is 20 K.

Table 12. Stream data for Example 4.

Stream	$T_s^{(in)}$ (K)	$T_s^{(out)}$ (K)	mc_p (kW/K)	$p_s^{(in)}$ (MPa)	$p_s^{(out)}$ (MPa)	h (kW/m²K)
S1	623	623	2	0.2	0.1	0.1
S2	593	593	4	0.2	0.1	0.1
S3	383	383	3	0.2	0.1	0.1
S4	323	323	3	0.1	0.2	0.1
S5	463	463	10	0.1	0.2	0.1
HU	673	673	-	-	-	1
CU	288	288	-	-	-	1

Table 13. Cost data for equipment and utilities for Example 4.

Reference	Equipment	*a*	*b*	*n*
Lin et al. [27]	COMP	-	51,104.85	0.62
	EXP	-	2585.47	0.81
	STHE	99,163	250.9	1.1577
Cost Parameter	$C_{el}^{(cons)}$	$C_{el}^{(prod)}$	C_{HU}	C_{CU}
$/kWy	455.05	364.03	337	100

For compressors and expanders, the equipment attribute is in kW; for heat exchangers, it is in m^2. The equipment cost is in $.

Figure 14 shows the optimal network design obtained using the proposed superstructure and optimization model. The S1 stream's expansion starts at the maximum temperature, producing 223.36 kW of work. The stream leaving the expander is a cold stream requiring heating with hot utilities, with the load equal to the work produced. The S2 stream changes identity, from first being a hot stream cooled before the expansion and then a cold stream after the expansion. The heating of the S2 stream leaving the expander is achieved in two heat exchangers, recovering 209.8 kW of heat in a match with the hot S5 stream and 467.66 kW with the hot S2 stream (self-recuperation). Additional heating (132.34 kW) is required to achieve the stream's target temperature. Stream S3 is expanded at the stream's initial temperature, producing 206.43 kW of work, and the heating of the stream after the expansion is done by heat recovery from the hot stream S5. Because the stream supply and target temperatures are the same, the heat required for heating (206.43 kW) is equal to the work produced by the expander. The S4 stream is not integrated with other streams.

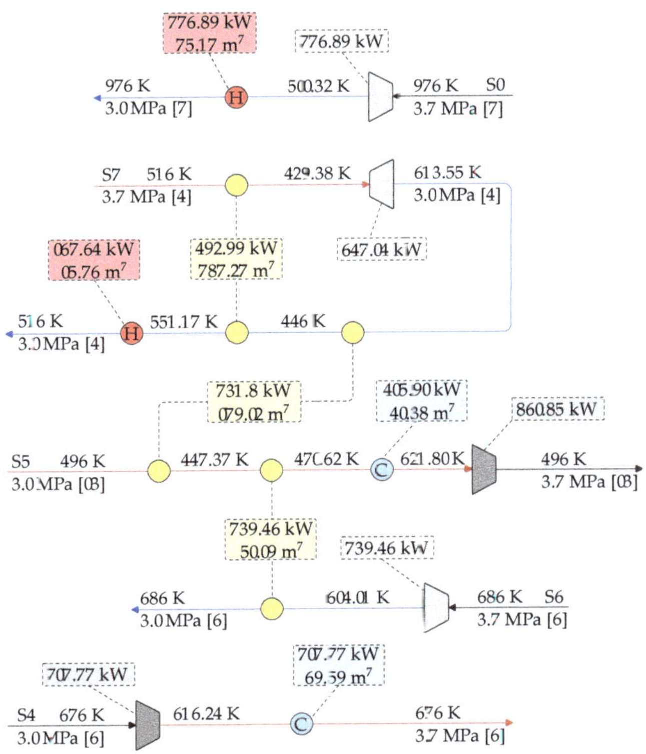

Figure 14. Optimal network design for Example 4.

The compression of the stream is performed at the stream's initial temperature, followed by cooling with cold utilities (212.22 kW). The S5 stream is identified as a hot stream being cooled in a series of two heat exchangers and a cooler to reduce the work consumption in the compressor (831.85 kW). The stream leaves the compressor exactly at the target temperature. Table 14 shows the key network performance indicators and network costs for Example 4. In addition, a comparison of the results with those in the literature is given. Lin et al. [27] presented a global optimization approach for a minimal WHEN network where the heat balance was based on enthalpy calculations using equations of state. The network performance indicators from Lin et al. [27] were thus recalculated using parameters from the optimal design and assuming a constant heat capacity flow rate. The optimal network design obtained by Lin et al. [27] is shown in Figure 15. The TAC for their network using the recalculated values (1549.028 $/y) given in Table 14 is only slightly different compared to the optimal value in their approach of 1544.845 $/y. The network design shown in Figure 15 consists of only one heat exchanger, with a heat recovery load of 206.22 kW. This increases the hot and cold utility consumption compared to the design reported in our work. The heat content of S2 is used to produce work and not for heat recovery; thus, more work is produced. Our design exhibits a higher CAPEX due to the higher HEN investment cost, but the OPEX is reduced compared to the work of Lin et al. [27].

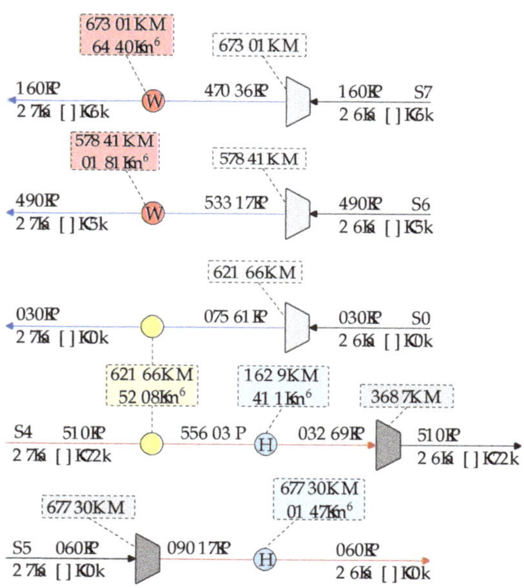

Figure 15. Optimal network design for Example 4 from the literature [27].

Table 14. Network performance indicators for Example 4.

Indicator	Unit	Yu et al. [34]	Lin et al. [27]	This Work
Number of COMP and EXP	-	7	5	5
Number of HE units	-	13	5	10
Hot utility demand	kW	40	635.92	356.2
Cold utility demand	kW	342.7	832.73	627.83
Compression work	kW	1127.3	1038.93	1044.07
Expansion work	kW	827.4	842.14	772.43
Total CAPEX	k$	9719.198	6029.103	6421.778
CAPEX	k$/y	1749.455	1085.239	1155.92
OPEX	k$/y	259.528	463.79	376.749
TAC	k$/y	2008.983	1549.028	1532.658

However, although the proposed approach used in Lin et al. [27] gives a global solution to their minimal WHEN superstructure, it is still not the global solution to the problem, since their superstructure may not include all possible options for work and heat integration. Using our approach, a set of local solutions can be obtained, as shown in Figure 16, and some of the solutions are practically the same as the solution obtained by Lin et al. [27]. It should also be mentioned that the solution from Yu et al. [34] in Table 14 is based on exergy rather than economy considerations.

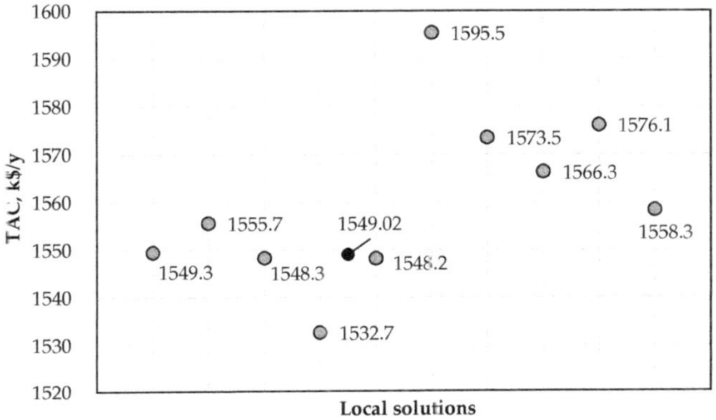

Figure 16. Local optimization solutions for Example 4.

4.5. Model Statistics

The model is solved on a computer with a 2.8 GHz processor with 4 cores and 16 GB of RAM. Table 15 shows the basic model statistics, including the model sizes and computational times for both steps of the proposed solution strategy. Statistics are presented for several cases to give a basic understanding of the model sizes and computational effort. The statistics for Example 1 are given for the model with the cost equation from Seider et al. [41] and those for Example 3 for the model with the cost equation from Towler and Sinnott [45]. The model statistics for Example 2 are given for an annualization factor of 0.1.

Table 15. Model statistics for studied examples.

	Example 1	Example 2	Example 3	Example 4
Step 1—NLP				
No. of equations	120	157	210	229
No. of continuous variables	114	153	202	221
CPU time, s	3.8	24	0.1	37.9
Step 2—MINLP				
No. of equations	353	486	773	713
No. of continuous variables	323	466	866	744
No. of discrete variables	23	31	50	45
CPU time, s	3.3	26.9	546	221

5. Conclusions

This work presents a mathematical programming approach for the simultaneous synthesis of work and heat exchange networks (WHENs). A new superstructure for work and heat integration is developed based on feasible thermodynamic paths, enabling the heating, cooling and pressure changes of defined process streams. An additional superstructure enables the HEN design of the overall problem. The proposed approach solves the optimization problem in two steps. The first step identifies hot and cold streams within the network and provides an initial solution for the second step of the proposed approach. In the second step, the WHEN is synthesized, minimizing the TAC of the network. The TAC of the network includes the OPEX (heating, cooling and electricity) and CAPEX (expanders, compressors and heat exchangers). To validate the methodology, four examples are studied, including additional analyses of the effects of the electricity cost, equipment cost and annualization factor on the network design. By varying the equipment depreciation periods, one can obtain and compare different designs concerning the network complexity, utility consumption and required CAPEX and OPEX. One important limitation of the methodology presented in this paper is the assumption that all streams behave as ideal gases. Both real gases and liquids should have been considered, adding pumps as new pressure change equipment. Another limitation is that the two-step approach cannot guarantee the global optimality of the networks developed. However, as argued in the paper, a simultaneous (one-step) approach with such a rich superstructure would be prohibitive from a computational point of view. Future work should consider handling both liquids and real gases, including phase change operations. Additional equipment should be added to the superstructure, such as pumps (to handle liquid streams) and single-shaft turbine compressors (SSTCs) to enable additional work integration opportunities. In addition, multiple thermal utilities with corresponding cost models should be considered. Finally, multi-objective optimization should be explored with economic and exergetic objective functions.

Author Contributions: Conceptualization, N.I. and T.G.; methodology, N.I.; software, N.I.; validation, N.I., C.F. and T.G.; formal analysis, N.I.; writing—original draft preparation, N.I., C.F. and T.G.; writing—review and editing, N.I., C.F. and T.G.; visualization, N.I.; supervision, T.G.; project administration, T.G. All authors have read and agreed to the published version of the manuscript.

Funding: This publication is funded by the HighEFF—Centre for an Energy Efficient and Competitive Industry for the Future, an 8-year Research Centre under the FME scheme (Centre for Environment-Friendly Energy Research, 257632).

Institutional Review Board Statement: Not applicable.

Informed Consent Statement: Not applicable.

Data Availability Statement: Data are contained within the article.

Acknowledgments: The authors gratefully acknowledge the financial support of the Research Council of Norway and user partners of HighEFF.

Conflicts of Interest: The authors declare no conflicts of interest. The funders had no role in the design of the study; in the collection, analyses, or interpretation of data; in the writing of the manuscript; or in the decision to publish the results.

Abbreviations

CAPEX	Capital Expenditures
COMP	Compressor
EMAT	Exchanger Minimum Approach Temperature
EXP	Expander
GAMS	General Algebraic Modeling System
GDP	Generalised Disjunctive Programming
HE	Heat Exchanger
HEN	Heat Exchanger Network
HRAT	Heat Recovery Approach Temperature
MINLP	Mixed Integer Nonlinear Programming
MP	Mathematical Programming
NLP	Nonlinear Programming
OPEX	Operational Expenditures
PA	Pinch Analysis
STHE	Shell and Tube Heat Exchangers
TAC	Total Annualized Cost
WHI	Work and Heat Integration
WHEN	Work and Heat Exchange Network

Indices	
i	Hot stream
j	Cold stream
k	Pressure change stage
s	Process stream
Sets	
K	Pressure change stages
S	Process streams
Subscripts/Superscripts	
bypass	Bypass
C	Cooler
comp	Compressor
cons	Consumed
cs	Cold stream/Cold supply
ct	Cold target
CU	Cold utility
ehs	Exchanger's hot stream
ecs	Exchanger's cold stream
exc	Exchanger
exp	Expander
H	Heater
hs	Hot stream/Hot supply
ht	Hot target
HU	Hot utility
IC	Inter-stage cooling
id	Ideal
IH	Inter-stage heating
in	Inlet
m	Mixer
max	Maximum
out	Outlet
prod	Produced
ub	Upper bound
valve	Expansion valve
Parameters	
a	Fixed cost for equipment, $
af	Equipment investment annualization factor, 1/y
b	Variable cost coefficient for equipment, $/attribute
BINT	Binary parameter denoting existence of bypass, [−]
C_{CU}	Cold utility cost, $/kWy
C_{HU}	Hot utility cost, $/kWy
C_{el}	Electricity cost, $/kWy
CINT	Binary parameter denoting existence of compression stages, [−]
dp	Depreciation period for investment, y
EINT	Binary parameter denoting existence of expansion stages, [−]
FBM	Bare module factor, [−]
h	Individual heat transfer coefficient, kW/(m^{2}K)
ir	Interest rate for investment, %/100
n	Cost exponent for equipment
κ	Isentropic expansion/compression coefficient, [−]
μ	Joule–Thompson expansion coefficient, K/MPa
η	Isentropic efficiency of compressor/expander, [−]
$T_s^{(in)}$	Process stream s supply temperature, K
$T_s^{(out)}$	Process stream s target temperature, K
$p_s^{(in)}$	Process stream s supply pressure, MPa
$p_s^{(out)}$	Process stream s target pressure, MPa
Γ	Upper bound for temperature driving force, K

Continuous variables

A	Heat exchange area, m^2
m, f	Heat capacity flow rate, kW/K
T	Temperature, K
p	Pressure, MPa
q	Heat load, kW
ΔTh	Temperature difference at hot end of heat exchanger, K
ΔTc	Temperature difference at cold end of heat exchanger, K
Binary variables	
z	Existence of equipment, -

Appendix A

Appendix A.1. Identification of Streams

The identification and ordering of hot and cold streams has been done according to the proposed WHI superstructure. To facilitate the understanding of the stream numbering, an extended superstructure for two process streams ($|S| = 2$) and three pressure change stages ($|K| = 3$) is shown in Figure A1. The numbering of the streams starts from the hot and cold streams related to the initial stream splitters for hot (HSS_s) and cold streams (CSS_s). The number of related hot/cold streams equals the number of streams s ($i = s$) and the connecting relations between the heat capacity flow rates and inlet/outlet temperatures are given by Equations (A1)–(A6).

$$f_i = m_s^{(hs)}, \quad i = s \tag{A1}$$

$$T_i^{(in)} = T_s^{(in)}, \quad i = s \tag{A2}$$

$$T_i^{(out)} = T_s^{(hs,out)}, \quad i = s \tag{A3}$$

$$f_j = m_s^{(cs)}, \quad j = s \tag{A4}$$

$$T_j^{(in)} = T_s^{(in)}, \quad j = s \tag{A5}$$

$$T_j^{(out)} = T_s^{(cs,out)}, \quad j = s \tag{A6}$$

The numbering of the streams continues with the hot/cold streams related to the hot target (HTM_s) and cold target (CTM_s) stream mixers. The number of related streams equals the number of streams s. However, the numbering starts from $|S|$, as given by Equations (A7)–(A12).

$$f_i = m_s^{(ht)}, \quad i = s + |S| \tag{A7}$$

$$T_i^{(in)} = T_s^{(ht,in)}, \quad i = s + |S| \tag{A8}$$

$$T_i^{(out)} = T_s^{(out)}, \quad i = s + |S| \tag{A9}$$

$$f_j = m_s^{(ct)}, \quad j = s + |S| \tag{A10}$$

$$T_j^{(in)} = T_s^{(ct,in)}, \quad j = s + |S| \tag{A11}$$

$$T_j^{(out)} = T_s^{(out)}, \quad j = s + |S| \tag{A12}$$

Cooling the streams before compression reduces work consumption. At the same time, compression increases the temperature of the stream. As a result, streams leaving the compression stages are defined as hot streams requiring inter-stage cooling. The numbering of hot streams continues from $2|S|$, and the number of hot streams is related to the number of compression stages $|K| - 1$, as described by the connecting Equations (A13)–(A15).

$$f_i = m_s^{(comp)}, \quad i = 2|S| + s(|K| - 1) - (|K| - 1 - k) \land k < |K| \land |K| > 1 \tag{A13}$$

$$T_i^{(in)} = T_{s,k}^{(IC,in)}, \quad i = 2|S| + s(|K| - 1) - (|K| - 1 - k) \land k < |K| \land |K| > 1 \tag{A14}$$

$$T_i^{(out)} = T_{s,k}^{(IC,out)}, \quad i = 2|S| + s(|K|-1) - (|K|-1-k) \wedge k < |K| \wedge |K| > 1 \tag{A15}$$

Heating of the streams before expansion increases the work production. At the same time, expansion decreases the temperature of the stream. As a result, streams leaving the expansion stages are defined as cold streams requiring inter-stage heating. The numbering of the cold streams continues from $2|S|$, and the number of cold streams is related to the number of expansion stages $|K|-1$, as described by the connecting Equations (A16)–(A18).

$$f_j = m_s^{(exp)}, \quad j = 2|S| + s(|K|-1) - (|K|-1-k) \wedge k < |K| \wedge |K| > 1 \tag{A16}$$

$$T_j^{(in)} = T_{s,k}^{(IH,in)}, \quad j = 2|S| + s(|K|-1) - (|K|-1-k) \wedge k < |K| \wedge |K| > 1 \tag{A17}$$

$$T_j^{(out)} = T_{s,k}^{(IH,out)}, \quad j = 2|S| + s(|K|-1) - (|K|-1-k) \wedge k < |K| \wedge |K| > 1 \tag{A18}$$

Appendix A.2. Additional Constraints

To facilitate the solution of the second MINLP model, the binary variables identifying the existence of heat exchangers, heaters and coolers are constrained to values 0 or 1 based on the existence of hot and cold streams within the first NLP model as follows:

$$z_{i,j} \leq \begin{cases} 1 & \text{if } ec_i > 0 \wedge ec_j > 0 \\ 0 & \text{else} \end{cases} \tag{A19}$$

$$z_i \leq \begin{cases} 1 & \text{if } ec_i > 0 \\ 0 & \text{else} \end{cases} \tag{A20}$$

$$z_j \leq \begin{cases} 1 & \text{if } ec_j > 0 \\ 0 & \text{else} \end{cases} \tag{A21}$$

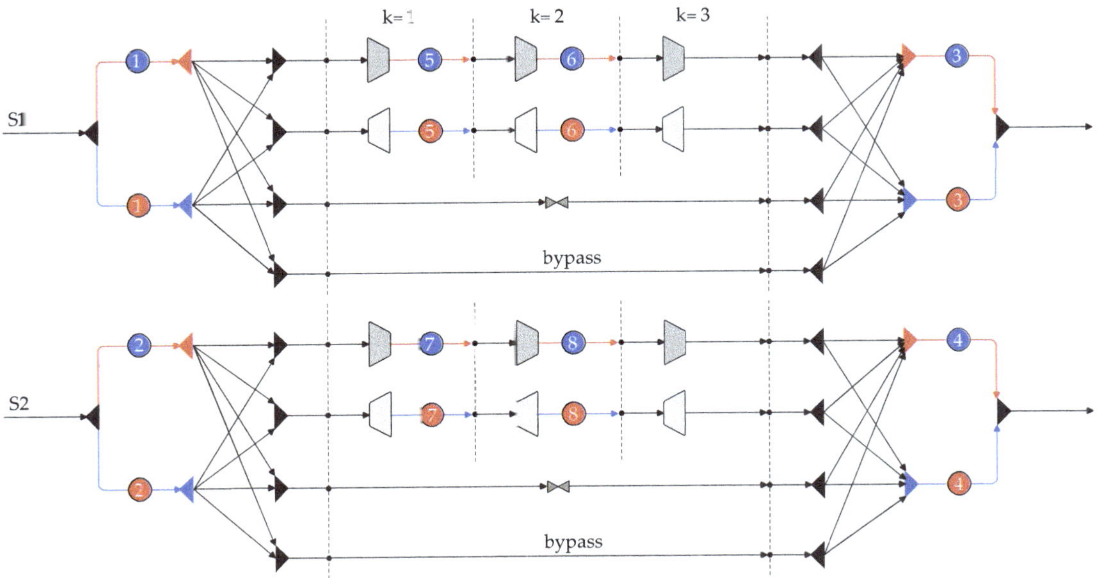

Figure A1. Representative WHI superstructure for two process streams and three pressure change stages.

References

1. Klemeš, J.J.; Kravanja, Z. Forty years of Heat Integration: Pinch Analysis (PA) and Mathematical Programming (MP). *Curr. Opin. Chem. Eng.* **2013**, *2*, 461–474. [CrossRef]
2. Fu, C.; Vikse, M.; Gundersen, T. Work and heat integration: An emerging research area. *Energy* **2018**, *158*, 796–806. [CrossRef]
3. Aspelund, A. Berstad, D.O.; Gundersen, T. An Extended Pinch Analysis and Design procedure utilizing pressure based exergy for subambient cooling. *Appl. Therm. Eng.* **2007**, *27*, 2633–2649. [CrossRef]
4. Fu, C.; Gundersen, T. Integrating compressors into heat exchanger networks above ambient temperature. *AIChE J.* **2015**, *61*, 3770–3785. [CrossRef]
5. Fu, C.; Gundersen, T. Integrating expanders into heat exchanger networks above ambient temperature. *AIChE J.* **2015**, *61*, 3404–3422. [CrossRef]
6. Fu, C.; Wang, X.; Gundersen, T. The importance of thermodynamic insight in Work and Heat Exchange Network Design. *Chem. Eng. Trans.* **2020**, *81*, 127–132. [CrossRef]
7. Wechsung, A.; Aspelund, A.; Gundersen, T.; Barton, P.I. Synthesis of heat exchanger networks at subambient conditions with compression and expansion of process streams. *AIChE J.* **2011**, *57*, 2090–2108. [CrossRef]
8. Huang, K.; Karimi, I.A. Work-heat exchanger network synthesis (WHENS). *Energy* **2016**, *113*, 1006–1017. [CrossRef]
9. Huang, K.F.; Al-mutairi, E.M.; Karimi, I.A. Heat exchanger network synthesis using a stagewise superstructure with non-isothermal mixing. *Chem. Eng. Sci.* **2012**, *73*, 30–43. [CrossRef]
10. Onishi, V.C.; Ravagnani, M.A.; Caballero, J.A. Simultaneous synthesis of work exchange networks with heat integration. *Chem. Eng. Sci.* **2014**, *112*, 87–107. [CrossRef]
11. Yee, T.F.; Grossmann, I.E. Simultaneous optimization models for heat integration—II. Heat exchanger network synthesis. *Comput. Chem. Eng.* **1990**, *14*, 1165–1184. [CrossRef]
12. Onishi, V.C. Ravagnani, M.A.; Caballero, J.A. MINLP Model for the Synthesis of Heat Exchanger Networks with Handling Pressure of Process Streams. *Comput. Aided Chem. Eng.* **2014**, *33*, 163–168. [CrossRef]
13. Pavão, L.V.; Costa, C.B.; Ravagnani, M.A. A new framework for work and heat exchange network synthesis and optimization. *Energy Convers. Manag.* **2019**, *183*, 617–632. [CrossRef]
14. Onishi, V.C.; Ravagnani, M.A.S.S.; Caballero, J.A. Simultaneous synthesis of heat exchanger networks with pressure recovery: Optimal integration between heat and work. *AIChE J.* **2014**, *60*, 893–908. [CrossRef]
15. Pavão, L.V.; Costa, C.B.B.; Ravagnani, M.A.S.S. An Enhanced Stage-wise Superstructure for Heat Exchanger Networks Synthesis with New Options for Heaters and Coolers Placement. *Ind. Eng. Chem. Res.* **2018**, *57*, 2560–2573. [CrossRef]
16. Pavão, L.V.; Caballero, J.A.; Ravagnani, M.A.; Costa, C.B. An extended method for work and heat integration considering practical operating constraints. *Energy Convers. Manag.* **2020**, *206*, 112469. [CrossRef]
17. Nair, S.K.; Nagesh Rao, H.; Karimi, I.A. Framework for work-heat exchange network synthesis (WHENS). *AIChE J.* **2018**, *64*, 2472–2485. [CrossRef]
18. Onishi, V.C.; Quirante, N.; Ravagnani, M.A.; Caballero, J.A. Optimal synthesis of work and heat exchangers networks considering unclassified process streams at sub and above-ambient conditions. *Appl. Energy* **2018**, *224*, 567–581. [CrossRef]
19. Duran, M.A.; Grossmann, I.E. Simultaneous optimization and heat integration of chemical processes. *Aiche J.* **1986**, *32*, 123–138. [CrossRef]
20. Quirante, N.; Caballero, J.A.; Grossmann, I.E. A novel disjunctive model for the simultaneous optimization and heat integration. *Comput. Chem. Eng.* **2017**, *96*, 149–168. [CrossRef]
21. Li, J.; Demirel, S.E.; Hasan, M.M.F. Building Block-Based Synthesis and Intensification of Work-Heat Exchanger Networks (WHENS). *Processes* **2019**, *7*, 23. [CrossRef]
22. Pavão, L.V.; Caballero, J.A.; Ravagnani, M.A.; Costa, C.B. A pinch-based method for defining pressure manipulation routes in work and heat exchange networks. *Renew. Sustain. Energy Rev.* **2020**, *131*, 109989. [CrossRef]
23. Linnhoff, B.; Ahmad, S. Cost optimum heat exchanger networks—1. Minimum energy and capital using simple models for capital cost. *Comput. Chem. Eng.* **1990**, *14*, 729–750. [CrossRef]
24. Yu, H.; Vikse, M.; Anantharaman, R.; Gundersen, T. Model reformulations for Work and Heat Exchange Network (WHEN) synthesis problems. *Comput. Chem. Eng.* **2019**, *125*, 89–97. [CrossRef]
25. Lin, Q.; Liao, Z.; Sun, J.; Jiang, B.; Wang, J.; Yang, Y. Targeting and Design of Work and Heat Exchange Networks. *Ind. Eng. Chem. Res.* **2020**, *59*, 12471–12486. [CrossRef]
26. Lin, Q.; Chang, C.; Liao, Z.; Sun, J.; Jiang, B.; Wang, J.; Yang, Y. Efficient Strategy for the Synthesis of Work and Heat Exchange Networks. *Ind. Eng. Chem. Res.* **2021**, *60*, 1756–1773. [CrossRef]
27. Lin, Q.; Liao, Z.; Bagajewicz, M.J. Globally optimal design of Minimal WHEN systems using enumeration. *AIChE J.* **2023**, *69*, e17878 [CrossRef]
28. Yang, R.; Zhuang, Y.; Zhang, L.; Du, J.; Shen, S. A thermo-economic multi-objective optimization model for simultaneous synthesis of heat exchanger networks including compressors. *Chem. Eng. Res. Des.* **2020**, *153*, 120–135. [CrossRef]
29. Zhuang, Y.; Yang, R.; Zhang, L.; Du, J.; Shen, S. Simultaneous synthesis of sub and above-ambient heat exchanger networks including expansion process based on an enhanced superstructure model. *Chin. J. Chem. Eng.* **2020**, *28*, 1344–1356. [CrossRef]
30. Zhuang, Y.; Xing, Y.; Zhang, L.; Liu, L.; Du, J.; Shen, S. An enhanced superstructure-based model for work-integrated heat exchange network considering inter-stage multiple utilities optimization. *Comput. Chem. Eng.* **2021**, *152*, 107388. [CrossRef]

31. Braccia, L.; Luppi, P.; Vallarella, A.J.; Zumoffen, D. Generalized simultaneous optimization model for synthesis of heat and work exchange networks. *Comput. Chem. Eng.* **2022**, *168*, 108036. [CrossRef]
32. Yu, H.; Fu, C.; Gundersen, T. Work Exchange Networks (WENs) and Work and Heat Exchange Networks (WHENs): A Review of the Current State of the Art. *Ind. Eng. Chem. Res.* **2020**, *59*, 507–525. [CrossRef]
33. Floudas, C.A.; Ciric, A.R.; Grossmann, I.E. Automatic synthesis of optimum heat exchanger network configurations. *AIChE J.* **1986**, *32*, 276–290. [CrossRef]
34. Yu, H.; Fu, C.; Vikse, M.; He, C.; Gundersen, T. Identifying optimal thermodynamic paths in work and heat exchange network synthesis. *AIChE J.* **2019**, *65*, 549–561. [CrossRef]
35. Biegler, L.T.; Grossmann, I.E.; Westerberg, A.W. *Systematic Methods for Chemical Process Design*; Prentice Hall, Inc., NJ. USA, 1997.
36. Chen, J. Comments on improvements on a replacement for the logarithmic mean. *Chem. Eng. Sci.* **1987**, *42*, 2488–2489. [CrossRef]
37. Chen, J. Logarithmic mean: Chen's approximation or explicit solution? *Comput. Chem. Eng.* **2019**, *120*, 1–3. [CrossRef]
38. Smith, R. *Chemical Process Design and Integration*, 2nd ed.; John Wiley & Sons, Inc.: Chichester, UK, 2016.
39. GAMS Development Corporation. *General Algebraic Modeling System (GAMS)*: Release 36.1.0; GAMS Development Corporation: Fairfax, VA, USA, 2022.
40. Systat Software Inc. *SigmaPlot v 14*; Systat Software Inc.: San Jose, CA, USA, 2017.
41. Seider, W.; Seader, J.; Lewin, D. *Product and Process Design Principles: Synthesis, Analysis and Evaluation*, 3rd ed.; John Wiley & Sons, Inc.: Hoboken, NJ, USA, 2008.
42. Peters, M.; Timmerhaus, K.; West, R. *Plant Design and Economics for Chemical Engineers*, 4th ed.; McGraw-Hill Chemical Engineering Series; McGraw-Hill, Inc.: New York, NY, USA, 2003.
43. Couper, J.R.; Penney, W.R.; Fair, J.R.; Walas, S.M. 21—Costs of Individual Equipment. In *Chemical Process Equipment*, 3rd ed.; Couper, J.R., Penney, W.R., Fair, J.R., Walas, S.M., Eds.; Butterworth-Heinemann: Boston, MA, USA, 2012; pp. 731–741 [CrossRef]
44. Woods, D.R. Appendix D: Capital Cost Guidelines. In *Rules of Thumb in Engineering Practice*; Wiley-VCH Verlag GmbH & Co.: KGaA, Germany, 2007; pp. 376–436. [CrossRef]
45. Towler, G.; Sinnott, R. Chapter 7—Capital cost estimating. In *Chemical Engineering Design*, 3rd ed.; Towler, G., Sinnott, R., Eds.; Butterworth-Heinemann: Cambridge, MA, USA, 2022; pp. 239–278. [CrossRef]
46. Turton, R.; Bailie, R.C.; Whiting, W.B.; Shaeiwitz, J.A.; Bhattacharyya, D. *Analysis, Synthesis, and Design of Chemical Processes*; Prentice-Hall International Series in Engineering; Prentice Hall: Upper Saddle River, NJ, USA, 2012.
47. Shah, N.M.; Rangaiah, G.P.; Hoadley, A.F. Multi-objective optimization of multi-stage gas-phase refrigeration systems. In *Multi-Objective Optimization: Techniques and Application in Chemical Engineering*; World Scientific Publishing Co. Pte. Ltd.: Singapore, 2017; pp. 247–290.
48. Fu, C.; Gundersen, T. Correct integration of compressors and expanders in above ambient heat exchanger networks. *Energy* **2016**, *116*, 1282–1293. [CrossRef]

energies

MDPI

Article

Advancing Industrial Process Electrification and Heat Pump Integration with New Exergy Pinch Analysis Targeting Techniques

Timothy Gordon Walmsley [1,*], Benjamin James Lincoln [1], Roger Padullés [2] and Donald John Cleland [3]

1 Ahuora—Centre for Smart Energy Systems, School of Engineering, University of Waikato, Hamilton 3240, New Zealand; b.lincoln@waikato.ac.nz
2 Department of Civil and Mechanical Engineering, Technical University of Denmark, Nils Koppels Allé, 2800 Kgs. Lyngby, Denmark; rpiso@dtu.dk
3 School of Food Technology and Natural Sciences, Massey University, Palmerston North 4474, New Zealand; d.cleland@massey.ac.nz
* Correspondence: tim.walmsley@waikato.ac.nz

Abstract: The process integration and electrification concept has significant potential to support the industrial transition to low- and net-zero-carbon process heating. This increasingly essential concept requires an expanded set of process analysis tools to fully comprehend the interplay of heat recovery and process electrification (e.g., heat pumping). In this paper, new Exergy Pinch Analysis tools and methods are proposed that can set lower bound work targets by acutely balancing process heat recovery and heat pumping. As part of the analysis, net energy and exergy load curves enable visualization of energy and exergy surpluses and deficits. As extensions to the grand composite curve in conventional Pinch Analysis, these curves enable examination of different pocket-cutting strategies, revealing their distinct impacts on heat, exergy, and work targets. Demonstrated via case studies on a spray dryer and an evaporator, the exergy analysis targets net shaft-work correctly. In the evaporator case study, the analysis points to the heat recovery pockets playing an essential role in reducing the work target by 25.7%. The findings offer substantial potential for improved industrial energy management, providing a robust framework for engineers to enhance industrial process and energy sustainability.

Keywords: process integration; pinch analysis; process electrification; heat pumps; exergy analysis; energy efficiency

1. Introduction

Industrial process electrification is an increasingly critical pathway to minimise on-site and lifecycle greenhouse gas (GHG) emissions as national and regional grids move toward low- and net-zero-carbon renewable electricity production. Process electrification involves the complex interplay of heat and power at the process level and is commonly achieved using closed-cycle heat pumps or directly integrating gas compressors and expanders into a process. New process technologies that use electricity can also replace old process technologies that rely more on steam heating. The appropriate adoption of electrification technologies requires supporting process integration methods, such as Pinch Analysis (PA) and its derivatives, to enable their effective application.

The development of PA in the 1970s has been extensively applied within the processing industry to improve heat recovery and design more efficient industrial systems [1]. Early in its development, PA focused on the efficient use of heat energy through heat exchanger network synthesis [2], appropriate heat recovery and utility placement, and multiple process heat integration, i.e., Total Site Heat Integration [3]. Case studies using PA resulted in energy-efficient heat exchanger networks with savings between 20 and 40% being historically common [4]. In conventional PA, the concepts of Heat Pinch and the appropriate placement principle for heat pumps are sufficient to analyse system performance and determine the best integration point for a heat pump using the grand composite curve (GCC). However, as electrical/work devices become a more common process utility, the need for deeper insights into the trade-offs between heat flow and temperature increases. As a result, exergy analysis becomes increasingly critical to design an efficient utility system that relies more on electricity than conventional fuels.

Exergy analysis adds a measure for work potential to different energy forms relative to the system's surrounding environment. It therefore provides a quantitative measure of process inefficiencies for differing energy streams [5]. The capability to determine the energy quantity of a given stream using exergy analysis therefore allows better insight into combined heat and work systems and how they should be designed and retrofitted when using existing process integration methods [6]. Exergy and electricity are closely related. Exergy represents the maximum potential for a system to perform useful work as it reaches equilibrium with the environment (dead state). The reverse is also true; exergy represents the minimum work required to increase the quality of energy in a direction away from the dead state conditions. Such work transfer can link to electricity generation or consumption, as the case may be. As a result, exergy has been applied as part of PA to understand the potential for electricity generation, consumption, and optimisation.

Exergy was first included in PA by Umeda et al. [7] using the Carnot factor ($\eta = 1 - T_L/T_H$). Feng and Zhu [8] refined the Carnot factor into a more accurate exergy-to-energy ratio, which they called the energy level. Anantharaman et al. [9] followed a similar approach to produce energy-level composite curves and an energy-level GCC. However, the energy level concept often produces nonlinear composite curves, complicating their interpretation. Aspelund et al. [10] presented a succinct Extended Pinch Analysis and Design (ExPAND) method focusing on sub-ambient processes where electricity-driven refrigeration systems are the primary utility. Their analysis showed that process modifications can be made to optimise the compression and expansion work to produce the required cooling duties. Marmolejo-Correa and Gundersen [11] introduced a new Exergy Pinch Analysis that mirrored conventional PA. Their Exergy Pinch Analysis used an exergetic temperature (a transformation of normal temperature given a dead state temperature) in place of temperature and temperature-based exergy in place of enthalpy (or heat flow). These improvements enabled direct targeting of exergy requirements, deficits, and destruction for the first time. Hamsani et al. [12] later extended the concept to an exergy GCC backed by an Exergy Problem Table Algorithm.

A common challenge with all the previous forms of Exergy Pinch Analysis is the difficulty in interpreting the results (e.g., graphs) in ways that add value beyond conventional heat PA. Marmolejo-Correa and Gundersen [11] attempted to set exergy targets, but these are still at least one step removed from a meaningful work (and electricity) target that can be used as a benchmark. Hamsani et al. [12] also proposed work targets; however, as will be demonstrated in this study, these targets assume all exergy surpluses are recovered through a heat engine and all exergy deficits are fulfilled by heat pumping from ambient temperature, which is inefficient in practice. As a result, new or extended process integration methods are needed to address the challenge and opportunity presented by process electrification.

PI&E aims to harness process integration to maximise system efficiency in the context of process electrification where systems rely on electricity as the primary utility (instead of heat). Process electrification includes a range of technology solutions, such as heat pumps, electrode boilers, and electricity-driven processes, that enable the transition from fossil fuels to electricity. This concept has found strategic value for low-temperature heat industries, e.g., dairy processing [13] and food processing [14]. Complete process electrification is now economically viable for many low-temperature processes using existing technologies. However, the efficient and economical integration of electrification technology requires systematic analysis to ensure appropriate placement, operation, and design. Lincoln et al. [15] presented an iterative approach to PI&E as shown in Figure 1. This method draws on many existing process integration tools and concepts, together with process simulation, to synthesise a systematic design method. A challenging part of the method is the application and interpretation of the Exergy Pinch Analysis step in ways that provide deeper insights than heat PA alone. This study looks in more depth at how to gain such insights through an improved exergy targeting method.

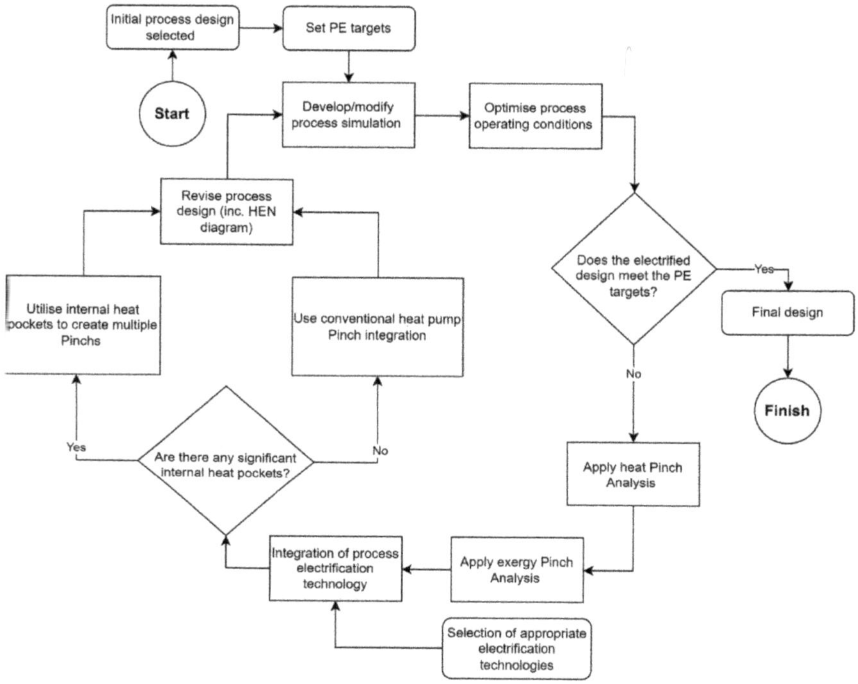

Figure 1. Process integration and electrification (PI&E) method proposed by Lincoln et al. [15].

This paper proposes a method to calculate a minimum (lower bound) net shaft-work target that takes advantage of heat recovery pockets. By setting a meaningful net shaft-work target, the interpretation of Exergy Pinch Analysis becomes more definitive, analogous to heat PA, and useful in PI&E studies. The study analyses two case studies: a spray dryer and an evaporator. The spray dryer is a relatively simple case where the process does not contain any heat recovery pockets and the heat and exergy Pinch Points occur at an equivalent temperature. The evaporator, in contrast, is a case where there are several heat recovery pockets, and the heat and exergy Pinch Points correspond to different temperatures. A systematic investigation of heat recovery pockets on the heat and exergy GCCs provides an improved understanding of the exergy, or shaft-work, trade-offs that are inherent with these pockets from the problem during targeting.

The organisation of the paper is as follows. In Section 2, the new method for exergy targeting is presented. Focus is given to setting targets that assume the thermodynamic potential within heat recovery pockets can be realised. The method produces two sets of exergy-based targets that correspond to lower and upper bound net shaft-work targets when integrating heat engines and heat pumps into the process. Section 3 then presents two case studies. The first is a spray dryer, a case without heat recovery pockets. The second is an evaporator, a case with heat recovery pockets. These two cases illustrate how the method can apply to both cases. In case two, a range of network shaft-work targets are calculated for different levels of heat recovery pocket utilisation. Following the results section, Section 4 discusses the key challenges in using the proposed method in practice and the implications for designing a combined heat and work exchanger network. Finally, Section 5 reiterates the most significant conclusions from the study.

2. Methods

In shifting towards process electrification for energy decarbonisation, exergy can be a useful metric to evaluate and compare the efficiency of different designs [16]. PA assists in guiding the design process, simulation provides a rating of a design, while exergy is used to determine the thermodynamic inefficiencies and potentials of unit operations and streams. All three analyses are critical to gaining the insights needed to improve design.

The exergy evaluation and targeting methods in this paper follow the principles of the ExPAND method [10] and its graphical representation of Marmolejo-Correa and Gundersen [11]. An extension by Hamsani et al. [12] details how an exergy GCC can determine accurate targets for the total exergy deficit and surplus of a process system. To better understand the context of this study, the reader is encouraged to review these latter two works. Figure 2 presents the new method developed as part of this study. It builds on the work of Hamsani et al. [12]. The method has been implemented in the OpenPinch Excel Workbook, which was created by the corresponding author. The interested reader can contact the corresponding author for more details.

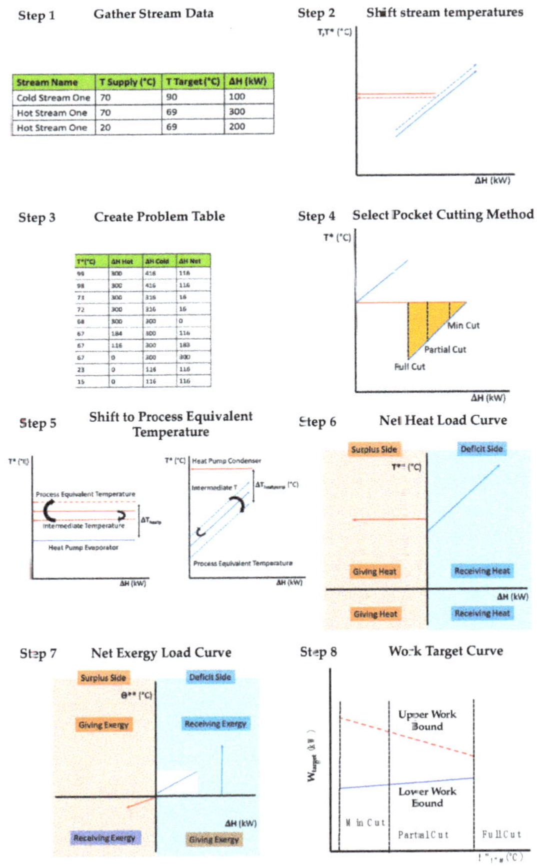

Figure 2. Graphical outline for Exergy Pinch Analysis targeting method. Note: T^* refers to shifted or intermediate temperatures, and T^{**} refers to utility temperatures.

2.1. Stream Data Extraction

The first step is common to all process integration studies. One needs to collect hot and cold stream data for a process, including supply and target temperatures, supply, target pressures (where pressure change is significant), and heat duty. Often extracting the stream data relies on a mass and energy balance or a process simulation to ensure the temperatures and pressures are accurate. This step, often non-trivial, can be one of the most time-consuming when conducting an industrial analysis.

2.2. Initial Stream Temperature Shifts and the Problem Table

Next is to calculate the problem table, which analyses a process from the perspective of defined temperature (or exergetic temperature) intervals. All process temperatures need to shift according to a minimum approach temperature contribution, ΔT_{cont} (or 0.5 ΔT_{min}), to an intermediate temperature scale, T^*. Hot streams (i) are shifted down in temperature, while cold streams (j) are shifted up in temperature.

$$T^*_{hot,i} = T_{hot,i} - \Delta T_{cont,i} \quad (1)$$

$$T^*_{cold,j} = T_{cold,j} + \Delta T_{cont,j} \quad (2)$$

The well-established Problem Table Algorithm (PTA) can then obtain heat recovery and utility targets. Its data also underpin the graphical plots of the process CCs and GCC. As this method is standard practice, the reader who is unfamiliar with this method is referred to the Handbook of Process Integration [17].

2.3. Heat Recovery Pocket Analysis and Cutting

Using the GCC, it is necessary to identify the portions of the heat pockets intended to be used for internal heat recovery or to be used in conjunction with heat pumps or heat engines. This step represents a deviation from previous works. Hamsani et al. [12] assumed heat recovery pockets would be used for internal heat recovery only, simplifying the exergy analysis but also eliminating the potential for thermodynamic efficiency gains. They did, however, calculate the avoidable exergy destruction that occurs due to heat recovery within a pocket at temperature differences above the pre-defined minimum approach temperatures. This gave rise to the question that this study seeks to answer: what is the benefit of utilising the thermodynamic potential of the heat recovery pockets in the context of process electrification?

Heat recovery pockets represent areas on a GCC with internal process heat recovery potential. Conventionally, extended PA methods, such as Total Site Heat Integration, remove all the pockets from the process-level GCCs before Total Site Analysis. This approach is generally well justified; however, it can also miss obvious and significant energy efficiency gains when the pocket involves large temperature differences. For example, Walmsley et al. [18] demonstrated the principle of assisted heat integration using pockets to recover additional heat and generate more shaft-work.

This study investigates how different pocket cutting strategies, which represent different internal heat exchanger networks, affect the exergy and work targets. The heat recovery pocket cutting strategies, illustrated in Figure 2, include:

1. Full Cut: this approach identifies and removes the full pocket from further analysis. This approach (although not referred to by this name) is commonly used in the Total Site Heat Integration method [3].
2. Min Cut: this approach leaves as much of the pocket in the stream data as possible for further analysis, i.e., $\Delta T_{min,cut} = \Delta T_{min,U}$. This approach determines the limits for exergy and work targets providing a thermodynamic benchmark to compare against.
3. Partial Cut: this approach identifies and removes the heat surplus and deficit segments of a pocket within a defined $\Delta T_{min,cut}$. This strategy removes part of the pocket depending on the selection of $\Delta T_{min,cut}$, which can be varied to understand its impact on the work targets.

A constraint for $\Delta T_{min,cut}$ is that it must be greater than $\Delta T_{min,U}$, the minimum approach temperature difference process and utility streams, i.e., $\Delta T_{min,cut} \geq \Delta T_{min,U}$. If $\Delta T_{min,cut} < \Delta T_{min,U}$, it indicates that transferring process heat to an intermediate utility and then to a process sink must be less thermodynamically efficient than direct process-to-process heat recovery. The Full-Cut approach sets $\Delta T_{min,cut}$ to the maximum ΔT of the largest pocket, which is identifiable on the GCC. The Partial-Cut approach is bounded by the Full-Cut and Min-Cut approaches, representing the continuum of values in between that could be explored further to understand the selection on heat and exergy targets.

2.4. Process Temperature Equivalent Shift

Given the pocket cutting strategy, each of the remaining GCC segments are temperature-shifted back to an effective process temperature scale, τ, in K using a common 0.5 $\Delta T_{min,U}$.

$$\tau_{hot} = T^*_{hot} + 0.5\,\Delta T_{min,U} + 273.15 \tag{3}$$

$$\tau_{cold} = T^*_{cold} - 0.5\,\Delta T_{min,U} + 273.15 \tag{4}$$

The rationale for the shift back is that Carnot efficiencies for heat engines and heat pumps are based on process temperatures (as opposed to working fluid temperatures, which would be the utility temperature scale). This shift back prepares the modified GCC to be converted into net load curves.

2.5. Net Heat and Exergy Load Curves

Net load curves are introduced to split the stream segments of the modified GCC into the heat sources (left of the y-axis) and heat sinks (right of the y-axis). These curves have the same styling as Total Site Profiles [19]. The net heat load curves (NHLCs) are based on a plot of temperature and enthalpy (or heat flow).

In previous works (e.g., [12]), an exergy GCC has been proposed and applied. However, this approach struggles to universally represent all types of problems. For problems that are entirely above or below the dead state temperature, the method is adequate. However, issues arise when streams cross the dead state temperature. Above ambient temperature, heat sources are also exergy sources, while below ambient temperature, heat sinks are exergy sources. This reversal of roles means it is difficult to draw and understand an exergy GCC with streams on both sides of the dead state on a single figure. One solution is the concept of net load curves that split up sources and sinks.

To determine the net exergy load curves (NXLCs), each heat load segment needs to be translated into exergetic temperatures and exergy flows. Using the data for the heat load segments, the supply (s) and target (t) exergetic temperatures, θ_s and θ_t, and their temperature-based exergy change, ΔX, can be determined using the following equations.

$$\theta = \tau - \tau_0\left(\ln\left(\frac{\tau}{\tau_0}\right) + 1\right) \tag{5}$$

$$\Delta X = CP(\theta_s - \theta_t) \tag{6}$$

where τ is the effective process stream temperature in K, τ_0 is the dead state temperature in K, and CP is the heat capacity flow rate (i.e., $\dot{m}c_p$). Note, the CP is the inverse slope of segments on the modified GCC plot. Equation (6) determines only the temperature-based component of exergy, which approximates the total thermomechanical exergy (temperature and pressure) for streams with minimal pressure-based exergy, as is the case with the present study. Given the stream data for these segments, the PTA can be applied (twice) to the GCC stream segments and converted into exergy flows and exergetic temperatures to obtain an exergy-based problem table and exergy load curves. This process is repeated twice, once for all the heat surplus segments of the GCC and a second time for all the heat deficit segments of the GCC. Although the input stream data differ, the exergy-based PTA is explained step by step by Hamsani et al. [12]. The NXLC shows the amounts of surplus exergy contained in the exergy sources and deficit exergy required by the exergy sinks. In terms of the exergy analysis in this work, a dead state of 15 °C (288.15 K) and atmospheric pressure (101.325 kPa), which correspond to typical ambient conditions, has been used in the exergy analysis.

For processes that cross the dead state temperature, i.e., ambient temperature, exergy sources and sinks reverse. Above ambient temperature, hot streams are exergy sources, while cold streams are exergy sinks. Below ambient temperature, hot streams are now exergy sinks (e.g., refrigeration) and cold streams are exergy sources. In this study, the x-axis of the NXLC plot divides above and below the dead state temperature. Note, the exergetic temperatures, above or below ambient temperature, are always positive. As a result, the y-axis, both up and down, displays positive exergetic temperatures, which is critical for calculating exergy differences. The left side of the y-axis is all the exergy sources while the right side of the axis is all the exergy sinks. Note, the NXLC are original to this study.

2.5. Net Shaft-Work Calculation

This final step of the method aims to identify and calculate a net shaft-work target. The NXLC, together with the concepts of a reversible heat engine and heat pump, provides a basis for determining net shaft-work targets. A reversible (Carnot) heat engine operating between temperature levels T_H and T_L achieves zero entropy generation and no exergy destruction, which means,

$$W_{HE} = W_H - W_L = X_H - X_L \tag{7}$$

where W_{HE} is the reversible work generated from the heat engine, W is the heat flow from T_H to T_L, and X is the exergy flow from T_H to T_L. To obtain a meaningful work value for an actual heat engine, Equation (8) can be multiplied by a Carnot efficiency, $\eta_{II,HE}$,

$$W_{HE,act} = \eta_{II,HE}(X_H - X_L) \tag{8}$$

A similar analysis of a reversible heat pump and dividing by a Carnot efficiency, $\eta_{II,HP}$, gives the following relationship for the actual work of a heat pump, $W_{HP,act}$:

$$W_{HP,act} = \frac{1}{\eta_{II,HP}}(X_H - X_L) \tag{9}$$

Hamsani et al. [12] applied these two equations to generate a net shaft-work target for a process,

$$W_{net(high)} = W_{comp} - W_{turb} \approx \frac{1}{\eta}X_{rej} - \eta X_{req},\ where\ \eta = \eta_{II,HP} = \eta_{II,HE}, \tag{10}$$

where W_{comp} is the work of compression, W_{turb} is the work of expansion, X_{rej} is the exergy rejection target (below the Pinch), and X_{req} is the exergy required target (above the Pinch). The Carnot efficiency, η_{II}, of heat engines and heat pumps often both sit around 50% [20], justifying the equivalence of the two efficiencies in Equations (8) and (9). Targets based on a Carnot efficiency of 50% are expected, therefore, to be achievable for practical systems; however, other values for η_{II} could be explored to understand its impact on the results in future studies.

In this study, the pockets on the GCC are cut using a defined strategy before the exergy analysis (another point of difference to Hamsani et al. [12]), which results in the exergy surplus and rejection targets being the same and the exergy deficit and required targets also being the same.

Equation (10) inherently assumes the exergy surplus (rejection) segments would generate work through a heat engine and the exergy deficit (required) segments would consume work using a heat pump from the specified dead state. As a result, Equation (10) may substantially overestimate the actual work of a well-designed system since a heat pump supplied by exergy surplus segments would be more efficient than applying a heat engine and then a heat pump in series.

An improved shaft-work target, representing a lower bound, would supply heat pump devices with as much of the exergy (and heat) surpluses (S) as possible to fulfil exergy (and heat) deficits (D), while any remaining exergy surplus segments would be recovered through a heat engine. As a result, this study defines γ as the fraction of heat surplus segments supplying heat pump devices (as opposed to heat engine devices). Given different γ values, net shaft-work targets can be determined using

$$W_{net} = \frac{1}{\eta_{II,HP}}(\sum \Delta X_D - \gamma \sum \Delta X_S) - \eta_{II,HE}(1-\gamma)\sum \Delta X_S \quad \text{where } 0 \le \gamma \le \min\left(\frac{\sum \Delta X_D}{\sum \Delta X_S}, 1\right), \tag{11}$$

where ΔX_D is an exergy deficit segment on the NXLC and ΔX_s is an exergy surplus segment on the NXLC.

The Full-Cut strategy would result in only exergy deficits above the Pinch and exergy surpluses below the Pinch, and Equation (9) is equivalent to Equation (8) when given $\gamma = 0$. In contrast, setting $\gamma = \min\left(\frac{\sum \Delta X_D}{\sum \Delta X_S}, 1\right)$ gives a lower bound net shaft-work target that represents an ideally designed system of heat pump and heat engine devices given the defined minimum approach temperatures, pocket cutting strategy, and Carnot efficiency.

3. Case Studies

3.1. Spray Dryer Case Study

This case study delves into the operation of a spray dryer employed in the production of milk powder. The primary function of the dryer is to transform milk concentrate, which is derived from evaporation lines, into a powdered state. This case study was initially presented by Liang et al. [21]. The stream data and the process flow diagram are found in Table 1 and Figure 3.

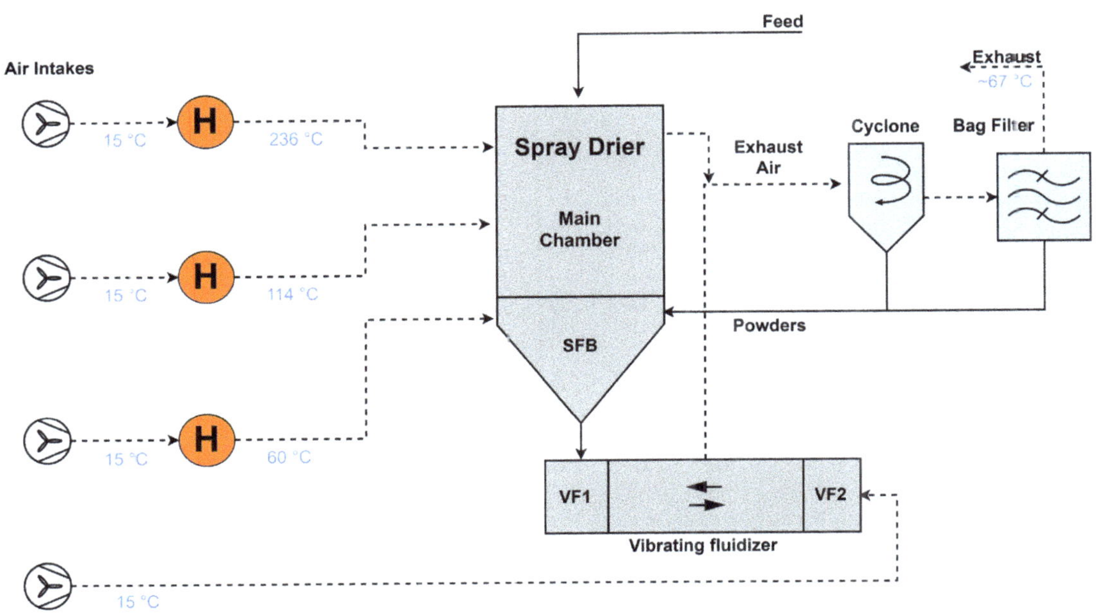

Figure 3. Flow diagram of the spray drying process.

Table 1. Stream data for the spray drying system.

Stream Name	T_s (°C)	T_t (°C)	ΔH (kW)	ΔT_{cont} (°C)
Exhaust 1 (above dew point)	67	40.3	1081	8
Exhaust 2 (below dew point)	40.3	15	5788	8
Vibrating Fluidised Bed	15	60	1741	8
Static Fluid Bed 1	60	86.5	781	8
Static Fluid Bed 2	86.5	114	814	8
Main Air Chamber	114	236	2695	5

In the drying process, the feed is initially atomized and exposed to the hot air from the main air chamber at 236 °C. Moving to the bottom of the chamber, a static fluid bed (SFB) is utilised for additional drying, using a separate air stream at 114 °C. Subsequently, the partially dried powder exits through a series of vibrating fluidised (VF) beds, which provide ample residence time to complete the drying process. The exhaust air streams from the main chamber and fluid beds merge before entering the cyclone and bag filter.

3.2. Milk Evaporator Case Study

The base case design, shown in Figure 4, is a two-effect mechanical vapour recompression (MVR) milk evaporator system with a milk heat treatment system derived from Walmsley et al. [22]. The original design was modified to include a second flash vessel immediately before milk enters the first effect to increase preheating efficiency [15].

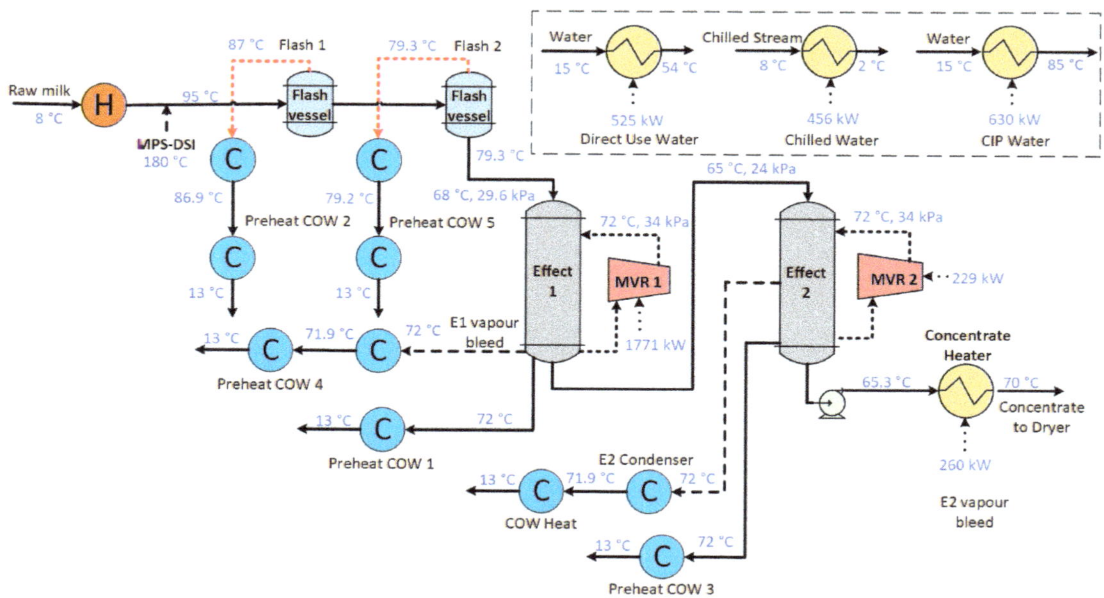

Figure 4. Milk evaporator process flow diagram.

The process takes 247 t/h of raw milk, which is sufficient to produce 30 t/h of milk powder product. The raw milk is heated from 8 °C to 95 °C using heat recovery, which is followed by Direct Steam Injection (DSI) to reach the desired temperature and to achieve rapid heating and low residence times to maintain high product quality. After preheating, two flash cooling steps produce low-pressure vapour that can be used for additional heat recovery. Milk then enters the evaporator with Effect 1, which has a mean temperature driving force of 4 °C, and Effect 2, which maintains a mean temperature driving force of 7 °C, with both MVRs upgrading vapour to 72 °C (saturation). The flash vessels and evaporators produce condensate and vapour streams that have heat recovery potential. Beyond the evaporation process, ancillary services, such as milk chilling and cleaning in place (CIP), require on average 1155 kW of additional heating duty and 456 kW of cooling duty. These hot and chilled water demands fluctuate with time and require heat storage to be integrated with the evaporation system.

In this study, the milk evaporation system is considered in isolation to define a fully electric milk evaporator system. The stream data are presented in Table 2. In the future, it is anticipated that placing the milk evaporator system in the context of Total Site Heat Integration may lead to additional performance gains [19].

Table 2. Stream data for the milk evaporator system.

Stream Name	T_s (°C)	T_t (°C)	ΔH (kW)	ΔT_{cont} (°C)
Chiller Water	8.0	4.0	456.0	0.5
CIP Water	15.0	84.0	630.0	2.5
Cow Heat	72.1	13.0	38.9	2.5
Direct Use	15.0	55.0	525.0	2.5
E1 Vapour Bleed	72.0	71.95	3920.5	0.5
E2 Condenser	72.0	71.95	364.1	0.5
HT Flash	87	86.95	2060.8	0.5
HT Flash 2	79.3	79.25	1988.8	0.5
Milk Concentrate	65.3	70.0	259.9	2.5
Preheat COW 1	72.0	13.0	10,697.8	2.5
Preheat COW 2	86.9	13.0	279.8	2.5
Preheat COW 3	72.0	13.0	412.1	2.5
Preheat COW 4	71.9	13.0	704.4	2.5
Preheat COW 5	79.2	13.0	240.3	2.5
Raw Milk	8.0	95.0	22,834.8	2.5

4. Results

The results comprise the analysis of two case studies: a milk spray dryer and a milk evaporator plant. The milk spray dryer case, with a simpler GCC, does not contain heat recovery pockets. As a result, it also serves as a good illustration of the basic method. The milk evaporator case study is more complicated due to the presence of heat recovery pockets on the GCC and that it includes streams on both sides of the dead state conditions. These case studies demonstrate how the method can be applied to a wide range of processes and sites.

4.1. Spray Dryer Analysis and Results

In the spray dryer case study, the exhaust air contains a large amount of excess heat that could be optionally cooled to ambient temperature and utilised for heat recovery and as a heat source for the heat pump. As shown in Figure 5, the GCC has no pockets, and indicates a hot utility target of 4.98 MW and a cold utility target of 3.99 MW.

In this case, the absence of heat pockets leads to NHLC (Figure 6A) without overlap, and the curves are equivalent to each of the regions of the GCC. This indicates that there is no further potential for heat recovery. Additionally, since the entire GCC is above the dead state temperature, the NHLC and NXLC (Figure 6B) appear similar in shape. However, the relative magnitudes of sources to sinks differ substantially. The sources close to the dead state contain minimal exergy, resulting in a change in proportions. Despite the lower temperature and therefore lower exergy content of the excess heat available, utilising this heat as a heat source for the heat pump system can still improve its performance and reduce the necessary work shaft for the electrification of the process.

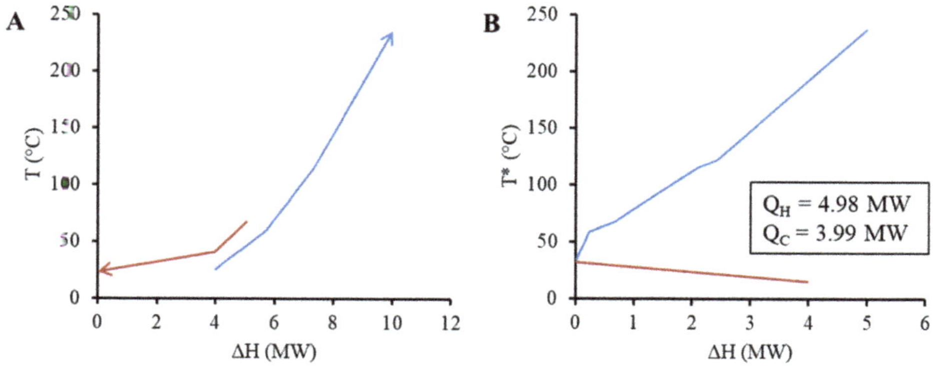

Figure 5. Composite curve, CC, (**A**) and grand composite curve, GCC, (**B**) for the milk powder spray dryer system. Note: T^* refers to shifted or intermediate temperature scale.

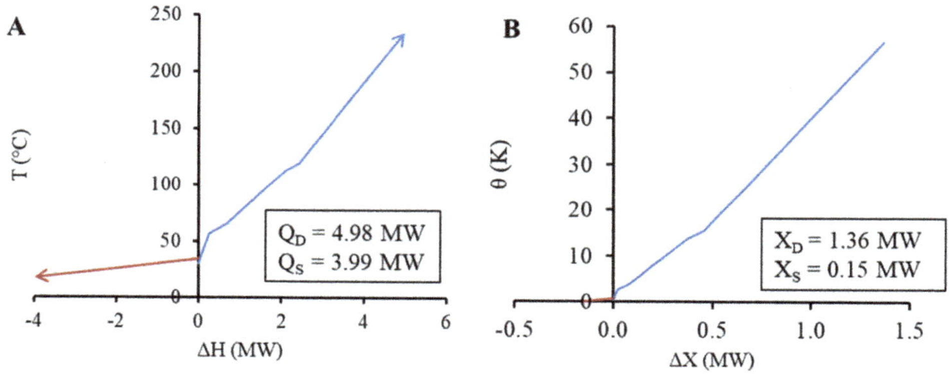

Figure 6. Net heat load curve (**A**) and net exergy load curve (**B**) for the milk powder spray dryer system.

Assuming an exergetic efficiency of 50%, the resulting work targets vary from 2.42 MW ($\gamma = 1$) to 2.65 MW ($\gamma = 0$) using Equation (11). The $\gamma = 0$ case assumes that the excess heat below the Pinch temperature is fully expanded to the dead state and the heat demand above the Pinch is supplied by a heat pump utilizing the dead state as a source. On the other hand, the $\gamma = 1$ case sets a lower bound that assumes that exergy is utilised, avoiding exergy destruction associated with expansion and compression to and from the dead state.

4.2. Evaporator Analysis

The evaporator system is significantly different from the spray dryer. As shown in Figure 7, this system contains three heat recovery pockets, where one is considerably larger than the other two. As a result, this subsection explores various strategies for cutting the heat recovery pockets and their impact on the utility demand.

4.2.1. The Full and Minimum Pocket Cutting Strategies

This paper introduces the concept of a $\Delta T_{min,cut}$ as part of the energy and exergy analysis. The "Full-Cut" pocket strategy assumes all pockets are used for internal process heat recovery (Figure 8). While this may save energy, there is a thermodynamic penalty that results in exergy destruction during the heat recovery process, especially in cases where the $\Delta T \gg \Delta T_{min}$ [12]. Figure 8A shows that the NHLCs do not overlap, indicating that the maximum heat recovery has been reached. The Pinch range is found between 5.5 and

10.5 °C, with most heating for the evaporator required between 90 and 100 °C. In this case, the Exergy Pinch in Figure 8B sits in the same corresponding exergetic temperature band.

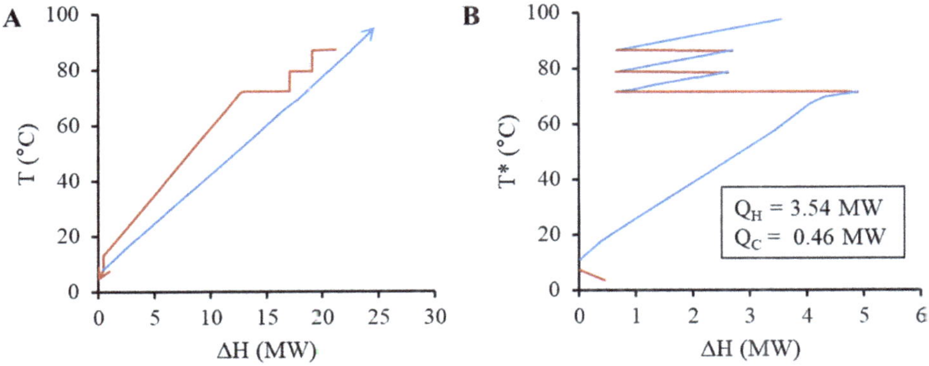

Figure 7. Composite curves (**A**) and GCC (**B**) for the milk evaporator system, where red segments identify hot streams and/or heat surpluses and blue segments identify cold streams and/or heat deficits. Note: T^* refers to shifted or intermediate temperature scale.

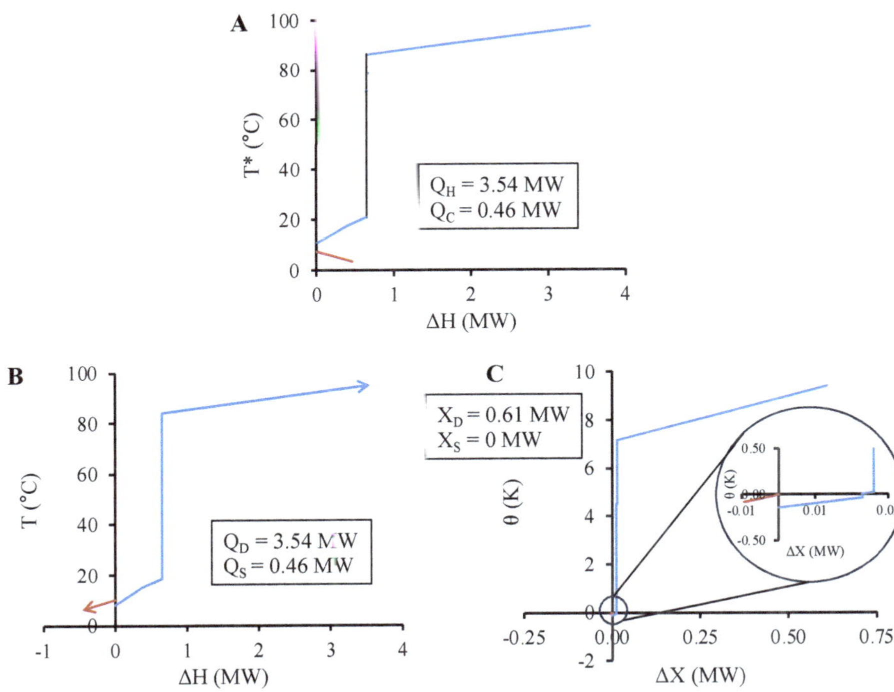

Figure 8. GCC (**A**), net heat load curve (**B**), and net exergy load curve (**C**) for the milk evaporator system based on the Full-Cut pocket strategy. Note: T^* refers to shifted or intermediate temperature scale. Note: T^* refers to shifted or intermediate temperature scale.

At the other extreme, the "Min-Cut" strategy (Figure 9) leaves as much of the pocket in the energy and exergy analysis as possible that may find benefit through indirect heat recovery or heat pumping schemes. In this study, $\Delta T_{min,U}$ = 5 °C, and this second shift accounts for the general need to transfer heat to and from an intermediate fluid, e.g., a heat pump refrigerant. This value also defines the lower limit of $\Delta T_{min,cut}$, i.e., 5 °C, as $\Delta T_{min,cut}$ is equal to or greater than $\Delta T_{min,U}$. In Figure 9A, the effect of $\Delta T_{min,cut}$ is visible as the vertical black lines in the 60–90 °C temperature range where three ends of pockets have been cut. In this case, there is a significant overlap on the temperatures of the NHLC, indicating potential for heat recovery. In Figure 9B, the temperatures of each segment are shifted and then converted to exergetic temperatures and exergy flow.

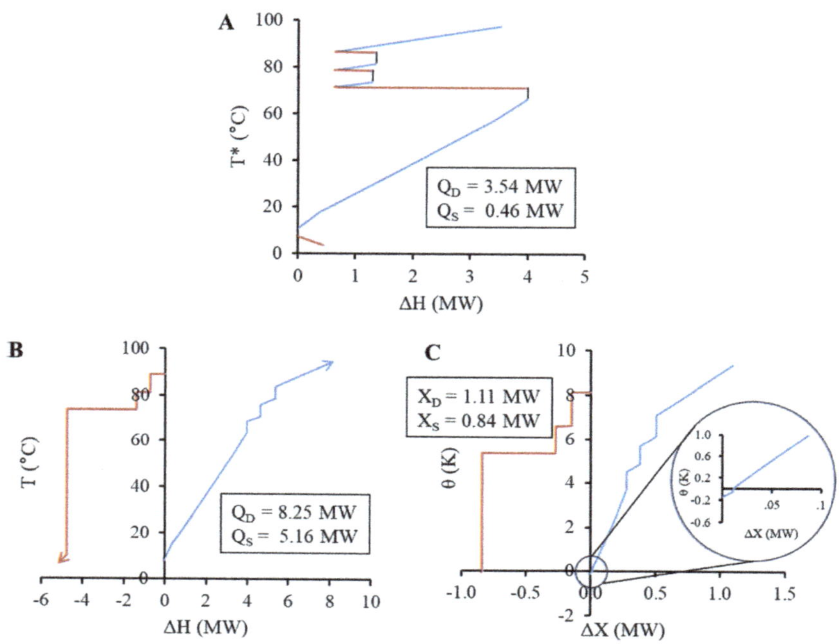

Figure 9. GCC (**A**), net heat load curve (**B**), and net exergy load curve (**C**) for the milk evaporator system based on the Min-Cut pocket strategy (5 °C cut). Note: T^* refers to shifted or intermediate temperature scale.

Pockets with a small amount of potential exergy destruction (X_{pd}) have limited opportunity to improve thermodynamic efficiency beyond internal heat recovery. Pockets with a large amount of exergy destruction can present additional opportunities to improve thermodynamic efficiency while also electrifying the process with electric technologies such as heat pumps. If the large bottom pocket is assumed to be recovered through heat exchange (Full Cut), the associated exergy destruction would reach 200 kW_x. The total impact of cutting versus minimally cutting the pockets is evident also in comparing the exergy targets as net values (deficits minus surpluses). Compared to the Full-Cut strategy, the Min-Cut strategy achieves a lower exergy deficit requirement above the Pinch and a lower surplus target below the Pinch in this case. This leads to, assuming an exergy efficiency of 50%, minimum shaft-work targets of 1.21 MW and 0.54 MW for the Full-Cut and Min-Cut strategies, respectively.

4.2.2. Partial-Cut Pocket Strategy

In between the extremes of the Full-Cut and Min-Cut strategies is the Partial-Cut strategy. Higher $\Delta T_{min,cut}$ values result in more of the pocket being removed and an inherent acceptance of more exergy destruction within the internal heat recovery network. As one further example, Figure 10 presents the results for the cutting pocket with a $\Delta T_{min,cut}$

of 20 °C. The exergy deficit is the same as the Min-Cut strategy, while the exergy surplus decreases from 0.168 MW (Min Cut) to 0.141 MW, where the gap of 0.027 MW is lost to exergy destruction during additional internal process heat recovery.

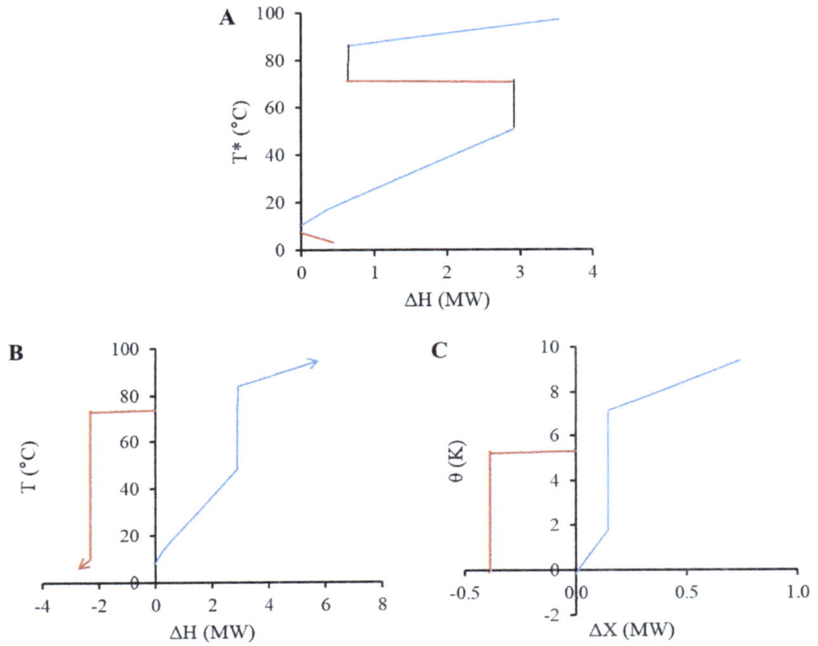

Figure 10. GCC (**A**), net heat load curve (**B**), and net exergy load curve (**C**) for the milk evaporator system based on the Partial-Cut pocket strategy (20 °C cut). Note: T^* refers to shifted or intermediate temperature scale.

4.2.3. Analysis of the Pocket-Cutting Strategy on Net Shaft-Work Targets

As mentioned, the variable $\Delta T_{min,cut}$ has specific lower and upper limits, representing the Min-Cut and Full-Cut approaches, respectively. This study has calculated targets within these limits to better understand the observed trends. Figure 11 shows the impact of $\Delta T_{min,cut}$ on the net shaft-work targets (i.e., red and blue lines). These targets are determined using Equation (9). For this case, the amount of exergy surpluses is negligible under the Min-Cut strategy, which results in the convergence of the two sets of work targets.

The label "simple design" in Figure 11 corresponds to a heat pump placement across the pinch temperature and would be in accordance with conventional pinch analysis strategies. First, the pockets are cut via heat integration, and the remaining heat utility is supplied by a single, high-temperature lift heat pump operating from ambient temperature to 105 °C. Due to the high-temperature lift, the resulting COP falls to 2.1.

The label "efficient design" corresponds to the design developed by Lincoln et al. [15]. The "efficient design" involved the placement of heat pumps around both Pinch regions to generate a practical process design that approaches the minimum work consumption (Figure 11). By leveraging smaller temperature lifts, the "efficient design" was able to raise the combined COP of the system to 3.17.

The practicality of a design is determined by many factors that are specific to a process but could include spatial layout, scheduling, and operability. Increasing the practicality generally acts to counter efficiency and therefore moves away from the minimum work target. For example, in the "efficient design", streams of different temperature levels ranging between 68 and 74 °C were combined, which increased the amount of exergy destruction in the system but reduced the number of heat exchangers required.

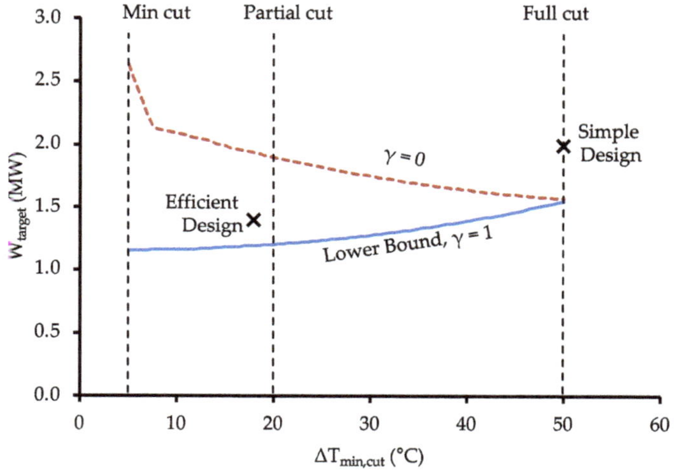

Figure 11. Net shaft-work targets for a range of $\Delta T_{min,cut}$ values, including both simple and efficient design points. The simple design and efficient design points are taken from the work of Lincoln et al. [15].

5. Interpretation and Implications of the Exergy Analysis on Design Decisions

It is important to recognise the value added by exergy PA, and particularly the lower bound work target, to conventional heat PA. The lower bound shaft-work target establishes a thermodynamic benchmark to compare the quality of a system's performance.

The net heat load curves and net exergy load curves are useful representations of the actual heat and exergy available as heat sources for heat pump integration. The inclusion of the heat pockets of the GCC with this method reduces the minimum exergy target, which highlights the importance of avoiding exergy destruction, which can sometimes be substantial for internal process (i.e., direct) heat recovery in the heat pockets. Integrating heat pumps in these heat pockets may appear counter to the conventional rules of Pinch Analysis because it increases the apparent heating and cooling utility targets. However, as shown, the net shaft-work requirement based on a practical level of exergy efficiency (50%) can decrease. To achieve the lower work target, multiple heat pumps need to be integrated into the processes.

The relationship between direct heat recovery on the heat pockets and the use of these pockets for heat pump integration can also be considered. Where pockets are cut out of the GCC (and therefore not present in the exergy analysis), the inherent assumption is that the heat source/sink segments will be matched as conventional process heat. If a pocket (or any portion of it) is left in the analysis, the exergy analysis assumes that a heat pump and/or heat engine could utilise its thermodynamic potential. By investigating different strategies for cutting the heat pockets, the trade-off between the solution with higher exergy efficiency (smaller $\Delta T_{min,cut}$ value) and the solution with simpler system design (larger $\Delta T_{min,cut}$ value) can be explored. The decision variable $\Delta T_{min,cut}$ encapsulates the general trade-off between capital and energy, i.e., simple inefficient design vs. complex efficient design. An efficient system design will consider the complexity of the system and the necessary investment, as well as the energy efficiency, resulting in a partial cut of the pocket.

Lastly, the presented method also highlights the importance of utilising excess heat as a heat source for the heat pumps rather than using a direct expansion device when setting exergy targets, as shown by the difference in the results with different gamma factors. The results show that the use of the available exergy as a heat source for heat pumping should be prioritised, as it avoids the additional exergy destruction of the expansion.

The use of the gamma factor avoids the application of the exergy efficiency factor for streams that are used as a heat source and are not expanded to the dead state temperature for minimum work targeting. However, when the available excess heat is found above the dead state temperature and is utilised by a heat pump, the exergy efficiency factor considered with this method does not represent the actual efficiency of the heat pump, since it is dependent on the dead state temperature. The use of a Carnot or Lorenz efficiency factor, which considers the source temperature instead of the dead state temperature, may be better fitted to represent the work targets for heat pump integration [23].

Net load curves underline the heat pockets' significance in establishing exergy targets and demonstrate the benefits of an efficient pocket-cut strategy when integrating multiple heat pumps. However, further investigation into optimally combining sink and source streams is necessary for an efficient system design.

6. Conclusions

This study focussed on refining the application of Exergy Pinch Analysis for targeting the efficient conversion of industrial process heating to renewable electricity through the utilisation of heat recovery pockets.

The investigations performed during the study led to the proposal of two concepts as additions to Exergy Pinch Analysis:

- An improved work targeting equation (Equation (11)), which includes a new gamma factor, that enables determination of a practical minimum work target. Gamma indicates the proportion of available surplus heat utilised by heat pumps while the remainder is expanded to the dead state. Future work can look further into the role of gamma and how its selection is best made for additional case studies where heat sinks are the limiting factor.
- The heat and exergy grand composite curves and the net energy and exergy load curves are used to visualise available energy and exergy within heat pockets, giving different perspectives and aiding targeted energy optimisation. Different pocket-cutting strategies can be investigated as a result. Such strategies significantly impact heat and exergy targets, showcasing various implications on system design.

These methods were applied to two case studies: a milk powder spray dryer system and a milk evaporator. In the case of the milk evaporator case study, it was shown that the lower boundary of the net shaft-work target escalates proportionally with the removal of more heat recovery pockets. The difference between the minimum work target (i.e., maximum utilisation of heat recovery pockets) and maximum work target (i.e., no utilisation of heat recovery pockets) is 400 kW, corresponding to 25.7%. Additionally, the case study illustrated scenarios where the heat and exergy analyses' Pinch points were manifested in distinct temperature ranges. The optimal solution implied placing heat pumps around both Pinch regions, defying the conventional rules of Pinch Analysis.

Author Contributions: Conceptualisation, T.G.W.; methodology, T.G.W. and B.J.L.; software T.G.W.; formal analysis, T.G.W., B.J.L. and R.P.; investigation, T.G.W., R.P., and B.J.L.; data curation, T.G.W., B.J.L. and R.P.; writing—original draft preparation, T.G.W.; writing—review and editing, D.J.C.; visualisation, T.G.W., B.J.L. and R.P.; funding acquisition, T.G.W. and D.J.C. All authors have read and agreed to the published version of the manuscript.

Funding: This research was funded by the programme "Ahuora: Delivering sustainable industry through smart process heat decarbonisation", an Advanced Energy Technology Platform, funded by the New Zealand Ministry of Business, Innovation and Employment.

Data Availability Statement: The data that support the findings of this study are available from the corresponding author, T.G.W., upon reasonable request.

Conflicts of Interest: The authors declare no conflicts of interest.

Abbreviations

Symbol	Description
CC	Composite Curve
CIP	Cleaning in Place
COP	Coefficient of Performance
CP	Heat Capacity
DSI	Direst Steam Injection
EPTA	Exergy Problem Table Algorithm
GCC	Grand Composite Curve
MVR	Mechanical Vapour Recompression
NHLC	Net Heat Load Curve
NXLC	Net Exergy Load Curve
PA	Pinch Analysis
PI&E	Process Integration and Electrification
PTA	Problem Table Algorithm
SFB	Static Fluidised Bed
T^*	Shifted Temperatures
VFB	Vibrating Fluidised Bed
W	Work
Υ	Fraction of heat surplus segments supplying heat pump deficits
ΔT_{cont}	Contribution to the approach temperature (°C)
ΔT_{min}	Minimum approach temperature
$\Delta T_{min,cut}$	Delta T to achieve a minimum pocket cut
$\Delta T_{min,U}$	Minimum utility approach temperature
ΔX	Change in exergy
η_{II}	Carnot efficiency
∂	Exegetic temperature
τ	Effective process temperature

References

1. Klemeš, J.J.; Varbanov, P.S.; Walmsley, T.G.; Jia, X. New Directions in the Implementation of Pinch Methodology (PM). *Renew. Sustain. Energy Rev.* **2018**, *98*, 439–468. [CrossRef]
2. Gundersen, T. 4—Heat Integration: Targets and Heat Exchanger Network Design. In *Handbook of Process Integration (PI)*; Klemeš, J.J., Ed.; Woodhead Publishing Series in Energy; Woodhead Publishing: Sawston, UK, 2013; pp. 129–167. ISBN 978-0-85709-593-0.
3. Dhole, V.R.; Linnhoff, B. Total Site Targets for Fuel, Co-Generation, Emissions, and Cooling. *Comput. Chem. Eng.* **1993**, *17*, S101–S109. [CrossRef]
4. Hackl, R.; Harvey, S. Total Site Analysis (TSA) and Exergy Analysis for Shaft Work and Associated Steam and Electricity Savings in Low Temperature Processes in Industrial Clusters. *Chem. Eng. Trans* **2012**, *29*, 73–78.
5. Kotas, T.J. *The Exergy Method of Thermal Plant Analysis*; Paragon Publishing: Trowbridge, UK, 2012; ISBN 978-1-908341-89-1.
6. Khaliq, A.; Kaushik, S.C. Thermodynamic Performance Evaluation of Combustion Gas Turbine Cogeneration System with Reheat. *Appl. Therm. Eng.* **2004**, *24*, 1785–1795. [CrossRef]
7. Umeda, T.; Harada, T.; Shiroko, K. A Thermodynamic Approach to the Synthesis of Heat Integration Systems in Chemical Processes. *Comput. Chem. Eng.* **1979**, *3*, 273–282. [CrossRef]
8. Feng, X.; Zhu, X.X. Combining Pinch and Exergy Analysis for Process Modifications. *Appl. Therm. Eng.* **1997**, *17*, 249–261. [CrossRef]
9. Anantharaman, R.; Abbas, O.S.; Gundersen, T. Energy Level Composite Curves—A New Graphical Methodology for the Integration of Energy Intensive Processes. *Appl. Therm. Eng.* **2006**, *26*, 1378–1384. [CrossRef]
10. Aspelund, A.; Berstad, D.O.; Gundersen, T. An Extended Pinch Analysis and Design Procedure Utilizing Pressure Based Exergy for Subambient Cooling. *Appl. Therm. Eng.* **2007**, *27*, 2633–2649. [CrossRef]
11. Marmolejo-Correa, D.; Gundersen, T. New Graphical Representation of Exergy Applied to Low Temperature Process Design. *Ind. Eng. Chem. Res.* **2013**, *52*, 7145–7156. [CrossRef]
12. Hamsani, M.N.; Walmsley, T.G.; Liew, P.Y.; Wan Alwi, S.R. Combined Pinch and Exergy Numerical Analysis for Low Temperature Heat Exchanger Network. *Energy* **2018**, *153*, 100–112. [CrossRef]
13. Bühler, F.; Zühlsdorf, B.; Nguyen, T.-V.; Elmegaard, B. A Comparative Assessment of Electrification Strategies for Industrial Sites: Case of Milk Powder Production. *Appl. Energy* **2019**, *250*, 1383–1401. [CrossRef]

14. Atuonwu, J.; Tassou, S. Decarbonisation of Food Manufacturing by the Electrification of Heat: A Review of Developments, Technology Options and Future Directions. *Trends Food Sci. Technol.* **2021**, *107*, 168–182. [CrossRef]
15. Lincoln, B.J.; Kong, L.; Pineda, A.M.; Walmsley, T.G. Process Integration and Electrification for Efficient Milk Evaporation Systems. *Energy* **2022**, *258*, 124885. [CrossRef]
16. Munir, M.T.; Young, B.R.; Yu, W. Can Exergy Be a Useful Tool for the Dairy Industry? In Proceedings of the 24th European Symposium on Computer Aided Process Engineering-ESCAPE 24, Budapest, Hungary, 18 June 2014; Elsevier B.V.: Amsterdam, The Netherlands, 2014; pp. 1129–1134.
17. Abd Manan, Z.; Alva-Argaez, A.; Arsenyeva, O.P.; Bandyopadhyay, S.; Berntsson, T.; Bogataj, M.; Bonhivers, J.-C.; Bulatov, I.; Chin, H.H.; Czamara, M.; et al. List of Contributors. In *Handbook of Process Integration (PI)*, 2nd ed.; Klemeš, J.J., Ed.; Woodhead Publishing Series in Energy; Woodhead Publishing: Sawston, UK, 2023; pp. xvii–xx. ISBN 978-0-12-823850-9.
18. Walmsley, T.G.; Atkins, M.J.; Tarighaleslami, A.H.; Liew, P.Y. Assisted Heat Transfer and Shaft Work Targets for Increased Total Site Heat Integration. *Chem. Eng. Trans.* **2016**, *52*, 403–408. [CrossRef]
19. Walmsley, T.G.; Atkins, M.J.; Ong, B.H.Y.; Klemeš, J.J.; Walmsley, M.R.W.; Varbanov, P.S. Total Site Heat Integration of Multi-Effect Evaporators with Vapour Recompression for Older Kraft Mills. *Chem. Eng. Trans.* **2017**, *61*, 265–270. [CrossRef]
20. Arpagaus, C.; Bless, F.; Uhlmann, M.; Schiffmann, J.; Bertsch, S.S. High Temperature Heat Pumps: Market Overview, State of the Art, Research Status, Refrigerants, and Application Potentials. *Energy* **2018**, *152*, 985–1010. [CrossRef]
21. Liang, J.; Andersen, M.P.; Solé, R.P.I.; Bellemo, L.; Bergamini, R.; Poulsen, J.L.; Zühlsdorf, B.; Schneider, P.; Jensen, J.K.; Markussen, W.B.; et al. Full Electrification Opportunities of Spray Dryers in Milk Powder Processes. In Proceedings of the ECOS 2022-The 35th International Conference on Efficiency, Cost, Optimization, Simulation and Environmental Impact of Energy Systems 2022, Copenhagen, Denmark, 3–7 July 2022.
22. Walmsley, T.G.; Atkins, M.J.; Walmsley, M.R.W.; Philipp, M.; Peesel, R.-H. Process and Utility Systems Integration and Optimisation for Ultra-Low Energy Milk Powder Production. *Energy* **2018**, *146*, 67–81. [CrossRef]
23. Lior, N.; Zhang, N. Energy, Exergy, and Second Law Performance Criteria. *Energy* **2007**, *32*, 281–296. [CrossRef]

energies

MDPI

Article

Optimal Dispatching Strategy for Textile-Based Virtual Power Plants Participating in GridLoad Interactions Driven by Energy Price

Tingyi Chai [1], Chang Liu [1], Yichuan Xu [1], Mengru Ding [2], Muyao Li [2], Hanyu Yang [2,*] and Xun Dou [2]

[1] Changzhou Power Supply Branch, State Grid Jiangsu Electric Power Co., Ltd., Changzhou 213004, China; chaity@js.sgcc.com.cn (T.C.); liuc9@js.sgcc.com.cn (C.L.); xuxtl@whu.edu.cn (Y.X.)

[2] College of Electrical Engineering and Control Science, Nanjing Tech University, Nanjing 211816, China; dmr@njtech.edu.cn (M.D.); limuyao@epri.sgcc.com.cn (M.L.); dxnjut@njtech.edu.cn (X.D.)

* Correspondence: hanyu93@njtech.edu.cn; Tel.: +86-13055176429

Abstract: The electricity consumption of the textile industry accounts for 2.12% of the total electricity consumption in society, making it one of the high-energy-consuming industries in China. The textile industry requires the use of a large amount of industrial steam at various temperatures during production processes, making its dispatch and operation more complex compared to conventional electricity–heat integrated energy systems. As an important demand-side management platform connecting the grid with distributed resources, a virtual power plant can aggregate textile industry users through an operator, regulating their energy consumption behavior and enhancing demand-side management efficiency. To effectively address the challenges in load regulation for textile industry users, this paper proposes a coordinated optimization dispatching method for electricity–steam virtual-based power plants focused on textile industrial parks. On one hand, targeting the impact of different energy prices on the energy usage behavior of textile industry users, an optimization dispatching model is established where the upper level consists of virtual power plant operators setting energy prices, and the lower level involves multiple textile industry users adjusting their purchase and sale strategies and changing their own energy usage behaviors accordingly. On the other hand, taking into account the energy consumption characteristics of steam, it is possible to optimize the production and storage behaviors of textile industry users during off-peak electricity periods in the power market. Through this electricity–steam optimization dispatching model, the virtual power plant operator's revenue is maximized while the operating costs for textile industry users are minimized. Case study analyses demonstrate that this strategy can effectively enhance the overall economic benefits of the virtual power plant.

Keywords: energy price; textile industry; virtual power plant; optimal dispatching; integrated energy system

1. Introduction

The textile industry, as a traditional pillar industry, has economic clusters whose scale approaches nearly 50% of the entire sector [1]. There is significant potential for energy conservation and consumption reduction within the textile industry [2]. Despite continuous enrichment of supply-side resources, the growing integration of distributed energy resources introduces output characteristics like randomness, intermittency, and fluctuations, which intensify challenges in regulating loads and allocating resources for textile industry users [3]. In this context, a virtual power plant (VPP) employs advanced information, communication, and control technologies to aggregate and optimize the management of distributed resources [4]. Acting as intermediaries between large-scale power systems and diverse distributed energy sources, a VPP is tailored to textile industrial parks to optimize energy dispatching by coordinating the energy consumption patterns of textile industry users. Furthermore, internally aggregated textile industry users and a virtual power plant operator (VPPO) engage in electricity markets through competitive bidding [5]. VPPO can additionally direct users to engage in demand response through strategic energy procurement and sales policies, fully capitalizing on the textile industry's diverse energy complementarity. This approach is critical for guiding industrial users towards efficient energy consumption and maximizing the overall benefits of a VPP.

A VPP aggregates generating units, renewable energy sources, energy storage systems, and various flexible loads, optimizing specific objective functions through coordinated control mechanisms [6]. To achieve supply–demand balance, many scholars have conducted research on the optimal dispatching problem of VPP. Reference [7] introduces a multi-objective economic optimization dispatching model for VPPs, which balances the interests of power suppliers and cooling/heating providers, enhances environmental performance, and ensures economically sustainable operations. Reference [8] develops a comprehensive energy optimization dispatching model for VPP considering electric–thermal conversion, minimizing integrated operational costs across multiple electricity markets. Reference [9] proposes a multi-agent internal optimization dispatching decision model for VPP considering risk preferences. It establishes a profit allocation model based on Nash–Harsanyi bargaining solutions, quantifying the actual contributions of distributed energy resources to VPP profits and enhancing member participation in markets. Reference [10] presents a method for solving the control boundary of industrial park-type VPP by considering power network constraints, achieving precise boundary solutions for VPP control in industrial parks. Reference [11] aggregates flexible electric–thermal loads in industrial production processes into a multi-energy VPP. It adjusts equipment operation states using measurement and control methods to provide frequency conversion services, meeting ancillary service market demands. Reference [12] proposes a two-stage robust resilience enhancement strategy considering thermal inertia for combined heat and power VPP, effectively improving their capability to withstand extreme disasters. Reference [13] establishes an assessment method for the aggregated response capability of VPP distributed resources. It optimizes and quantifies multi-dimensional indicators of aggregated resources and analyzes indicator weights to obtain an overall response capability score for the VPP. Reference [14] introduces an optimized dispatching strategy for a multi-energy complementary VPP by considering renewable energy integration. It constructs an operational mechanism and optimization dispatching model covering generation, grid, load, and storage, significantly enhancing renewable energy utilization and enabling low-carbon economic operations of VPP.

Currently, there are existing studies on the interaction behaviors between VPPO and users. Reference [15] constructs a Stackelberg game model involving VPPO, energy suppliers, and users, aimed at optimizing VPPO pricing strategies, energy supply-side dispatch plans, and user demand response. Reference [16] proposes a Stackelberg game pricing model between VPPO and electric vehicle (EV) users, aiming for mutual benefits and optimizing operational revenue. Reference [17] develops a tripartite electricity pricing game involving distributed energy resources, VPPO, and users to achieve Nash equilibrium in revenue for all parties. Reference [18] introduces a Stackelberg game model where VPPO engage in coordinated charging management of electric vehicles, using reasonable electricity pricing to guide orderly charging and coordinate various distributed resources in the electricity market. Reference [19] investigates strategies for VPPO managing controllable loads of industrial users, exploring methods to reduce interaction costs and optimize overall VPP performance. Reference [20] integrates carbon trading with demand response strategies to develop a Stackelberg game model between operators and users, leveraging demand-side resources for emission reduction and response capabilities. Reference [21] constructs a multi-energy dynamic pricing model for operators by setting price ranges, maximizing their own revenue while ensuring maximal benefits for users. Reference [22] proposes a bi-level optimization model for a VPP including electric vehicles, balancing the interests of VPPO and EV users in a principal-agent game framework.

The abovementioned studies primarily focus on the optimal dispatching strategy of specific electric–thermal integrated systems from a macro level regardless of its industrial characteristics. Each industry has its own production process with different coupling forms of multiple energy flows [23,24]. Considering this research gap, this paper takes the textile industry, which is listed as one of top six high-density industries, as an example. As the textile industry is usually developing in clusters, we introduce a textile-based VPPO to dominate multiple textile users in the industrial park. The main contributions of this paper are summarized as:

1. This paper first establishes a bi-level dispatching strategy for VPP targeting textile industrial parks to specially address the coupling characteristics of electricity and steam.
2. Considering the impact on the user behavior brought by the energy prices, this strategy is modeled as where VPPO aims to maximize its own revenue in the upper level, while multiple textile industry users aim to minimize total operational costs in the lower level, finally resulting in an economic increase of the overall system.
3. Considering that the textile industry users utilize electrically-driven industrial steam boilers under decarbonization, the storage-like characteristics of the steam accumulator (SA) is specially addressed.

2. Bi-Level Dispatching Framework for VPP

As shown in Figure 1, the overall framework of the VPP includes the VPPO and the textile industry user system.

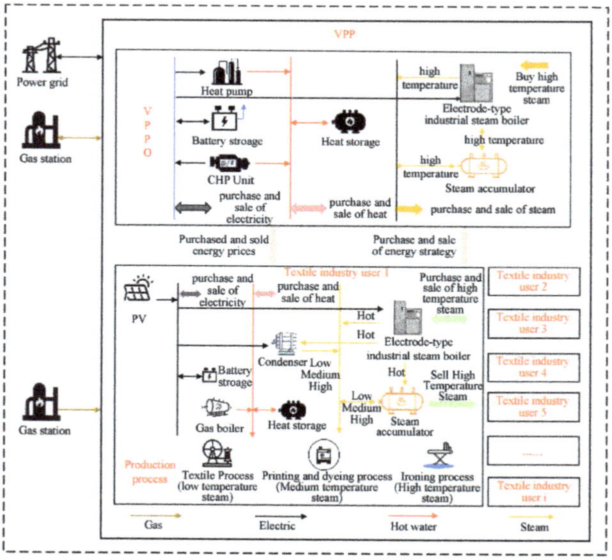

Figure 1. The overall structure of the VPP.

The VPPO structure consists of heat pumps, battery storage systems, combined heat and power units (CHP), electrode industrial steam boilers, and steam accumulators. VPPO has certain power and thermal capabilities, and can also provide a unified steam supply to textile industry users using electrode boilers and steam accumulators. When VPPO supply is insufficient, distributed supply is provided by textile industry users. These users have demand for electricity, heat, gas, and steam. When the industrial user's own system can meet both electricity and heating demands, and the external gas station can fulfill their natural gas needs, there is no requirement for interaction with the VPPO. Only when the user's own system fails to satisfy their energy demands do they engage with the VPPO to balance energy supply and demand.

In this framework, textile industry users initially prefer to use their own distributed photovoltaic (PV) systems to meet their electricity demands. When PV generation and discharge from their own energy storage systems cannot meet the users' electricity needs, users can engage in electricity transactions with VPPO to balance energy supply and demand. When there is surplus PV generation, users can store the excess electricity in their own energy storage systems and choose to sell the surplus electricity to VPPO.

Based on the conventional electricity purchasing strategies between VPPO and users, this study further explores the purchasing strategies for thermal energy and steam. Industrial users can purchase natural gas either from external gas stations or through interactions with the VPPO. Additionally, users can sell excess natural gas when there is an oversupply in the market. Users purchase natural gas to produce thermal energy using gas boilers, and engage in thermal energy transactions with VPPO. Users can sell surplus thermal energy when there is excess supply, or purchase additional thermal energy when demand increases. Additionally, users sell high-temperature steam through electrode industrial steam boilers and SA, while simultaneously purchasing high-temperature steam that enters the outlet of the electrode industrial steam boiler. Steam transacted between VPPO and textile industry users is consistently high-temperature steam. The energy purchase and sale prices, as well as the energy trading strategies between the VPPO and industrial users, are based on real-time interactions, with dynamic adjustments made on an hourly basis.

3. The Bi-Level Optimal Dispatching Model for VPPO and Users

3.1. The VPPO-Level Optimal Dispatching Model

3.1.1. VPPO-Level Objective Function

The VPPO, as the upper-level leader, specifies various energy prices for electricity, heat, and steam purchases and sales in each time period, guiding the textile industry users in their strategies for purchasing and selling multiple energy sources of electricity, heat, and steam, with the maximization of total revenue as the objective function. The VPPO's total revenue includes revenue from the purchase and sale of multiple energy sources between the VPPO and users, revenue from transactions between the VPPO and the external grid, and the VPPO's gas purchase costs, as Equation (1) shows:

$$\max F_{\text{VPPO}} = \sum_{t=1}^{T} (F_{\text{e},t} + F_{\text{h},t} + F_{\text{s},t} + F_{\text{e},t}^{\text{grid}} - C_{\text{g},t}^{\text{VPPO}}) \tag{1}$$

In Equation (1), F_{VPPO} represents the total revenue of the VPPO, $F_{\text{e},t}$, $F_{\text{h},t}$, $F_{\text{s},t}$, $F_{\text{e},t}^{\text{grid}}$ and $C_{\text{g},t}^{\text{VPPO}}$, respectively, represent the revenue from VPPO's purchase and sale of electricity, heat, and steam with users, revenue from transactions between VPPO and the external grid, and VPPO's gas purchase costs, in yuan.

$$F_{\text{e},t} = (\lambda_{\text{e},t}^{\text{su}} P_{\text{e},t}^{\text{su}} - \lambda_{\text{e},t}^{\text{bu}} P_{\text{e},t}^{\text{bu}})\Delta t \tag{2}$$

$$F_{\text{h},t} = (\lambda_{\text{h},t}^{\text{su}} P_{\text{h},t}^{\text{su}} - \lambda_{\text{h},t}^{\text{bu}} P_{\text{h},t}^{\text{bu}})\Delta t \tag{3}$$

$$F_{\text{s},t} = (\lambda_{\text{s},t}^{\text{su}} Q_{\text{s},t}^{\text{su}} - \lambda_{\text{s},t}^{\text{bu}} Q_{\text{s},t}^{\text{bu}})\Delta t \tag{4}$$

$$F_{\text{e},t}^{\text{grid}} = (\lambda_{\text{e},t}^{\text{sg}} P_{\text{e},t}^{\text{sg}} - \lambda_{\text{e},t}^{\text{bg}} P_{\text{e},t}^{\text{bg}})\Delta t \tag{5}$$

$$C_{\text{g},t}^{\text{VPPO}} = \lambda^{\text{gas}} V_t^{\text{VPPO}} \tag{6}$$

In Equations (2)–(6), $\lambda_{\text{e},t}^{\text{bu}}$, $\lambda_{\text{e},t}^{\text{su}}$, $\lambda_{\text{h},t}^{\text{bu}}$ and $\lambda_{\text{h},t}^{\text{su}}$, respectively, represent the electricity and heat prices for VPPO's purchase and sale with users at time t, yuan/kWh; $\lambda_{\text{e},t}^{\text{bg}}$ and $\lambda_{\text{e},t}^{\text{sg}}$ are the electricity grid prices for VPPO's purchase and sale at time t, yuan/kWh. $\lambda_{\text{s},t}^{\text{bu}}$ and $\lambda_{\text{s},t}^{\text{su}}$ are the prices of steam purchased and sold by VPPO to users at time t, yuan/kJ; $P_{\text{e},t}^{\text{bu}}$,

$P_{e,t}^{su}$, $P_{h,t}^{bu}$ and $P_{h,t}^{su}$, respectively, represent the electricity and thermal power purchased and sold by VPPO to users at time t, kW; $Q_{s,t}^{bu}$ and $Q_{s,t}^{su}$ are steam power purchased and sold by VPPO to users at time t, kJ/h; $P_{e,t}^{bg}$ and $P_{e,t}^{sg}$ are electricity grid power purchased and sold by VPPO at time, (kW); λ^{gas} and V_t^{VPPO} are the price of natural gas at time t(yuan/m^3) and amount of gas purchased (m^3).

The electricity, thermal power, and steam energy purchased and sold to users by VPPO can be represented as:

$$\begin{cases} P_{e,t}^{su} = \sum\limits_{i \in B} P_{e,i,t}^{ub} \\ P_{e,t}^{bu} = \sum\limits_{i \in S} P_{e,i,t}^{us} \end{cases} \tag{7}$$

$$\begin{cases} P_{h,t}^{su} = \sum\limits_{i \in B} P_{h,i,t}^{ub} \\ P_{h,t}^{bu} = \sum\limits_{i \in S} P_{h,i,t}^{us} \end{cases} \tag{8}$$

$$\begin{cases} Q_{s,t}^{su} = \sum\limits_{i \in B} M_{s,i,t}^{ub} h_{s,i,t}^{ub} = \sum\limits_{i \in B} Q_{s,i,t}^{ub} \\ Q_{s,t}^{bu} = \sum\limits_{i \in S} M_{s,i,t}^{us} h_{s,i,t}^{us} = \sum\limits_{i \in S} Q_{s,i,t}^{us} \end{cases} \tag{9}$$

In Equations (7)–(9), $i \in B$ represents the set of users purchasing electricity, heat, and steam; $i \in S$ represents the sold of users purchasing electricity, heat, and steam; $P_{e,i,t}^{ub}$, $P_{e,i,t}^{us}$, $P_{h,i,t}^{ub}$ and $P_{h,i,t}^{us}$, respectively, represent the electricity and heat power purchased and sold by user i to VPPO at time t, kW; $M_{s,i,t}^{ub}$, $h_{s,i,t}^{ub}$, $M_{s,i,t}^{us}$ and $h_{s,i,t}^{us}$, respectively, represent the steam flow rate and steam enthalpy purchased and sold by user i to VPPO at time t. $Q_{s,i,t}^{ub}$ and $Q_{s,i,t}^{ub}$, respectively, represent the steam power purchased and sold by user i to VPPO at time t, kJ/h.

3.1.2. Energy Price Constraint

The energy price constraints for VPPO's sale of electricity, heat, and steam to textile industry users are shown in Equations (10)–(12). To prevent sustained high prices for the multi-energy sold by VPPO, average price constraints are set for VPPO's sale prices of electricity, heat, and steam, respectively [25].

$$\begin{cases} \lambda_e^{min} \le \lambda_{e,t}^{bu} \le \lambda_{e,t}^{sg} \\ \lambda_{e,t}^{bu} \le \lambda_{e,t}^{su} \le \lambda_{e,t}^{bg} \\ \sum\limits_{t=1}^{T} \lambda_{e,t}^{su} \le 24\lambda_{e,ave}^{su} \end{cases} \tag{10}$$

$$\begin{cases} \lambda_h^{min} \le \lambda_{h,t}^{bu} \le \lambda_h^{max} \\ \lambda_h^{min} \le \lambda_{h,t}^{su} \le \lambda_h^{max} \\ \sum\limits_{t=1}^{T} \lambda_{h,t}^{su} \le 24\lambda_{h,ave}^{su} \end{cases} \tag{11}$$

$$\begin{cases} \lambda_s^{min} \le \lambda_{s,t}^{bu} \le \lambda_s^{max} \\ \lambda_s^{min} \le \lambda_{s,t}^{su} \le \lambda_s^{max} \\ \sum\limits_{t=1}^{T} \lambda_{s,t}^{su} \le 24\lambda_{s,ave}^{su} \end{cases} \tag{12}$$

The minimum price at which VPPO purchases electricity from textile industry users is λ_e^{min}, while the grid selling price serves as the maximum price at which VPPO sells electricity to textile industry users; the upper and lower limits of VPPO's buying and selling prices for thermal energy and steam to textile industry users are λ_h^{max}, λ_h^{min}, λ_s^{max}

and $\lambda_{\mathrm{s}}^{\min}$; and the average prices at which VPPO sells electricity, heat, and steam are $\lambda_{\mathrm{e,ave}}^{\mathrm{su}}$, $\lambda_{\mathrm{h,ave}}^{\mathrm{su}}$, and $\lambda_{\mathrm{s,ave}}^{\mathrm{su}}$.

3.1.3. VPPO's Constraints on Multi-Energy Transaction Volumes with Users

In Equations (13)–(15), $P_{\mathrm{e,max}}^{\mathrm{grid}}$ represents the maximum electricity volume at which VPPO can transact with the grid in each time period; $P_{\mathrm{h,max}}^{\mathrm{VPPO}}$ represents the maximum heat volume VPPO can transact with users in each time period; $Q_{\mathrm{s,max}}^{\mathrm{VPPO}}$ represents the maximum steam volume VPPO can transact with users in each time period.

$$\begin{cases} 0 \leq P_{\mathrm{e},t}^{\mathrm{su}} \leq P_{\mathrm{e,max}}^{\mathrm{grid}} \\ 0 \leq P_{\mathrm{e},t}^{\mathrm{bu}} \leq P_{\mathrm{e,max}}^{\mathrm{grid}} \end{cases} \tag{13}$$

$$\begin{cases} 0 \leq P_{\mathrm{h},t}^{\mathrm{su}} \leq P_{\mathrm{h,max}}^{\mathrm{VPPO}} \\ 0 \leq P_{\mathrm{h},t}^{\mathrm{bu}} \leq P_{\mathrm{h,max}}^{\mathrm{VPPO}} \end{cases} \tag{14}$$

$$\begin{cases} 0 \leq Q_{\mathrm{s},t}^{\mathrm{su}} \leq Q_{\mathrm{s,max}}^{\mathrm{IPO}} \\ 0 \leq Q_{\mathrm{s},t}^{\mathrm{bu}} \leq Q_{\mathrm{s,max}}^{\mathrm{IPO}} \end{cases} \tag{15}$$

3.1.4. VPPO-Level Power Balance Constraint

The upper-level model centers around the VPPO, with electrical energy inputs including purchases from the public grid, electricity generation from CHP, discharges from electric energy storage, and purchases from textile industry users. Electrical energy outputs consist of electricity sales to the grid, electricity consumption by heat pumps, electricity consumption by electrode industrial steam boilers, charging of electric energy storage, and electricity sales to textile industry users. VPPO's thermal energy inputs include heat production from heat pumps, heat production from cogeneration units, heat release from thermal energy storage, and purchases of heat from textile industry users. Thermal energy outputs include charging of thermal energy storage and thermal energy sales to textile industry users. VPPO's steam energy inputs comprise steam production from electrode industrial steam boilers, steam release from SA, and steam purchases from textile industry users. Steam energy outputs include steam charging into SA and steam sales to textile industry users. The power balance is shown in Equations (16)–(18):

$$P_{\mathrm{e},t}^{\mathrm{bg}} + P_{\mathrm{e},t}^{\mathrm{CHP}} + P_{\mathrm{e},t}^{\mathrm{VPPOdis}} + P_{\mathrm{e},t}^{\mathrm{bu}} = P_{\mathrm{e},t}^{\mathrm{HP}} + P_{\mathrm{e},t}^{\mathrm{ESB\_VPPO}} + P_{\mathrm{e},t}^{\mathrm{VPPOcha}} + P_{\mathrm{e},t}^{\mathrm{su}} + P_{\mathrm{e},t}^{\mathrm{sg}} \tag{16}$$

$$P_{\mathrm{h},t}^{\mathrm{HP}} + P_{\mathrm{h},t}^{\mathrm{CHP}} + P_{\mathrm{h},t}^{\mathrm{VPPOdis}} + P_{\mathrm{h},t}^{\mathrm{bu}} = P_{\mathrm{h},t}^{\mathrm{VPPOcha}} + P_{\mathrm{h},t}^{\mathrm{su}} \tag{17}$$

$$Q_{t}^{\mathrm{ESB\_VPPO}} + Q_{\mathrm{s},t}^{\mathrm{bu}} + M_{t}^{\mathrm{VPPOSAdis}} h^{\mathrm{VPPOSAdis}} \Delta t = Q_{\mathrm{s},t}^{\mathrm{su}} + M_{t}^{\mathrm{VPPOSAcha}} h^{\mathrm{VPPOSAcha}} \Delta t \tag{18}$$

3.2. *User-Level Optimal Dispatching Model*

3.2.1. User-Level Objective Function

The user-level optimization and dispatching model minimizes the daily operational costs of textile industry users as its objective function.

$$\min C_{\mathrm{u}} = \sum_{i=1}^{N} C_{\mathrm{e},i} + C_{\mathrm{h},i} + C_{\mathrm{s},i} + C_{\mathrm{g},i}^{\mathrm{u}} \tag{19}$$

$$C_{\mathrm{e},i} = \sum_{t=1}^{T} (\lambda_{\mathrm{e},t}^{\mathrm{su}} P_{\mathrm{e},i,t}^{\mathrm{ub}} - \lambda_{\mathrm{e},t}^{\mathrm{bu}} P_{\mathrm{e},i,t}^{\mathrm{us}}) \tag{20}$$

$$C_{\mathrm{h},i} = \sum_{t=1}^{T} (\lambda_{\mathrm{h},t}^{\mathrm{su}} P_{\mathrm{h},i,t}^{\mathrm{ub}} - \lambda_{\mathrm{h},t}^{\mathrm{bu}} P_{\mathrm{h},i,t}^{\mathrm{us}}) \tag{21}$$

$$C_{s,i}=\sum_{t=1}^{T}(\lambda_{s,t}^{su}Q_{s,i,t}^{ub}-\lambda_{s,t}^{bu}Q_{s,i,t}^{us}) \tag{22}$$

$$C_{g,i}^{u}=\sum_{t=1}^{T}\lambda^{gas}V_{i,t}^{GB} \tag{23}$$

In Equations (19)–(23): C_u represents the total cost of textile industry users; $C_{e,i}$, $C_{h,i}$, $C_{s,i}$ and $C_{g,i}^{u}$, respectively, represent the costs in yuan for textile industry user i to purchase or sell VPPO electricity, heat, steam, and natural gas.

3.2.2. Constraints on Purchasing and Selling Electricity, Heat, and Steam Power

Textile industry users cannot be in a buying and selling state simultaneously. There are maximum transaction volume constraints for buying and selling electricity, heat, and steam.

$$\begin{cases} 0\leq P_{e,i,t}^{us}\leq \omega_{e,i,t}P_{e,max} \\ 0\leq P_{e,i,t}^{ub}\leq (1-\omega_{e,i,t})P_{e,max} \end{cases} \tag{24}$$

$$\begin{cases} 0\leq P_{h,i,t}^{us}\leq \omega_{h,i,t}P_{h,max} \\ 0\leq P_{h,i,t}^{ub}\leq (1-\omega_{h,i,t})P_{h,max} \end{cases} \tag{25}$$

$$\begin{cases} 0\leq Q_{s,i,t}^{us}\leq \omega_{s,i,t}Q_{s,max} \\ 0\leq Q_{s,i,t}^{ub}\leq (1-\omega_{s,i,t})Q_{s,max} \end{cases} \tag{26}$$

In Equations (24)–(26): $\omega_{e,i,t}$, $\omega_{h,i,t}$, $\omega_{s,i,t}$ are binary variables ranging from 0 to 1, representing the buying and selling statuses of textile industry user i for electricity, heat, and steam during time period t. When $\omega_{e,i,t}=1$, the user sells electricity. Otherwise, the user buys electricity. When $\omega_{h,i,t}=1$, the user sells heat. Otherwise, the user buys heat. When $\omega_{s,i,t}=1$, the user sells steam. Otherwise, the user buys steam. The maximum values for buying and selling electricity, heat, and steam between VPPO and users are $P_{e,max}$ for electricity, $P_{h,max}$ for heat, and $Q_{s,max}$ for steam.

Considering the presence of nonlinear terms in Equations (23) and (24), a linearization of these equations is performed.

$$\begin{cases} P_{e,i,t}^{us}\geq 0 \\ P_{e,i,t}^{us}\leq x \\ x\geq P_{e,max}-(1-\omega_{e,i,t})M \\ P_{e,i,t}^{us}\leq P_{e,max} \\ P_{e,i,t}^{us}\leq \omega_{e,i,t}M \\ P_{e,i,t}^{us}\geq -\omega_{e,i,t}M \end{cases} \quad \begin{cases} P_{e,i,t}^{ub}\geq 0 \\ P_{e,i,t}^{ub}-P_{e,max}\leq y \\ y\geq -P_{e,max}-(1-\omega_{e,i,t})M \\ P_{e,i,t}^{us}\leq -P_{e,max} \\ P_{e,i,t}^{us}\leq \omega_{e,i,t}M \\ P_{e,i,t}^{us}\geq -\omega_{e,i,t}M \end{cases} \tag{27}$$

$$\begin{cases} P_{h,i,t}^{us}\geq 0 \\ P_{h,i,t}^{us}\leq \alpha \\ \alpha\geq P_{h,max}-(1-\omega_{h,i,t})M \\ P_{h,i,t}^{us}\leq P_{h,max} \\ P_{h,i,t}^{us}\leq \omega_{h,i,t}M \\ P_{h,i,t}^{us}\geq -\omega_{h,i,t}M \end{cases} \quad \begin{cases} P_{h,i,t}^{ub}\geq 0 \\ P_{h,i,t}^{ub}-P_{h,max}\leq \beta \\ \beta\geq -P_{h,max}-(1-\omega_{h,i,t})M \\ P_{h,i,t}^{us}\leq -P_{h,max} \\ P_{h,i,t}^{us}\leq \omega_{h,i,t}M \\ P_{h,i,t}^{us}\geq -\omega_{h,i,t}M \end{cases} \tag{28}$$

$$\begin{cases} Q^{\text{us}}_{s,i,t} \geq 0 \\ Q^{\text{us}}_{s,i,t} \leq \delta \\ \delta \geq Q_{\text{s,max}} - (1-\omega_{s,i,t})M \\ Q^{\text{us}}_{s,i,t} \leq Q_{\text{s,max}} \\ Q^{\text{us}}_{s,i,t} \leq \omega_{s,i,t}M \\ Q^{\text{us}}_{s,i,t} \geq -\omega_{s,i,t}M \end{cases} \quad \begin{cases} Q^{\text{ub}}_{s,i,t} \geq 0 \\ Q^{\text{ub}}_{s,i,t} - Q_{\text{s,max}} \leq \varepsilon \\ \varepsilon \geq -Q_{\text{s,max}} - (1-\omega_{s,i,t})M \\ Q^{\text{ub}}_{s,i,t} \leq -Q_{\text{s,max}} \\ Q^{\text{ub}}_{s,i,t} \leq \omega_{s,i,t}M \\ Q^{\text{ub}}_{s,i,t} \geq -\omega_{s,i,t}M \end{cases} \tag{29}$$

In Equations (27)–(29): M is a sufficiently large number, and $x, y, \alpha, \beta, \delta, \varepsilon$ are continuous variables.

3.2.3. User-Level Power Balance Constraint

The lower-level model primarily consists of textile industry users. Electrical energy input includes distributed photovoltaic generation $P^{\text{PV}}_{i,t}$, energy storage discharge, and purchasing from VPPO. Electrical energy output includes selling electricity to VPPO, base electrical load $P^{\text{base}}_{\text{e},i,t}$, energy storage charging, electric boiler consumption for industrial steam, condenser consumption, and electricity demands of various production processes. Thermal energy input for textile industry users includes purchasing heat from VPPO, heating supplied by gas boilers, and heat release from thermal energy storage. Thermal energy output includes selling heat to VPPO, base thermal load $P^{\text{base}}_{\text{h},i,t}$, thermal energy storage charging, and thermal requirements of various production processes. Steam energy input for textile industry users includes steam output provided by condensers to various processes, steam output directly supplied to ironing processes by electric industrial steam boilers, and steam energy release provided to various processes by SA. Steam energy output corresponds to steam requirements of various production processes.

$$\begin{aligned} P^{\text{PV}}_{i,t} + P^{\text{ub}}_{\text{e},i,t} + P^{\text{Udis}}_{\text{e},i,t} &= P^{\text{base}}_{\text{e},i,t} + P^{\text{us}}_{\text{e},i,t} + P^{\text{Ucha}}_{\text{e},i,t} + P^{\text{ESB}}_{i,t} + \sum_{m=1}^{N_{\text{te1}}} P^{\text{CON\_te1}}_{i,m,t} + \sum_{n=1}^{N_{\text{pd2}}} P^{\text{CON\_pd2}}_{i,n,t} \\ &+ \sum_{l=1}^{N_{\text{ir3}}} P^{\text{CON\_ir3}}_{i,l,t} + D^{\text{e\_te1}}_{i} \sum_{m=1}^{N_{\text{te1}}} v^{\text{te1}}_{i,m,t} + D^{\text{e\_pd2}}_{i} \sum_{n=1}^{N_{\text{pd2}}} v^{\text{pd2}}_{i,n,t} + D^{\text{e\_ir3}}_{i} \sum_{l=1}^{N_{\text{ir3}}} v^{\text{ir3}}_{i,l,t} \end{aligned} \tag{30}$$

$$P^{\text{ub}}_{\text{h},i,t} + P^{\text{Udis}}_{\text{h},i,t} + Q^{\text{GB}}_{i,t} = P^{\text{base}}_{\text{h},i,t} + P^{\text{us}}_{\text{h},i,t} + P^{\text{Ucha}}_{\text{h},i,t} + D^{\text{h\_te1}}_{i} \sum_{m=1}^{N_{\text{te1}}} v^{\text{te1}}_{i,m,t} + D^{\text{h\_pd2}}_{i} \sum_{n=1}^{N_{\text{pd2}}} v^{\text{pd2}}_{i,n,t} + D^{\text{h\_ir3}}_{i} \sum_{l=1}^{N_{\text{ir3}}} v^{\text{ir3}}_{i,l,t} \tag{31}$$

$$\begin{aligned} &\sum_{m=1}^{N_{\text{te1}}} (Q^{\text{CONdis\_te1}}_{i,m,t} + M^{\text{SAdis\_te1}}_{i,m,t} h^{\text{SAdis\_te1}}_{i,m,t} \Delta t) + \sum_{n=1}^{N_{\text{pd2}}} (Q^{\text{CONdis\_pd2}}_{i,n,t} + M^{\text{SAdis\_pd2}}_{i,n,t} h^{\text{SAdis\_pd2}}_{i,n,t} \Delta t) + \\ &\sum_{l=1}^{N_{\text{ir3}}} (Q^{\text{CONdis\_ir3}}_{i,l,t} + M^{\text{SAdis\_ir3}}_{i,l,t} h^{\text{SAdis\_ir3}}_{i,l,t} \Delta t + Q^{\text{ESB\_ir3}}_{i,l,t}) = D^{\text{s\_te1}}_{i} \sum_{m=1}^{N_{\text{te1}}} v^{\text{te1}}_{i,m,t} + D^{\text{s\_pd2}}_{i} \sum_{n=1}^{N_{\text{pd2}}} v^{\text{pd2}}_{i,n,t} + D^{\text{s\_ir3}}_{i} \sum_{l=1}^{N_{\text{ir3}}} v^{\text{ir3}}_{i,l,t} \end{aligned} \tag{32}$$

In Equations (27)–(29): $D^{\text{e\_te1}}_{i}$, $D^{\text{e\_}pd2}_{i}$, $D^{\text{e\_ir3}}_{i}$ represent the electricity demand of production processes for user i per unit time; $D^{\text{e\_te1}}_{i}$, $D^{\text{e\_pd2}}_{i}$, and $D^{\text{e\_ir3}}_{i}$, respectively, represent the heat demand of production processes for user i per unit time; $D^{\text{s\_te1}}_{i}$, $D^{\text{s\_pd2}}_{i}$, and $D^{\text{s\_ir3}}_{i}$, respectively, represent the steam demand of production processes for user i per unit time.

3.3. Solution Process

Based on the EMP algorithm package, utilizing the BARON solver within the GAMS environment, the Bi-level model described above is solved. The solving process of the textile-based virtual power plant constructed in this chapter to participate in the optimal dispatching strategy of grid-load interaction driven by energy price is shown in Figure 2.

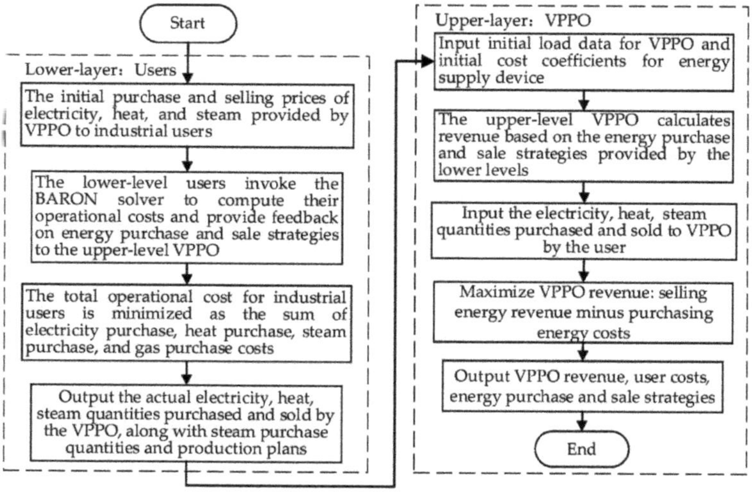

Figure 2. Solving process.

Initially, input the electric, heat, and steam purchase and sale prices provided by VPPO to textile industry users. These initial energy prices enter the lower-level model with the objective of minimizing total operational costs for textile industry users. The model optimizes the actual purchase and sale strategies of electric, heat, and steam from VPPO, along with user gas purchase strategies and production plans. The optimized strategies for electric, heat, and steam purchases from VPPO are then transmitted to the upper-level model. The upper-level model aims to maximize VPPO revenue by optimizing the energy conversion equipment and dispatching plans. It updates the purchase and sale prices of electric, heat, and steam provided to textile industry users and feeds this information back to the lower-level model as model parameters. This iterative process continues to search for the optimal solution, ultimately determining the purchase and sale strategies of electric, heat, and steam that maximize VPPO revenue while minimizing user costs.

4. Constraints on the Operation of Various Types of Equipment

4.1. Modeling of Multi-Energy Coupling Devices by VPPO

4.1.1. Electrode Industrial Steam Boilers

The electrical–steam energy conversion constraints of electrode industrial steam boilers and the constraints on the high-temperature steam flow output from electrode industrial steam boilers are as follows [26].

$$Q_t^{\mathrm{ESB}} = 3600\eta^{\mathrm{ESB}} P_t^{\mathrm{ESB}} \Delta t \tag{33}$$

$$Q_t^{\mathrm{ESB}} = M_t^{\mathrm{ESB}} h^{\mathrm{ESB}} \Delta t \tag{34}$$

In Equations (30) and (31), The electrical–steam energy conversion constraints of electrode industrial steam boilers include the output steam energy Q_t^{ESB}, input electrical power P_t^{ESB} and thermal efficiency η^{ESB}; M_t^{ESB} represents the steam flow output from the electrode industrial steam boiler at time t; h^{ESB} represents the enthalpy value of the steam output from the electrode industrial steam boiler.

4.1.2. Steam Accumulator

VPPO and users can only buy and sell high-temperature steam, and the steam released by the SA is also high-temperature steam.

$$\begin{cases} 0 \leq M_t^{\text{VPPOSAcha}} \leq \omega_t^{\text{VPPOcha}} M_{\max}^{\text{VPPOSAcha}} \\ 0 \leq M_t^{\text{VPPOSAdis}} \leq \omega_t^{\text{VPPOdis}} M_{\max}^{\text{VPPOSAdis}} \\ \omega_t^{\text{VPPOcha}}, \omega_t^{\text{VPPOdis}} \in \{0,1\} \\ S_{\min}^{\text{VPPOSA}} \leq S_t^{\text{VPPOSA}} \leq S_{\max}^{\text{VPPOSA}} \\ S_1^{\text{VPPOSA}} = S_T^{\text{VPPOSA}} \\ S_t^{\text{VPPOSA}} = S_{t-1}^{\text{VPPOSA}} + M_t^{\text{VPPOSAcha}} h^{\text{VPPOSAcha}} \Delta t - M_t^{\text{VPPOSAdis}} h^{\text{VPPOSAdis}} \Delta t \end{cases} \tag{35}$$

In Equation (32), $\omega_t^{\text{VPPOcha}}$ and $\omega_t^{\text{VPPOdis}}$ are both binary variables ranging from 0 to 1. When $\omega_t^{\text{VPPOcha}} = 1$, SA injects steam; otherwise, it does not inject steam. When $\omega_t^{\text{VPPOdis}} = 1$, SA releases steam; otherwise, it does not release steam. $M_{\max}^{\text{VPPOSAcha}}$, $M_{\max}^{\text{VPPOSAdis}}$ are maximum steam flow rates for SA's steam charging and discharging; At time t, the remaining capacity S_t^{VPPOSA} is constrained by $S_{\min}^{\text{VPPOSA}}$ and $S_{\max}^{\text{VPPOSA}}$. The remaining capacity of S_1^{VPPOSA} at the initial time is equal to the remaining capacity of S_T^{VPPOSA} at the end of the optimization period; $M_t^{\text{VPPOSAcha}}$, $M_t^{\text{VPPOSAdis}}$, respectively, represent the steam inflow and outflow rates of electrode industrial steam boiler SA at time t; $h^{\text{VPPOSAcha}}$, $h^{\text{VPPOSAdis}}$, respectively, represent the enthalpy values of high-temperature steam intake and discharge for SA. Constraints on energy storage and thermal storage equipment are described here, while constraints on heat pump units and combined heat and power units are not discussed in detail.

4.2. Multi-Energy Coupling Equipment Model for Textile Industry Users

4.2.1. Electrode Industrial Steam Boiler for Users

Part of the steam sold by users is provided by electrode industrial steam boilers, and all steam purchased by users enters the outlet of electrode industrial steam boilers. Therefore, the modeling of electrode industrial steam boilers for textile industry users is as follows.

$$Q_{i,t}^{\text{ESB}} = 3600\eta^{\text{ESB}} P_{i,t}^{\text{ESB}} \Delta t \tag{36}$$

$$Q_{i,t}^{\text{ESB}} = M_{i,t}^{\text{ESB}} h^{\text{ESB}} \Delta t \tag{37}$$

$$Q_{i,t}^{\text{ESB}} + Q_{s,i,t}^{\text{ub}} = Q_{i,t}^{\text{CONcha}} + Q_{i,t}^{\text{SAcha}} + \sum_{l=1}^{N_{\text{ir3}}} Q_{i,l,t}^{\text{ESB\_ir3}} + Q_{i,t}^{\text{ESB}\rightarrow\text{VPPO}} \tag{38}$$

$$Q_{i,t}^{\text{SAcha}} = M_{i,t}^{\text{SAcha}} h^{\text{SAcha}} \Delta t \tag{39}$$

$$Q_{i,t}^{\text{CONcha}} = M_{i,t}^{\text{CONcha}} h^{\text{CONcha}} \Delta t \tag{40}$$

$$Q_{i,l,t}^{\text{ESB\_ir3}} = M_{i,l,t}^{\text{ESB\_ir3}} h^{\text{ESB\_ir3}} \Delta t \tag{41}$$

$$Q_{i,t}^{\text{ESB}\rightarrow\text{VPPO}} = M_{i,t}^{\text{ESB}\rightarrow\text{VPPO}} h^{\text{ESB}\rightarrow\text{VPPO}} \Delta t \tag{42}$$

$$0 \leq P_{i,t}^{\text{ESB}} \leq P_{i,\max}^{\text{ESB}} \tag{43}$$

$$0 \leq M_{i,t}^{\text{ESB}} \leq M_{i,\max}^{\text{ESB}} \tag{44}$$

In Equations (33)–(41), At time t, the steam output energy of electrode industrial steam boilers $Q_{i,t}^{\text{ESB}}$ depends on the electrical power input $P_{i,t}^{\text{ESB}}$; $M_{i,t}^{\text{ESB}}$ represents the steam flow rate outputted by user i's electrode industrial steam boiler at time t; The steam energy purchased at the outlet of electrode industrial steam boilers by user i is $Q_{s,i,t}^{\text{ub}}$, which represents the steam energy purchased by user i; The steam energy sold by electrode industrial steam boilers $Q_{i,t}^{\text{ESB}\rightarrow\text{VPPO}}$ depends on the steam flow rate $M_{i,t}^{\text{ESB}\rightarrow\text{VPPO}}$. Since the outlet high-temperature steam temperature of electrode industrial steam boilers is constant, $h^{\mathrm{ESB}\rightarrow\mathrm{VPPO}}$ remains constant; $Q_{i,t}^{\mathrm{CONcha}}$, $M_{i,t}^{\mathrm{CONcha}}$, respectively, represent the steam energy input to the condenser by user i's electrode industrial steam boiler and the steam flow rate at time t; the steam energy input to the SA by user i's electrode industrial steam boiler and the steam flow rate at time t are $Q_{i,t}^{\mathrm{SAcha}}$ and $M_{i,t}^{\mathrm{SAcha}}$; $Q_{i,l,t}^{\mathrm{ESB\_ir3}}$, $M_{i,l,t}^{\mathrm{ESB\_ir3}}$ represent the steam energy and steam flow rate directly delivered by user i's electrode industrial steam boiler to the l-th ironing process production line at time t; $P_{i,\max}^{\mathrm{ESB}}$ is the upper limit of the input electrical power for user i's electrode industrial steam boiler, and $M_{i,\max}^{\mathrm{ESB}}$ is the upper limit of the steam flow rate output for user i's electrode industrial steam boiler.

4.2.2. User's Condenser

The energy balance constraints on condenser input, the relationship between the steam energy delivered by each process's condenser and the electrical power consumed by the condenser, the output steam energy balance constraints, the enthalpy (temperature) constraints on steam delivered to each process, and the constraints on output steam flow rate can be expressed as:

$$Q_t^{\mathrm{CON}} = Q_t^{\mathrm{CONcha}} + 3600\left(\sum_{m=1}^{N_{\mathrm{te1}}} P_{m,t}^{\mathrm{CON\_te1}} + \sum_{n=1}^{N_{\mathrm{pd2}}} P_{n,t}^{\mathrm{CON\_pd2}} + \sum_{l=1}^{N_{\mathrm{ir3}}} P_{l,t}^{\mathrm{CON\_ir3}}\right)\Delta t \tag{45}$$

$$\begin{cases} P_{m,t}^{\mathrm{CON\_te1}}\Delta t = Q_{m,t}^{\mathrm{CONdis\_te1}}\left(\frac{\eta_{\mathrm{e}}^{\mathrm{CON}}}{\eta_{\mathrm{s}}^{\mathrm{CON}}}\right)/3600 \\ P_{n,t}^{\mathrm{CON\_pd2}}\Delta t = Q_{n,t}^{\mathrm{CONdis\_pd2}}\left(\frac{\eta_{\mathrm{e}}^{\mathrm{CON}}}{\eta_{\mathrm{s}}^{\mathrm{CON}}}\right)/3600 \\ P_{l,t}^{\mathrm{CON\_ir3}}\Delta t = Q_{l,t}^{\mathrm{CONdis\_ir3}}\left(\frac{\eta_{\mathrm{e}}^{\mathrm{CON}}}{\eta_{\mathrm{s}}^{\mathrm{CON}}}\right)/3600 \end{cases} \tag{46}$$

$$Q_t^{\mathrm{CON}}\frac{\eta_{\mathrm{s}}^{\mathrm{CON}}}{1+\eta_{\mathrm{e}}^{\mathrm{CON}}} = \sum_{m=1}^{N_{\mathrm{te1}}} Q_{m,t}^{\mathrm{CONdis\_te1}} + \sum_{n=1}^{N_{\mathrm{pd2}}} Q_{n,t}^{\mathrm{CONdis\_pd2}} + \sum_{l=1}^{N_{\mathrm{ir3}}} Q_{l,t}^{\mathrm{CONdis\_ir3}} \tag{47}$$

$$\begin{cases} Q_{m,t}^{\mathrm{CONdis\_te1}} = M_{m,t}^{\mathrm{CON\_te1}} h_{m,t}^{\mathrm{CON\_te1}}\Delta t \\ Q_{n,t}^{\mathrm{CONdis\_pd2}} = M_{n,t}^{\mathrm{CON\_pd2}} h_{n,t}^{\mathrm{CON\_pd2}}\Delta t \\ Q_{l,t}^{\mathrm{CONdis\_ir3}} = M_{l,t}^{\mathrm{CON\_ir3}} h_{l,t}^{\mathrm{CON\_ir3}}\Delta t \end{cases} \tag{48}$$

$$\sum_{m=1}^{N_{\mathrm{te1}}} M_{m,t}^{\mathrm{CON\_te1}} + \sum_{n=1}^{N_{\mathrm{pd2}}} M_{n,t}^{\mathrm{CON\_pd2}} + \sum_{l=1}^{N_{\mathrm{ir3}}} M_{l,t}^{\mathrm{CON\_ir3}} \le M_{\max}^{\mathrm{CONdis}} \tag{49}$$

In Equations (42)–(46), The production lines of spinning, dyeing/printing, and ironing processes are denoted by m, n, and l; N_{te1}, N_{pd2}, and N_{ir3} represent the quantities of production lines for spinning, dyeing/printing, and ironing processes, respectively; Q_t^{CON} represents the total energy input of the condenser at time t, kJ; Q_t^{CONcha} represents the steam energy input of the condenser at time t; $P_{m,t}^{\mathrm{CON\_te1}}$, $P_{n,t}^{\mathrm{CON\_pd2}}$, and $P_{l,t}^{\mathrm{CON\_ir3}}$ represent the electrical power required by the condenser at time t to supply steam to the production lines of spinning, dyeing/printing, and ironing processes; $\eta_{\mathrm{e}}^{\mathrm{CON}}$, $\eta_{\mathrm{s}}^{\mathrm{CON}}$ represent the power consumption efficiency and heating efficiency of the condenser; $Q_{m,t}^{\mathrm{CONdis\_te1}}$, $Q_{n,t}^{\mathrm{CONdis\_pd2}}$, $Q_{l,t}^{\mathrm{CONdis\_ir3}}$, $M_{m,t}^{\mathrm{CON\_te1}}$, $M_{n,t}^{\mathrm{CON\_pd2}}$, $M_{l,t}^{\mathrm{CON\_ir3}}$, $h_{m,t}^{\mathrm{CON\_te1}}$, $h_{n,t}^{\mathrm{CON\_pd2}}$, and $h_{l,t}^{\mathrm{CON\_ir3}}$, respectively, represent the steam energy, steam flow rate, and steam enthalpy delivered by the condenser to each production line of various processes at time t; the condenser's maximum output steam flow rate is $M_{\max}^{\mathrm{CONdis}}$.

4.2.3. User's Steam Accumulator

Part of the steam sold by textile industry users is provided by SA; therefore, the SA modeling for textile industry user i can be represented as:

$$\begin{cases} 0 \leq M_{i,t}^{\mathrm{SAcha}} \leq \omega_{i,t}^{\mathrm{cha}} M_{i,\max}^{\mathrm{SAcha}} \\ 0 \leq \sum\limits_{m=1}^{N_{\mathrm{te1}}} M_{i,m,t}^{\mathrm{SAdis\_te1}} + \sum\limits_{n=1}^{N_{\mathrm{pd2}}} M_{i,n,t}^{\mathrm{SAdis\_pd2}} + \sum\limits_{l=1}^{N_{\mathrm{ir3}}} M_{i,l,t}^{\mathrm{SAdis\_ir3}} \leq \omega_{i,t}^{\mathrm{dis}} M_{i,\max}^{\mathrm{SAdis}} \\ \omega_{i,t}^{\mathrm{cha}}, \omega_{i,t}^{\mathrm{dis}} \in \{0,1\} \\ S_{i,\min}^{\mathrm{SA}} \leq S_{i,t}^{\mathrm{SA}} \leq S_{i,\max}^{\mathrm{SA}} \\ S_{i,1}^{\mathrm{SA}} = S_{i,T}^{\mathrm{SA}} \\ S_{i,t}^{\mathrm{SA}} = S_{i,t-1}^{\mathrm{SA}} + M_{i,t}^{\mathrm{SAcha}} h^{\mathrm{SAcha}} \Delta t - M_{i,t}^{\mathrm{SA}\to\mathrm{VPPO}} h^{\mathrm{SA}\to\mathrm{VPPO}} \Delta t - \\ (\sum\limits_{m=1}^{N_{\mathrm{te1}}} M_{i,m,t}^{\mathrm{SAdis\_te1}} h_{i,m,t}^{\mathrm{SAdis\_te1}} + \sum\limits_{n=1}^{N_{\mathrm{pd2}}} M_{i,n,t}^{\mathrm{SAdis\_pd2}} h_{i,n,t}^{\mathrm{SAdis\_pd2}} + \sum\limits_{l=1}^{N_{\mathrm{ir3}}} M_{i,l,t}^{\mathrm{SAdis\_ir3}} h_{i,l,t}^{\mathrm{SAdis\_ir3}}) \Delta t \end{cases} \tag{50}$$

In Equation (47): Variables $\omega_{i,t}^{\mathrm{cha}}$ and $\omega_{i,t}^{\mathrm{dis}}$ are both binary (0–1) variables. When $\omega_{i,t}^{\mathrm{cha}} = 1$, the SA charges steam for textile industry user i; otherwise, it does not charge steam. When $\omega_{i,t}^{\mathrm{dis}} = 1$, the SA discharges steam for textile industry user i; otherwise, it does not discharge steam; The maximum charging and discharging steam flow rates of the SA for textile industry user i are $M_{i,\max}^{\mathrm{SAcha}}$ and $M_{i,\max}^{\mathrm{SAdis}}$; $M_{i,m,t}^{\mathrm{SAdis\_te1}}$, $M_{i,n,t}^{\mathrm{SAdis\_pd2}}$, $M_{i,l,t}^{\mathrm{SAdis\_ir3}}$, $h_{i,m,t}^{\mathrm{SAdis\_te1}}$, $h_{i,n,t}^{\mathrm{SAdis\_pd2}}$ and $h_{i,l,t}^{\mathrm{SAdis\_ir3}}$ represent the steam flow rate and steam enthalpy delivered by the SA to each process and production line of textile industry user i at time t; At time t, the remaining capacity $S_{i,t}^{\mathrm{SA}}$ for textile industry user i should be equal to its initial value; $S_{i,\min}^{\mathrm{SA}}$, $S_{i,\max}^{\mathrm{SA}}$, respectively, denote the minimum and maximum values of the remaining capacity of the SA for textile industry user i; $S_{i,1}^{\mathrm{SA}}$, $S_{i,T}^{\mathrm{SA}}$ represent the remaining capacity of SA for textile industry user i at the initial time and at the end of the optimization period; $M_{i,t}^{\mathrm{SA}\to\mathrm{VPPO}}$, $h^{\mathrm{SA}\to\mathrm{VPPO}}$, respectively, represent the steam flow rate and steam enthalpy sold by the SA of textile industry user i at time t. Since the user sells isothermal steam, $h^{\mathrm{SA}\to\mathrm{VPPO}}$ remains constant [27].

$$Q_{i,t}^{\mathrm{SA}\to\mathrm{VPPO}} = M_{i,t}^{\mathrm{SA}\to\mathrm{VPPO}} h^{\mathrm{SA}\to\mathrm{VPPO}} \Delta t \tag{51}$$

$$Q_{s,i,t}^{\mathrm{us}} = Q_{i,t}^{\mathrm{ESB}\to\mathrm{VPPO}} + Q_{i,t}^{\mathrm{SA}\to\mathrm{VPPO}} \tag{52}$$

The total steam energy sold by textile industry $Q_{s,i,t}^{\mathrm{us}}$ is the sum of the steam energy $Q_{i,t}^{\mathrm{SA}\to\mathrm{VPPO}}$ sold by the SA and the steam flow rate $Q_{i,t}^{\mathrm{ESB}\to\mathrm{VPPO}}$.

Modeling the steam temperature boundary of the condensing unit for textile industry user i:

$$\begin{cases} 0.95 T^{\mathrm{s\_te1}} v_{m,t}^{\mathrm{te1}} \leq T_{m,t}^{\mathrm{CON\_te1}} \leq 1.05 T^{\mathrm{s\_te1}} v_{m,t}^{\mathrm{te1}} \\ 0.95 T^{\mathrm{s\_pd2}} v_{n,t}^{\mathrm{pd2}} \leq T_{n,t}^{\mathrm{CON\_pd2}} \leq 1.05 T^{\mathrm{s\_pd2}} v_{n,t}^{\mathrm{pd2}} \\ 0.95 T^{\mathrm{s\_ir3}} v_{l,t}^{\mathrm{ir3}} \leq T_{l,t}^{\mathrm{CON\_ir3}} \leq 1.05 T^{\mathrm{s\_ir3}} v_{l,t}^{\mathrm{ir3}} \end{cases} \tag{53}$$

$$\begin{cases} P_{m,t}^{\mathrm{CON\_te1}} = e^{a - \frac{b}{T_{m,t}^{\mathrm{CON\_te1}} + c}} \\ P_{n,t}^{\mathrm{CON\_pd2}} = e^{a - \frac{b}{T_{n,t}^{\mathrm{CON\_pd2}} + c}} \\ P_{l,t}^{\mathrm{CON\_ir3}} = e^{a - \frac{b}{T_{l,t}^{\mathrm{CON\_ir3}} + c}} \end{cases} \tag{54}$$

$$\begin{cases} h_{m,t}^{\text{CON\_te1}} = k_1(k_2 T_{m,t}^{\text{CON\_te1}} - k_3) P_{m,t}^{\text{CON\_te1}} k_4 + (k_5 T_{m,t}^{\text{CON\_te1}} + k_6) \\ h_{n,t}^{\text{CON\_pd2}} = k_1(k_2 T_{n,t}^{\text{CON\_pd2}} - k_3) P_{n,t}^{\text{CON\_pd2}} k_4 + (k_5 T_{n,t}^{\text{CON\_pd2}} + k_6) \\ h_{l,t}^{\text{CON\_ir3}} = k_1(k_2 T_{l,t}^{\text{CON\_ir3}} - k_3) P_{l,t}^{\text{CON\_ir3}} k_4 + (k_5 T_{l,t}^{\text{CON\_ir3}} + k_6) \end{cases} \quad (55)$$

The modeling of the steam temperature boundary delivered by SA to each process is similar to the steam temperature boundary modeling of the condensing unit for textile industry user i mentioned above. Constraints on the operation of the gas boiler are not reiterated here.

5. Result Analysis

5.1. Basic Data

To validate the effectiveness of the optimal dispatching strategy for textile-based virtual power plants participating in grid-load interactions driven by energy price proposed in this paper, this section selects daily operational tasks and production equipment parameters from actual statistics of a textile enterprise in Jintan, Changzhou. Utilizing the EMP algorithm package and BARON solver within the GAMS environment, this study conducts simulation-based optimization analysis on the model. The bi-level model operates on a time scale of 1 h, with an overall scheduling horizon of 24 h. The bi-level model operates on a time scale of 1 h, with an overall scheduling horizon of 24 h. Time-of-use energy prices are shown in Table 1. VPPO sells electricity to the external grid at 50% of the purchase price and purchases natural gas at a rate of 3.7 yuan/m^3, with a lower heating value of 9.97 kWh/m^3.

Table 1. The time interval division of time-of-use energy price.

Time Interval	Time Range
Peak period	11:00~19:00
off-peak period	10:00, 20:00~22:00
valley period	1:00~9:00, 23:00~24:00

To validate the effectiveness of the bi-level dispatching strategy, two scenarios are compared based on various energy price settings. The scenarios are as follows:

S1: VPPO sells electricity, heat, and steam energy to users at time-of-use energy prices;

S2: VPPO sells electricity, heat, and steam energy to users at real-time energy prices.

In scenario S1, multiple energy prices are based on time-of-use energy prices, with specific time intervals as detailed in Table 1. Given the widespread implementation of time-of-use electricity pricing, time-of-use energy prices extend this concept by dividing multiple energy prices into peak, off-peak, and valley periods to guide user energy consumption behavior. In scenario S2, multiple energy prices are based on real-time energy prices, where energy prices generally vary at each moment in time.

5.2. Strategy Results Analysis

The profits of VPPO within industrial park under different scenarios, total daily operating costs for textile industry users, and overall economic costs of the industrial park are shown in Table 2.

Table 2. Comparison of VPPO profits, total operating costs of industrial users and overall economic cost of VPP under two scenarios.

Scenario	VPPO Earnings/(yuan)	Total Operating Costs for Textile Industry Users/(yuan)	Overall Economic Costs of the VPP/(yuan)
S1	5 471.167 1	88 800.828 7	83 329.661 6
S2	10 258.255 5	90 145.685 1	79 887.429 6

Comparing the VPPO earnings, total daily operating costs for textile industry users, and overall economic costs of the VPP between scenarios S1 and S2 as depicted in Table 3, the VPPO earnings in scenario S2 increased by 87.5% compared to scenario S1. The total operating costs for textile industry users rose by 1.51%, while the overall economic costs of the VPP decreased by 4.13% in scenario S2. This is attributed to the fact that in scenario S2, the VPPO multi-energy prices were generally higher than in S1, leading to increased VPPO earnings but also contributing to higher total operating costs for textile industry users. Additionally, in scenario S2, textile industry users reduced their energy purchases during peak pricing periods and increased purchases during off-peak periods, resulting in lower overall economic costs for the VPP. This demonstrates that the energy pricing strategy proposed in this paper effectively enhances the overall economic performance of VPP targeted at industrial parks, thereby improving energy utilization efficiency.

Table 3. Comparison of daily operating costs of industrial users under different scenarios.

Scenario	Cost for User 1	Cost for User 2	Cost for User 3
S1	22 075.684 1	30 818.277 9	35 906.866 7
S2	22 218.254 9	31 329.119 6	36 598.310 5
cost volatility	−0.646%	−1.658%	−1.926%

Comparing the operating costs of three textile industry users under scenarios S1 and S2 as shown in Table 3, the operating costs for the three textile industry users in scenario S2 increased by 0.646%, 1.658%, and 1.926%, respectively, compared to scenario S1. This is due to Textile Industry User 1 having the highest output from distributed photovoltaics and concurrently the highest SA capacity, resulting in the smallest increase in operating costs for User 1 among the three. Users 2 and 3 have similar levels of distributed photovoltaic output, with User 2 having slightly higher SA capacity than User 3, contributing to slightly better economic performance for User 2 than User 3. This illustrates that the installed capacity of new energy sources and SA capacity significantly impact the operating costs of textile industry users, with higher capacities leading to lower operating costs.

5.3. Analysis of Energy Trading Outcomes

In scenarios S1 and S2, the pricing results of VPPO for textile industry users are depicted in Figures 3 and 4.

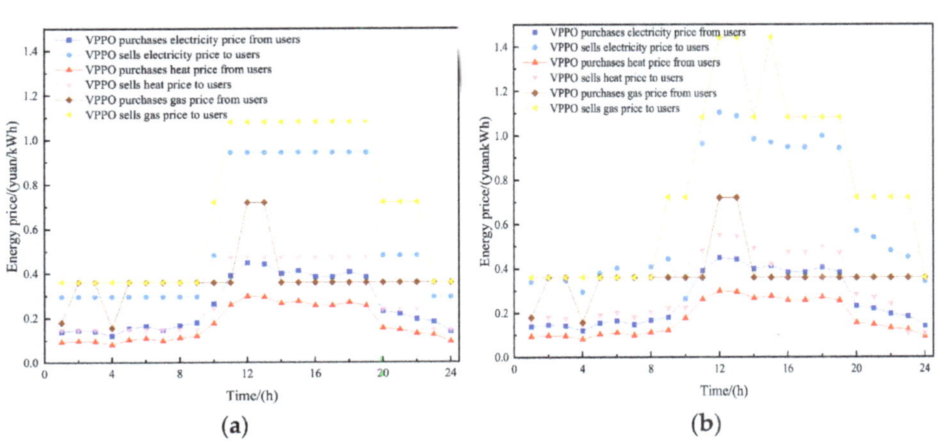

Figure 3. VPPO energy price under scenario S1 and S2. (**a**) S1 scenario VPPO trades energy prices. (**b**) S2 scenario VPPO trades energy prices.

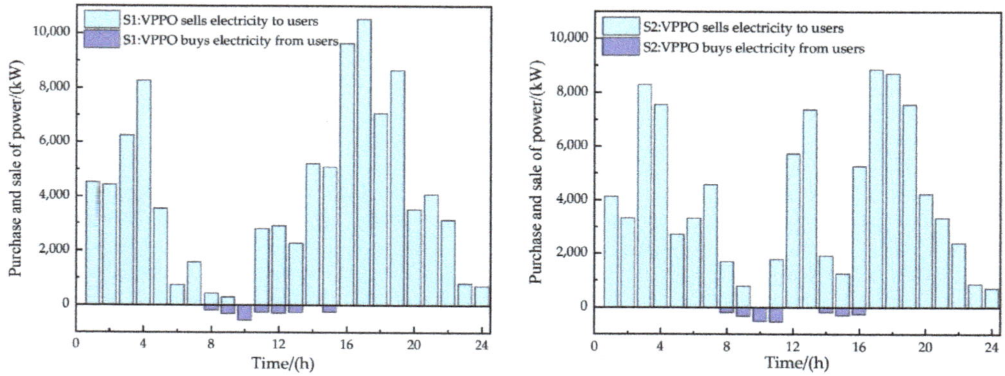

(**a**) (**b**)

Figure 4. Purchase and sale electricity strategy of VPPO under scenario S1 and S2. (**a**) S1 scenario VPPO trades electrical power. (**b**) S2 scenario VPPO trades electrical power.

Figures 4 and 5, respectively, depict the electricity trading patterns between VPPO and textile industry users in scenarios S1 and S2. From the figures, it is evident that the textile industry users predominantly sell electricity during the high solar output period from 8:00 to 16:00. During the peak period from 11:00 to 19:00, compared to scenario S1, VPPO in scenario S2 sell electricity at higher prices, resulting in lower sales volume by VPPOs and lower electricity purchase by users. Textile industry User 3, with limited SA capacity, purchases significant amounts of electricity to produce steam during the 11:00 to 13:00 period to meet production demands. In contrast, during the off-peak period from 1:00 to 5:00, VPPO in scenario S2 sell electricity at lower prices compared to S1, leading to higher electricity purchases by users and lower sales volume by VPPO. Textile industry User 1 purchases substantial amounts of electricity during this period for production purposes. These observations demonstrate that the proposed bi-level optimization dispatching method under time-of-use energy pricing effectively encourages adjustments in electricity purchasing plans for textile industry users.

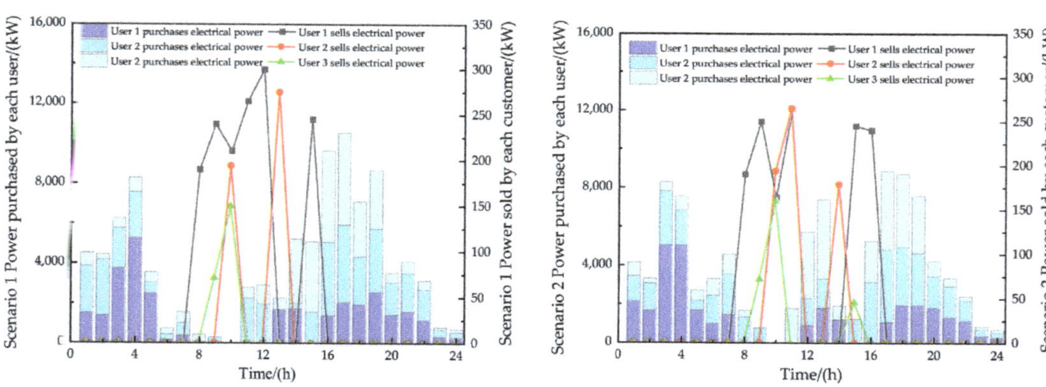

(**a**) (**b**)

Figure 5. Purchase and sale electricity strategy of users under scenario S1 and S2. (**a**) S1 scenario user trades electrical power. (**b**) S2 scenario user trades electrical power.

Figures 6 and 7 depict the heat trading patterns between VPPO and textile industry users in scenarios S1 and S2, respectively. A comparison reveals that there is minimal variation in the total heat sold by VPPO between scenarios S1 and S2. However, VPPO in scenario S2 shows significantly reduced heat purchases. This is because during peak periods in scenario S2, electricity prices are higher, prompting VPPO to utilize CHP units to generate electricity while producing substantial amounts of heat. Consequently, VPPO in scenario S2 exhibit lower heat purchase volumes. Additionally, based on the electricity selling prices by VPPOs as depicted in Figure 5, which influences the timing of industrial production schedules, the heat purchase volumes for textile industry users 1 and 3 in scenario S2, as shown in Figure 7, also shift compared to scenario S1, concentrating during off-peak hours.

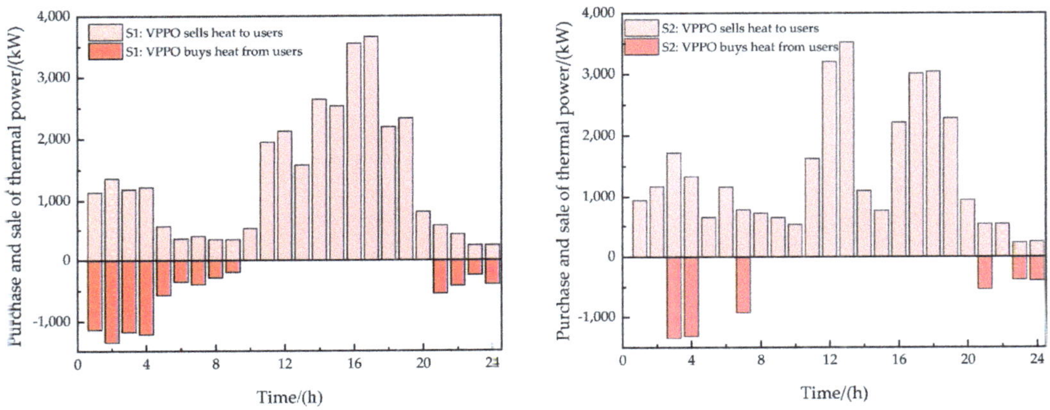

(a) (b)

Figure 6. Purchase and sale heat strategy of VPPO under scenario S1 and S2. (**a**) S1 scenario VPPO trades thermal power. (**b**) S2 scenario VPPO trades thermal power.

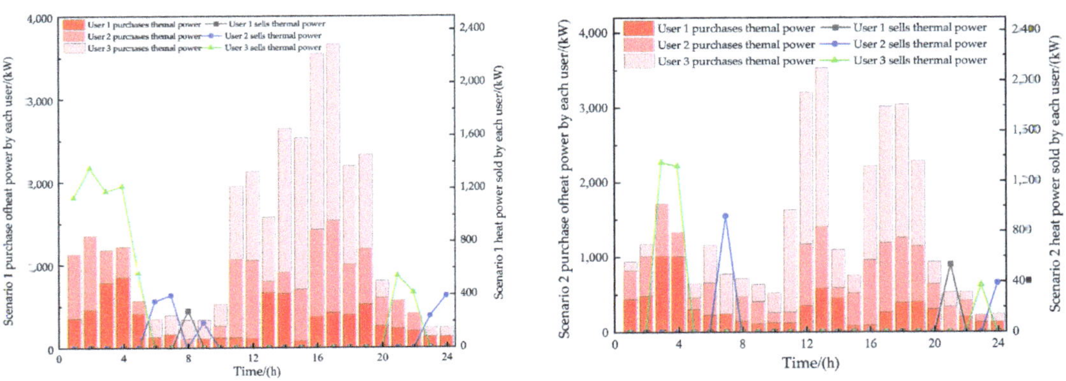

(a) (b)

Figure 7. Purchase and sale heat strategy of users under scenario S1 and S2. (**a**) S1 scenario users trade thermal power. (**b**) S2 scenario users trade thermal power.

Figures 8 and 9, respectively, illustrate the steam trading patterns between VPPO and textile industry users in scenarios S1 and S2. Comparing Figures 8 and 9, during the

11 00 to 19:00 period in scenario S2, VPPO significantly reduce their steam sales compared to scenario S1. This reduction is attributed to the increase in steam prices during this period due to the surge in electricity prices, leading to decreased steam purchases by textile industry users. Additionally, the steam sales volume increases for textile industry User 1 and User 2 in scenario S2. This is because User 2 has a smaller SA capacity, while User 1 has a larger SA capacity, allowing them to increase their economic benefits through the proposed strategies.

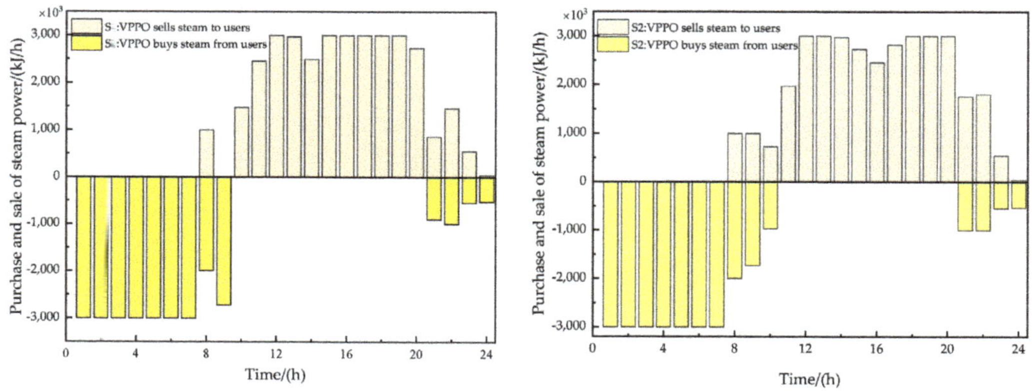

(a) (b)

Figure 8. Purchase and sale steam strategy of VPPO under scenario S1 and S2. (**a**) S1 scenario VPPO trades steam power. (**b**) S2 scenario VPPO trades steam power.

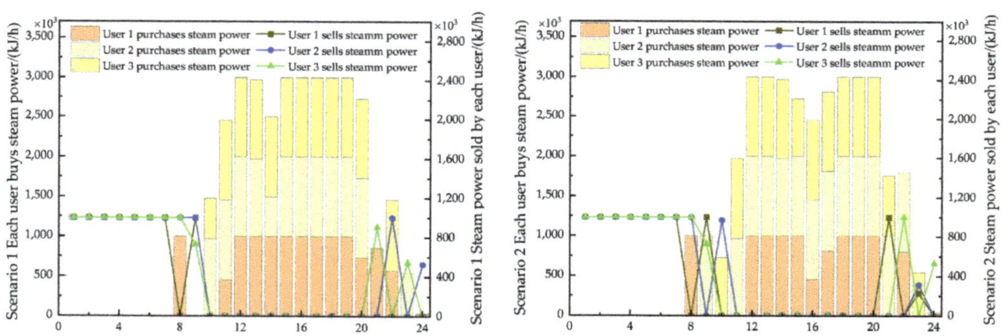

(a) (b)

Figure 9. Purchase and sale steam strategy of VPPO under scenario S1 and S2. (**a**) S1 scenario users trade steam power. (**b**) S2 scenario users trade steam power.

5.4. Analysis of Optimal Dispatching Results

In scenario S2, the optimized dispatching results for VPPO regarding electricity, heat, and steam are depicted in Figure 10. From the figure, it is evident that VPPO achieve multi-energy power balance constraints for electricity, heat, and steam.

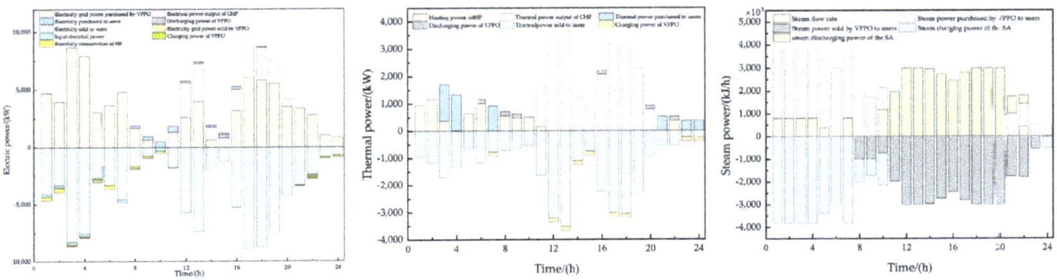

(a) Electrical Power (b) Thermal Power (c) Steam Power

Figure 10. Multi-energy balance diagram of VPPO.

Photovoltaic generation is the primary source of internal electricity within the VPPO. As shown in Figure 10 for electric power, VPPO primarily sources electricity from the grid and CHP units. During off-peak electricity price periods, VPPO purchases electricity from the grid, while during peak periods, it utilizes gas and the CHP units to supplement power supply. VPPO's electricity consumption mainly involves selling electricity to textile industry users, with its electrode industrial steam boilers and heat pumps predominantly operating during off-peak electricity price periods. Energy storage systems charge during off-peak periods and discharge during peak periods, thereby enhancing VPPO's electricity revenue. As depicted in Figure 10 for heat power, VPPO's heat sources primarily involve heat pumps for heating during off-peak electricity price periods and purchasing heat for textile industry users. During peak electricity price periods, energy conversion from gas to heat is facilitated through CHP units as an energy substitution measure. Regarding steam power, as shown in Figure 10, VPPO utilizes electrode industrial steam boilers to produce steam during off-peak electricity price periods. Simultaneously, it purchases steam from textile industry users, storing the obtained steam in SA. During peak electricity price periods, steam is released from SA and sold to textile industry users. The operational status of SA for each textile industry user is illustrated in Figure 11.

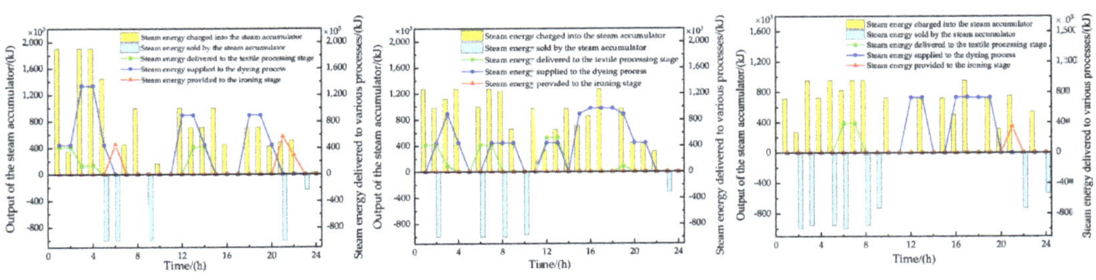

(a) Industrial user 1 (b) Industrial user 2 (c) Industrial user 3

Figure 11. SA operation of each user.

6. Conclusions

This paper proposes an optimal dispatching strategy for textile-based virtual power plants participating in grid-load interactions driven by energy price. The results are as follows:

(1) In the bi-level optimization strategy proposed in this paper, compared to VPPO selling electricity at peak, flat, and off-peak multi-energy prices, VPPO' profits increased by 87.5% under real-time energy prices. Total operating costs for textile industry users

increased by 1.51%. Meanwhile, the overall economic costs of the VPP decreased by 4.13%, effectively enhancing the overall economic viability of the VPP.

(2) Under the guidance of real-time energy prices set by the VPP, users adjusted their energy consumption behaviors by shifting energy purchase times to off-peak periods. This adjustment facilitated production planning and the transfer of multi-energy loads, resulting in smoother fluctuations in demand for textile industry users.

(3) The operating costs of textile industry users are related to the capacity of SA. Textile industry user 1 has the highest SA capacity, while user 3 has the lowest. Under real-time energy prices, the operating costs for these three textile industry users increased by 0.646%, 1.658%, and 1.926%, respectively. This indicates that the size of SA capacity affects the economic viability of textile industry users; higher SA capacity leads to lower operating costs for textile industry users.

Author Contributions: Supervision, T.C. and C.L.; methodology, M.L.; data curation, Y.X.; writing—original draft preparation, M.D. and M.L.; writing—review and editing, H.Y. and X.D. All authors have read and agreed to the published version of the manuscript.

Funding: This research was funded by State Grid Changzhou Power Supply Company, "Research on the Economic and Multi-Scenario Applications of Electric-Thermal Synergy Control Technology Considering Steam Equivalent Energy Storage Characteristics", grant number SGJSCZ00KJJS2312263.

Data Availability Statement: The data presented in this study are available on request from the corresponding author due to legal and privacy reasons.

Conflicts of Interest: Authors Tingyi Chai, Chang Liu and Yichuan Xu were employed by the company Changzhou Power Supply Branch, State Grid Jiangsu Electric Power Co., Ltd. The remaining authors declare that the research was conducted in the absence of any commercial or financial relationships that could be construed as a potential conflict of interest.

References

1. Xu, C.; Guan, Y. Cluster Convergence: Pioneering the Future -Symposium on Advancing the Modernization of Textile Industry Clusters. *Text. Appar. Wkly.* **2024**, *10*, 21.
2. Liu, J.; Peng, J.; Deng, J.; Liu, S. Two-stage Robust Optimization-based Economic Dispatch of Virtual Power Plants Considering Cogeneration. *J. Syst. Simul.* **2023**, *35*, 1799–1813.
3. Bhuiyan, E.A.; Hossain, M.Z.; Muyeen, S.M. Towards Next Generation Virtual Power Plant: Technology Review and Frameworks. *Renew. Sustain. Energy Rev.* **2021**, *150*, 111358. [CrossRef]
4. Gao, C. Conceptual reflection on virtual power plants. *Distrib. Util.* **2023**, *40*, 2–6.
5. Xie, B.; Zhang, L.; Hao, P. Real-time Pricing Demand Response Mechanism of Virtual Power Plant Based on Stackelberg Game. *Chin. J. Manag. Sci.* **2024**, 1–18. [CrossRef]
6. Tao, W.; Ai, Q.; Li, X. Research Status and Prospects of Collaborative Dispatching and Market Trading of Virtual Power Plants. *South Power Syst. Technol.* **2024**, 1–15. Available online: http://kns.cnki.net/kcms/detail/44.1643.tk.20240318.1018.008.html (accessed on 27 September 2024).
7. Hao, J.; Huang, T.; Xu, Q.; Sun, Y. Virtual power plant multi-objective optimal dispatching strategy with multiple energy suppliers. *Distrib. Util.* **2023**, *40*, 32–42.
8. Li, B.; Gu, C.; Du, J.; Tian, K.; Chen, S. Virtual power plant resource dispatching optimization strategy considering electrothermal conversion. *Distrib. Util.* **2023**, *40*, 16–21.
9. Yu, H.; Lu, W.; Tang, L.; Zhang, Y.; Zou, X. Economic dispatch and profit distribution strategy for multi-agent virtual power plants considering risk preference. *Integr. Intell. Energy* **2024**, *46*, 66–77.
10. Liao, S.; He, C.; Li, L.; Xu, J.; Sun, Y. Solution Method of Regulatory Boundary for Industrial Park Type Virtual Power Plant Considering Power Network Constraints. *Autom. Electr. Power Syst.* **2024**, *48*, 66–75.
11. Wang, J.; Shen, X.; Xu, Y. Ancillary service for frequency regulation based on multi-energy virtual power plant aggregating factory load. In Proceedings of the 11th IET International Conference on Advances in Power System Control, Operation and Management (APSCOM 2018), Hong Kong, China, 11–15 November 2018.
12. Yu, S.; Zhang, J.; Yuan, Z.; Fang, F. Resilience Enhancement Strategy of Combined Heat and Power-Virtual Power Plant Considering Thermal Inertia. *Power Gener. Technol.* **2023**, *44*, 758–768.
13. Jiang, Z.; Zhang, F.; Hu, F.; Sun, Y.; Jiang, W. Evaluation method of aggregated responsiveness of distributed resources in virtual power plant. *Electr. Power Eng. Technol.* **2022**, *41*, 39–49.
14. Zhang, N.; Zhu, H.; Yang, L.; Hu, C. Optimal Dispatching Strategy of Multi-Energy Complementary Virtual Power Plant Considering Renewable Energy Consumption. *Power Gener. Technol.* **2023**, *44*, 625–633.

15. Wang, J.; Niu, Y.; Chen, Y.; Du, M. Master-slave game optimal dispatching of virtual power plant. *Electr. Power Autom. Equip.* **2023**, *43*, 235–242.
16. Li, Q.; Zhu, D.; Huang, D.; Wu, S.; Yang, Y. Stackelberg game pricing strategy between virtual power plant operators and electric vehicle users. *Electr. Power Eng. Technol.* **2022**, *41*, 183–191.
17. Cang, H.; Liu, H. Economic research on virtual power plant source-load coordination based on electricity price linkage. *Mod. Electron. Tech.* **2022**, *45*, 149–155.
18. Qiang, L.; Jiale, T.; Ke, Z. Pricing Strategy for a Virtual Power Plant Operator with Electric Vehicle Users Based on the Stackelberg Game. *World Electr. Veh. J.* **2023**, *14*, 72. [CrossRef]
19. Li, Q.; Zhou, Y.C.; Wei, F.C. Multi-time scale dispatching for virtual power plants: Integrating the flexibility of power generation and multi-user loads while considering the capacity. *Appl. Energy* **2024**, *362*, 122980. [CrossRef]
20. Zhang, Y. On Low-Carbon Optimization Operation of Integrated Energy Systems Considering Hydrogen Energy Coupling and Master-Slave. Master's Thesis, Nanchang University, Nanchang, China, 2023.
21. Ehsan, H.-F.; Saeed, H.; Seifeddine, B.-E. Virtual Power Plant Operational Strategies: Models, Markets, Optimization, Challenges, and Opportunities. *Sustainability* **2022**, *14*, 12486. [CrossRef]
22. Hua, Y.; Wang, Y.; Bai, H.; Han, D.; Pu, F. Interval master-slave game optimal dispatching of virtual power plant with electric vehicle. *Therm. Power Gener.* **2022**, *51*, 163–170.
23. Tang, F. Optimal Dispatching of Combined Heat and Power of Virtual Power Plant Based on Hybrid Energy Storage. Master's Thesis, North China Electric Power University, Beijing, China, 2021.
24. Liang, A.; Zeng, S.; Ren, J.; Zhang, X.; Cao, F. Intelligent Operation Dynamic Characteristics of Heat Pump System in Integrated Electric-thermal Cooperative Grid Based on Game Optimization Algorithm. *J. Refrig.* **2024**, *45*, 109–111.
25. Sun, W.; Wu, J.; Zhang, Q. Coordinated optimization of a distribution network and multi-integrated energy microgrid based on a double-layer game. *Power Syst. Prot. Control* **2024**, *52*, 26–38.
26. Maddah, H.; Sadeghzadeh, M.; Ahmadi, M.H.; Kumar, R.; Shamshirband, S. Modeling and Efficiency Optimization of Steam Boilers by Employing Neural Networks and Response-Surface Method (RSM). *Mathematics* **2019**, *7*, 629. [CrossRef]
27. Bai, Z.; Miao, C.; Wang, W.; Sun, F.; Zhang, P. The Value of Thermal Storage Technology in Integrated Energy System. *Distrib. Util.* **2019**, *36*, 20–26.

energies

MDPI

Article

Multi-Objective Optimization Design for Cold-Region Office Buildings Balancing Outdoor Thermal Comfort and Building Energy Consumption

Fei Guo, Shiyu Miao, Sheng Xu, Mingxuan Luo, Jing Dong * and Hongchi Zhang *

School of Architecture and Fine Art, Dalian University of Technology, Dalian 116024, China; guofei@dlut.edu.cn (F.G.); miaoshiyu@mail.dlut.edu.cn (S.M.); xusheng@mail.dlut.edu.cn (S.X.); 32216047@mail.dlut.edu.cn (M.L.)

* Correspondence: jdong@dlut.edu.cn (J.D.); zhanghc@dlut.edu.cn (H.Z.)

Abstract: Performance parameters and generative design applications have redefined the human–machine collaborative relationship, challenging traditional architectural design paradigms and guiding the architectural design process toward a performance-based design transformation. This study proposes a multi-objective optimization (MOO) design approach based on performance simulation, utilizing the Grasshopper-EvoMass multi-objective optimization platform. The Non-dominated Sorting Genetic Algorithm II (NSGA-II) is applied to coordinate two performance metrics—outdoor thermal comfort and building energy loads—for the multi-objective optimization of architectural design. The results indicate that (1) a performance-based multi-objective optimization design workflow is established. Compared to the baseline design, the optimized building form shows a significant improvement in performance. The Pareto optimal solutions, under 2022 meteorological conditions, demonstrate an annual energy efficiency improvement of 16.55%, and the outdoor thermal neutrality ratio increases by 1.11%. These results suggest that the optimization approach effectively balances building energy loads and outdoor thermal comfort. (2) A total of 1500 building form solutions were generated, from which 16 optimal solutions were selected through the Pareto front method. The resulting Pareto optimal building layouts provide multiple feasible form configurations for the early-stage design phase.

Keywords: multi-objective optimization; building energy consumption; outdoor thermal comfort; office buildings

1. Introduction

The rapid urbanization process is profoundly altering the urban climate, with extreme weather events becoming more frequent, the urban heat island effect intensifying, and carbon emissions rising globally [1]. Poor environmental conditions pose significant threats to public health. Against the backdrop of climate and energy concerns, scholars worldwide have begun to focus on urban form studies based on multiple performance indicators [2]. In the cold regions of northeast China, where winters are long and harsh, thermal comfort significantly impacts the actual experience. Office buildings need to strike a balance between functional demands, sustainability, and comfort. Multi-objective optimization has become an effective tool [3], capable of simultaneously balancing various performance metrics and adapting flexibly to external meteorological conditions.

In recent years, generative design has developed rapidly. The application of performance parameters and generative design has redefined the human–machine collaborative relationship, breaking through traditional architectural design paradigms and guiding the architectural design process toward a performance-based design transformation [4]. As an iterative algorithm that coordinates multiple parameters, multi-objective optimization is suitable for meeting the need to coordinate various goals in architectural design, and it is crucial when balancing conflicting objectives [5–7]. This study selects a typical office building in Dalian as the research sample and analyzes the influence patterns of multiple factors through multi-objective optimization of architectural forms. The research aims to explore strategies for performance-based design in cold regions and provide methods and theoretical references for architects in the early-stage design of building forms.

Integrating multi-objective optimization methods into the conceptual phase of architectural design helps address the complex goals of achieving optimal energy efficiency while enhancing thermal comfort. Current research on the application of multi-objective optimization in architectural design can be categorized into three main areas. (1) From the perspective of green buildings, research on multi-objective optimization design coordinates multiple performance indicators (such as acoustics, lighting, and thermal properties in building physics). Over the past decade, researchers in building performance have focused on the role of artificial intelligence technologies in shaping architectural forms, conducting extensive work in underlying algorithms, theoretical methods, and engineering practices [5,8–10]. Architectural designers have applied multi-objective optimization to various performance metrics. In 2016, Delgarm N. et al. [11] optimized building energy consumption and indoor thermal comfort using the Multi-objective Artificial Bee Colony (MOABC) algorithm. In 2021, Deb K. et al. [12] optimized materials for residential buildings in cold regions, targeting building life cycle assessment (LCA) costs and CO2 emissions. In 2022, Jingjin Li et al. [13] conducted multi-objective optimization for a residential area in Nanjing, using three volumetric combinations (horizontal, vertical, and hybrid) as variables, targeting the floor area ratio and solar radiation. In 2024, Xian Zhang et al. [14] applied the MOPSO algorithm to optimize indoor spaces, with objectives including visual effects and energy efficiency. (2) From the perspective of urban design, multi-objective optimization is applied at various scales (building, block, and city levels). Multi-objective optimization is employed across multiple design scales (city—block—building—room—component). In 2022, S. Mirzabeigi et al. [15] proposed a multi-objective optimization framework for urban block-scale design using parametric design and energy consumption simulations. In 2024, Maksoud A. et al. [16] studied urban block optimization, coordinating solar radiation intensity with flood resilience. In 2024, Abdul Mateen Khan et al. [17] used design–build simulations combined with machine learning methods (LGBM and LIME) to predict and optimize the energy efficiency of building units, achieving energy savings and reduced carbon emissions through multi-objective optimization (MOO). In 2024, Juan Gamero Salinas et al. [18] explored multi-objective optimization to balance the minimization of indoor overheating time and the maximization of useful daylight, focusing on multiple rooms within a building. (3) From the perspective of architectural form, research focuses on generating combinations of forms based on multiple building performance indicators through multi-objective optimization. In 2022, Hankun Lin et al. [19] developed a coupled simulation method using tools such as Ansys Fluent, Ladybug, and Honeybee to simulate the shading and wind effects of green facades. Regarding architectural form exploration, research has primarily concentrated on the early design phase of overall building form exploration. In 2023, Yu Li et al. [20] applied the multi-material bi-directional evolutionary structural optimization (BESO) method to design the "Xiong'an Wing" large cantilever core structure, improving the practical application of multi-material BESO methods in large-scale building designs. In 2024, Younes Noorollahi et al. [21] demonstrated the significant role of climate conditions in optimizing building energy use by studying design parameters such as building orientation, shading parameters, insulation thickness, temperature set points, window-to-wall ratio, and roof insulation thickness.

The evaluation systems for the built environment are rich and diverse, with outdoor thermal comfort and building energy load both being criteria for assessing the built environment. For cold regions, the outdoor environment in winter is harsh and often uncomfortable, necessitating improvements in the built environment. Moreover, in response to the societal call for low carbon emissions, office buildings should consider how to reduce operational energy loads during the design phase. It is noteworthy that outdoor thermal comfort and building energy load are interrelated, and with a reasonable architectural layout, it is possible to coordinate these two evaluation indicators simultaneously. Therefore, this study uses outdoor thermal comfort and building energy load as two evaluation indicators, and based on these, formulates objective functions. It employs multi-objective optimization methods to explore architectural forms.

Based on the above background, current research on the application of multi-objective optimization in architectural design still has several limitations. (1) Limited scope of building types for optimization. Research primarily focuses on residential buildings, given their higher frequency of use. However, there is a lack of multi-objective optimization design studies for office buildings, especially office parks. (2) Optimization objectives need further refinement. There are few studies that standardize the treatment of outdoor thermal comfort and energy load. Existing research often uses default program methods to balance various performance metrics, applying trade-off factors to standardize multiple performance indicators, thus enabling better coordination of optimization objectives. (3) Few integrated workflows for multi-objective optimization and generative design. Current multi-objective optimization research typically focuses on idealized models, leading to results that are combinations of permutations rather than generating new design solutions. Parametric generative design addresses this gap, and integrating generative design with multi-objective optimization can better guide architects in early-stage form generation.

This study aims to investigate multi-objective optimization design for office parks in cold regions, focusing on coordinating annual energy loads and outdoor thermal comfort, with an emphasis on form changes and optimization potential under multi-objective optimization conditions. The main objectives of the research are as follows. (1) To establish a broadly applicable performance-based simulation workflow (focusing on building energy loads and outdoor thermal comfort) to create a quantitative evaluation system for the built environment. (2) To perform climate and form adaptability settings for urban areas in cold regions (modifying meteorological data based on the urban heat island effect), thereby enhancing the accuracy of performance simulation workflows. (3) To explore the possibilities of architectural forms within a plot by combining multi-objective optimization with generative design under constraints such as building form parameters (floor area ratio, number of stories, enclosure degree, etc.).

The remainder of this paper is structured as follows. Section 2 describes the research methods, including performance indicator simulation programs, generative design parameter settings, and data analysis and visualization techniques. Section 3 presents the research results, focusing on the exploration of the multi-objective optimization process and the various form outcomes generated by generative design. Section 4 discusses the results and the potential for performance improvement through optimization, as well as the limitations of the study. Section 5 provides conclusions and future outlook.

2. Methodology

This study combines generative design and multi-objective optimization methods to create a design workflow, focusing on optimization directions such as outdoor thermal comfort and building energy consumption, as shown in Figure 1. The research spans three stages, from subjective perception to performance improvement in the built environment: (1) perception detection; (2) analysis, diagnosis, and management; (3) quality efficiency improvement. Furthermore, the research involves five specific steps: (1) subjective perception; (2) urban modeling; (3) building performance simulation; (4) block scale building form generation; and (5) multi-objective genetic optimization.

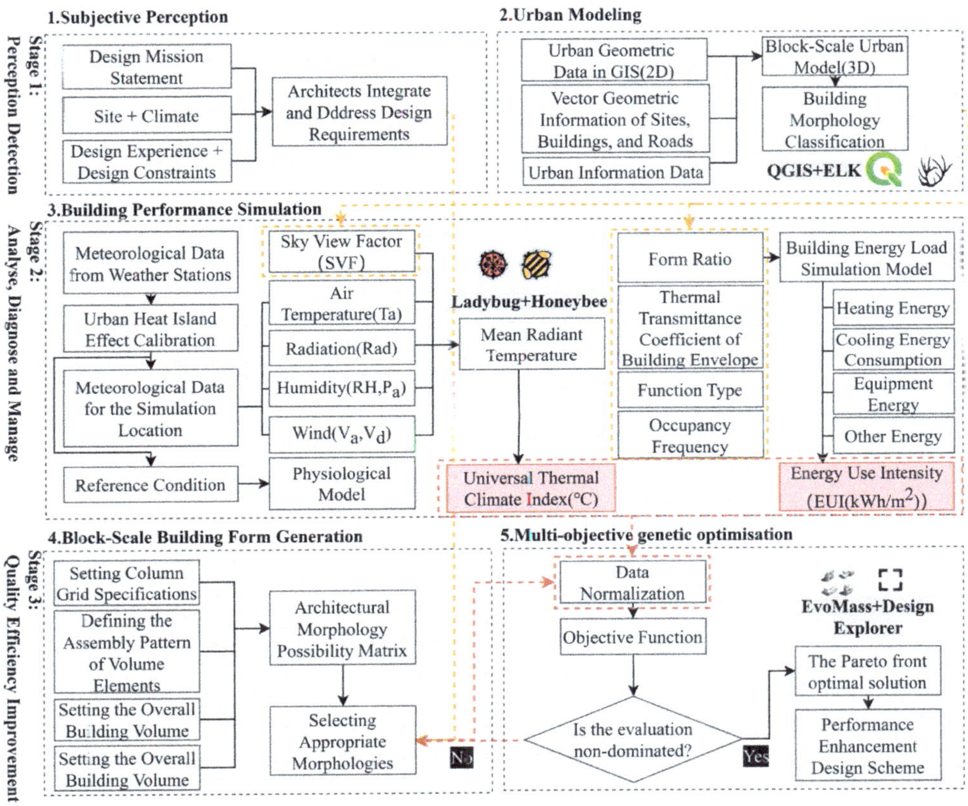

Figure 1. Technology roadmap.

In the morphological generation process, parameters such as enclosure ratio, building height, floor area, floor area ratio, and orientation offset angle are selected. Rhino and EvoMass are used as modeling and parametric form generation platforms, while Ladybug Tools serves as the performance simulation platform, simulating outdoor thermal comfort and annual building energy consumption [22,23]. The Universal Thermal Climate Index (UTCI) is used as the performance metric for outdoor thermal comfort, supplemented by secondary indicators such as Predicted Mean Vote (PMV), Mean Radiant Temperature (TMRT), Sky View Factor (SVF), and Air Temperature (Ta). Energy Use Intensity (EUI) is used as the performance metric for building energy consumption, with sub-metrics including heating energy consumption (Heat Generation), air conditioning and refrigeration (Chiller), lighting energy consumption (Lighting), other electrical equipment (Equipment), and total energy load (Total Load) [24,25]. Using the generative design program [26–28], multiple form indicators are selected as optimization targets, iterating and optimizing within a multi-objective optimization cycle.

2.1. *Research Location*

Dalian (latitude 38.91° N, longitude 121.61° E) is a major city and sub-provincial municipality in China, with a permanent population of 7.539 million and a total area of 12,574 km^2. Located at the southernmost point of northeast China, Dalian consists of 10 districts and counties, including Zhongshan District, Shahekou District, Xigang District, Ganjingzi District, Pulandian District, Changhai County, Jinzhou District, Zhuanghe City, Wafangdian City, and Lushunkou District. The city has a temperate monsoonal climate with maritime characteristics, featuring mild winters and warm summers with distinct seasons. The average annual temperature is 10.5 °C, with annual precipitation ranging between 550 and 950 mm and total annual sunshine duration between 2500 and 2800 h. In recent years, the occurrence of extreme weather events in China has increased [29], with both summer heatwaves and harsh winter cold significantly impacting urban thermal perception, inevitably increasing the use of air conditioning, heating, and other electrical equipment.

The study site is located in the Huaxin Industrial Park, Ganjingzi District, Dalian, as shown in Figure 2. The designated plot spans 126,000 m^2. The existing building on-site is an L-shaped office building with built structures on the south and east sides and open spaces to the north and west. An urban road on the south side connects the site with its surroundings. To the southwest and north of the site are office and residential buildings, respectively, arranged as linear standalone blocks. A digital model of the existing buildings and site conditions was created using the Grasshopper-Elk urban modeling workflow. This model was integrated with a multi-objective optimization design method to optimize the form of office buildings within the site.

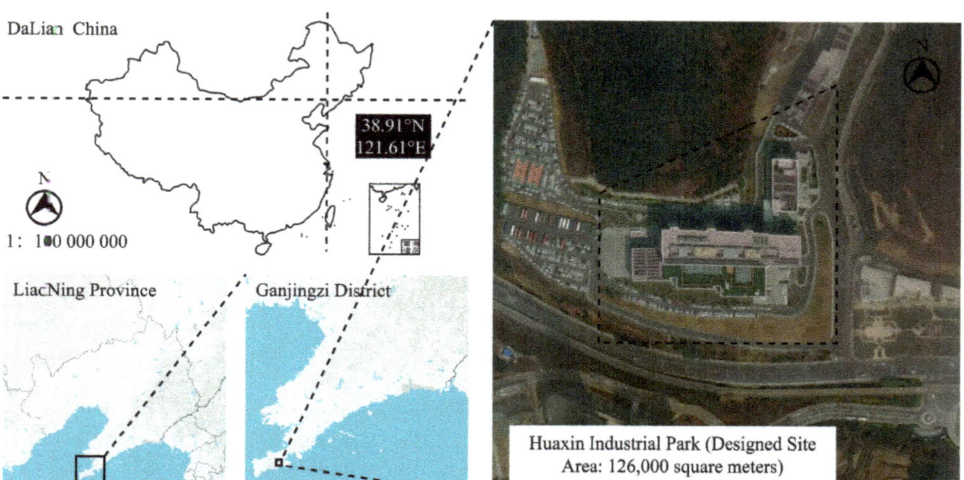

Figure 2. Research site location and existing building satellite map. Note: The base map is a standard map, and there are no modifications to the map. The map review number is GS (2016) 1552.

This section offers context regarding the research location, climate, and site-specific characteristics, emphasizing the practical challenges faced by the design process. The detailed modeling workflow, combining urban-scale modeling with form optimization, provides a foundation for generating adaptive and high-performance design solutions tailored to Dalian's climate and urban environment.

2.2. Energy Load Simulation

In its broad definition, building energy consumption refers to the total energy used throughout the entire life cycle of a building, including the production and transportation of building materials, the construction phase, the operational phase, and the demolition phase. In a narrower sense, building energy consumption specifically refers to the energy used during the operational phase, encompassing heating and air conditioning, lighting, and equipment energy use. The primary energy sources include water, electricity, natural gas, coal, and biomass. This study focuses on passive design strategies, which are significantly influenced by building morphology [30] and are directly controlled by architects. Active systems, such as heating, ventilation, and air conditioning (HVAC) system design [31], are beyond the scope of this research. The building energy consumption simulation only considers the operational phase [32,33], excluding the energy required for construction and demolition.

To simplify the building energy simulation calculations [34–36], the building envelope is modeled as a zero-thickness plane with a certain thermal resistance. This approach aligns with the guidelines outlined in the "General Specification for Building Energy Efficiency and Renewable Energy Utilization (GB55015-2021 [37])" for the cold A and B zones in China. The thermal transmittance of the building envelope has a significant impact on heating energy consumption, as heat loss due to the temperature difference between indoor and outdoor spaces during the heating season is dominant. The window-to-wall ratios for south-, north-, east-, and west-facing walls are set at 0.55, 0.35, and 0.40, respectively. The thermal transmittance values for external walls, roofs, and floors are set to 0.35 W/m^2·K, 0.25 W/m^2·K, and 0.35 W/m^2·K, respectively, with heating energy consumption for cold regions calculated at 82 MJ/(m^2·a). The hourly occupancy rates are shown in Table 1.

Table 1. Room occupancy rates (in percent).

Building Type		Time											
		1	2	3	4	5	6	7	8	9	10	11	12
Office Building	Weekdays	0	0	0	0	0	0	10	50	95	95	95	80
	Holidays	0	0	0	0	0	0	0	0	0	0	0	0
		13	14	15	16	17	18	19	20	21	22	23	24
	Weekdays	80	95	95	95	95	30	30	0	0	0	0	0
	Holidays	0	0	0	0	0	0	0	0	0	0	0	0

The building energy consumption simulation utilizes the Honeybee 0.0.69 plugin from Ladybug Tools 1.6.0, which allows Rhino 7.0 models to be imported into EnergyPlus 9.5.0 and OpenStudio 3.4.0 for energy analysis. The steps for the building energy simulation experiment are as follows. (1) Model validation: Import the model into Grasshopper 2.0 to check for and eliminate any overlapping surfaces. Once the model is verified, import it into Honeybee, assigning material properties to components such as external walls, roofs, and floors based on predefined parameters. (2) Setting basic parameters: Define the schedule based on the functional zoning requirements and input the local climate data for Dalian, provided by the national meteorological service. (3) Building the energy model: Construct the Building Energy Model (BEM) within Honeybee. The model primarily consists of HB Room and Context, where HB Room includes elements such as HB Object, Construction Set, Program, and Conditioned space definitions. (4) Generating energy simulation results: Simulate to output annual Energy Use Intensity (EUI), heating energy consumption, and cooling energy consumption. The energy use is calculated in units of kWh/m^2. (5) Visualization of simulation results: Organize the energy consumption data, make necessary adjustments, and use Origin software 2023 to visualize the experimental results.

2.3. Outdoor Thermal Comfort Simulation

Thermal comfort, also known as thermal satisfaction or thermal comfort level, refers to the human body's feedback on the comfort level of the thermal environment [38]. Research on thermal comfort initially began in the fields of indoor environmental design and heating, ventilation, and air conditioning (HVAC), later expanding to outdoor spaces in landscape architecture, specifically referred to as outdoor thermal comfort. Comfort is defined as the degree of satisfaction individuals feel, both physiologically and psychologically, with the objective physical environment, which includes factors such as thermal conditions, lighting, acoustics, and air quality [39]. When subjective evaluations are involved, other factors like cognition, psychology, and personal habits must also be considered [40,41].

Thermal comfort describes a state where individuals feel satisfied with the thermal conditions of their surroundings, which is influenced by factors such as air temperature, humidity, wind speed, and solar radiation. In this study, the term "comfort" specifically refers to thermal comfort, one of the earliest research topics in the field of building science. Thermal comfort is affected by multiple factors, including environmental conditions, human factors, and climatic influences.

Simulations of outdoor thermal comfort and building energy consumption were performed using the Rhino-Grasshopper platform [42–44], incorporating plugins such as Ladybug, Dragonfly, and Honeybee [45]. The thermal environment simulation results are quantified through several indices, including the Universal Thermal Climate Index (UTCI), Predicted Mean Vote (PMV), and Mean Radiant Temperature (TMRT). The research process follows these key steps. (1) Sky View Factor (SVF) Simulation: The Sky View Factor (SVF) simulates the extent to which a person is exposed to the sky, used to assess the natural impact of sunlight on individuals in outdoor environments. In building design, evaluating SVF is a crucial parameter for assessing building morphology. This index is calculated by considering factors such as building orientation, height, and angle. The input is defined through "LB Human to Sky Relation", which sets the measurement points and site locations, while the output is generated by "LB Outdoor Solar TMRT", which calculates solar irradiance under shadow conditions in outdoor environments. (2) Mean Radiant Temperature (TMRT) Simulation: Mean Radiant Temperature (TMRT), also referred to as ambient or environmental temperature, is the temperature at which the total radiant energy emitted by all surrounding surfaces is uniformly distributed at a given time. TMRT is crucial for describing the overall thermal radiation environment surrounding an individual. Although it represents an idealized temperature value, TMRT plays a significant role in both human thermal comfort and building energy consumption. During winter, people typically adjust indoor temperatures to maintain comfort, and the level of TMRT can influence heat loss and retention. In summer, elevated TMRT values can increase building energy consumption, as higher radiant temperatures require more cooling energy from air conditioning systems to maintain comfortable indoor conditions.

$$T_{mrt} = \left[(T_g + 273)^4 + \frac{1.10 \times 10^8 V_a^{0.6}}{\varepsilon D^{0.4}} (T_g - T_a) \right]^{\frac{1}{4}} - 273 \quad (1)$$

The "LB Outdoor Solar TMRT" program in the Ladybug software 0.0.69 utilizes the "Solar Cal" model from ASHRAE-55 [46] to calculate both long-wave and short-wave solar radiation. This model estimates the Mean Radiant Temperature (TMRT) using the sky view factor. The output provides hourly simulated values of TMRT. In addition to these calculations, the simulation can be refined by incorporating parameters such as human activity level, clothing insulation, metabolic rate, and ground surface reflectivity, as shown in Formula (1). (3) Outdoor Thermal Comfort Simulation: The Universal Thermal Climate Index (UTCI) is a metric that evaluates and describes human comfort based on physiological and psychological responses to the thermal environment. Influencing factors include temperature, humidity, airflow speed, and radiant temperature, along with the body's physiological and psychological reactions to these environmental conditions. The UTCI assumes a walking activity with a metabolic rate of 2.4 met, and it automatically adjusts clothing insulation based on outdoor temperatures. Although originally designed for indoor environments, the UTCI provides architects with a quantifiable method to analyze user comfort, allowing them to design and modify environments to improve overall comfort and health.

The UTCI has several advantages. (1) It is applicable under various thermal exchange conditions; (2) it can be used across a wide range of climates, seasons, and scales; (3) it supports interdisciplinary research across fields such as meteorology, geography, and architecture, as shown in Formula (2).

$$UTCI = T_a + offset(T_a, T_{mrt}, V_a, RH) \tag{2}$$

The "LB UTCI Comfort" module for UTCI calculation takes four key climate parameters as inputs: air temperature, mean radiant temperature, relative humidity, and wind speed. The output generates a thermal map of the simulated site, displaying hourly UTCI values and the corresponding outdoor thermal comfort zones. This process allows for the selection of grid points within the site as simulation points for multiple UTCI calculations, with the results visualized to depict the distribution of outdoor thermal comfort across the site.

2.4. Standardization Processing and Objective Function

Building energy consumption and outdoor thermal comfort are two performance indices with different dimensions, making them challenging to optimize as single-objective problems through direct target function settings [47]. This necessitates data preprocessing [48]. In this study, dimensionless data processing is employed, as demonstrated in Formula (3). In this equation, the greater the difference between a performance index value and its maximum, the smaller the resulting standardized value. The standardized value indicates the relative magnitude of a specific value within the total range of values, which lies between 0 and 1. For building energy consumption, the objective function is to minimize the total building energy consumption. For outdoor thermal comfort, the neutral thermal temperature serves as the reference point, comparing individual values to this neutral temperature. The objective function aims to minimize the deviation from the neutral temperature, thus making the simulated site temperatures as close as possible to the neutral thermal zone.

$$x_i' = \frac{maxx_i - x_i}{maxx_i} \tag{3}$$

The designated objective function needs to articulate a formulaic expression that simultaneously optimizes for both comfort and energy consumption. Through the normalization of indices, it is possible to balance two types of data that differ in units and have significant numerical disparities. The traditional method of balancing uses Formula (4), which employs a balancing factor w to weigh the two performance indices. For instance, if the weight for building energy consumption is set at w = 0.6, then the balancing factor for thermal comfort should be 0.4. However, this method is susceptible to subjectivity in setting the weight factors, potentially leading to suboptimal results constrained by personal biases, often resulting in a locally optimal solution (LOS). In this study, the weight factor w is set to 0.5.

This research adopts the Pareto optimization approach to enhance solution finding, effectively avoiding local optima. The core of Pareto optimization lies in identifying the optimal frontier. Points on this frontier are Pareto optimal solutions. If a testing scheme's two-dimensional performance scatter plot shows no dominated solutions, the scheme is considered a Pareto optimal solution. All such optimal solutions form a scattered set, which, when connected, creates the Pareto frontier. This method minimizes subjective interference in the optimization process. The criteria for forming a Pareto optimal solution involve setting objective functions to simultaneously minimize both indices—building energy consumption and thermal comfort—as shown in Formula (5). Performance indices are evaluated using a fitness metric, which represents the normalized result of the performance simulation.

The designated objective function needs to articulate a formulaic expression that simultaneously optimizes for both comfort and energy consumption. Through the normalization of indices, it is possible to balance two types of data that differ in units and have significant numerical disparities. The traditional method of performance simulation.

$$minize\ Total\ Fitness = wFitness\ of\ Load + (1 - w)Fitness\ of\ Comfort \tag{4}$$

$$minize(Total\ Fitness) = minize\begin{pmatrix} Fitness\ of\ Load \\ Fitness\ of\ Comfort \end{pmatrix} \tag{5}$$

2.5. Multi-Objective Optimization

In practical design engineering projects, the challenge often lies in simultaneously coordinating multiple project objectives. Architectural performance goals such as thermal comfort and energy consumption require differing approaches to building form, and historically, balancing these objectives has heavily depended on the experience and intuition of architects. However, with the advancement of quantitative science, multi-objective optimization methods have emerged to address the deficiencies in coordinating multiple building performance indices. The non-dominated set of solutions in multi-objective optimization is considered as Pareto optimal solutions, with the optimization direction aimed at minimizing each objective, thus optimizing multiple objectives $F(x)$ simultaneously. The formula is as follows:

$$\begin{array}{ll} minimize & F(x) = (f_1(x),\ f_2(x),\ \ldots,\ f_m(x))^T \\ subject\ to & x \in \Omega \end{array} \tag{6}$$

According to Formula (6), each $F_i(x)$ represents an individual objective function associated with different performance criteria of the building project. This approach facilitates a systematic exploration of architectural solutions, enabling the integration of multiple performance metrics without undue compromise. The adoption of multi-objective optimization allows architects and engineers to derive balanced solutions that better meet the complex demands of modern building projects.

Outdoor thermal comfort and building energy consumption are critical determinants of architectural form variations, with the thermal comfort index subdivided into several metrics such as Sky View Factor (SVF), Mean Radiant Temperature (MRT), Universal Thermal Climate Index (UTCI), and Predicted Mean Vote (PMV), denoted as Index1, Index2, Index3, Index4, respectively. Moreover, the building layout is influenced by foundational architectural parameters, with adjustable building parameters, including Width, Length, Orientation, and Capacity Ratio. In this setting, architectural form parameters serve as independent variables, which are freely generated through the EvoMass system based on morphological parameters [49]. Performance parameters are treated as dependent variables for evaluating generated designs [50].

Architectural volume generation is facilitated through the Additive Form Massing Component, based on a summary of morphological parameters for office buildings in Dalian, as discussed in Chapter 4. Key morphological settings include an 8 × 8 column grid system with 8 m spacing, building heights ranging from 6 to 10 stories, and a floor height of 4 m. Additionally, building orientation uses true south as a reference, with a variation angle range from −15° to 15° and a granularity of 1°.

For optimizing outdoor thermal comfort, the annual average UTCI value is used as a quantitative index to ensure that the site surrounding the building volume remains as close to thermal neutrality as possible. For energy consumption optimization, the energy load per unit area serves as a quantitative index, with initial energy consumption parameters set to simulate the annual energy consumption of the building form. To balance these performance goals, the research employs the Pareto optimal solution method to weigh the importance of multiple design objectives, optimizing them simultaneously. To simplify calculations, the UTCI simulation resolution is set at 50 m × 50 m, with averages from each generated site scenario used as quantitative metrics.

The optimization design process is implemented through the Steady State Island Evolutionary Algorithm, with five islands optimizing concurrently, each containing 15 individuals. A total of 1400 optimization iterations are conducted to ensure adequate evolution.

3. Results

3.1. Multi-Objective Optimization Result Analysis

Figure 3 illustrates the distribution of all generated samples, with the horizontal and vertical axes representing the standardized values of outdoor thermal comfort and building energy consumption, respectively. The overall trend in the optimization scatterplot shows the samples converging toward the minimal values of the two performance indices [51], indicating that the multi-objective optimization method successfully addresses the aim of balancing multiple performance indices [52,53]. Additionally, most optimized design samples exhibit performance indicators superior to existing buildings. However, since actual design processes involve multiple factors, there are numerous other influences to consider beyond just outdoor thermal comfort and building energy consumption.

Significantly, building orientation has a notable impact on performance-based design. From the solutions on the Pareto frontier, it is evident that south or southeast orientations are optimal. A south-facing orientation allows more direct sunlight during the day, reducing the need for artificial lighting and thereby lowering energy consumption. Conversely, buildings facing southeast create more shadows over the courtyard, maintaining more of the site in a thermally neutral comfort zone, thus improving the thermal environment.

A total of 1500 multi-objective optimization schemes were generated, with the distribution of optimization samples from six islands, as shown in Figure 3. The figure combines the individual and aggregate relationships. The y-axis represents the standardized result of outdoor thermal comfort, while the x-axis represents the standardized total annual building energy consumption. The black line represents the Pareto frontier's optimal solutions, indicating the best outcomes in the optimization direction. The objective of the optimization design is to minimize the standardized indices, with points closer to the origin representing better outcomes. Since multiple islands can produce multiple optimal solutions, the research consolidates these island-specific optima into an optimal solution set, providing design tendencies and optimization suggestions for the early stages of architectural form design [54].

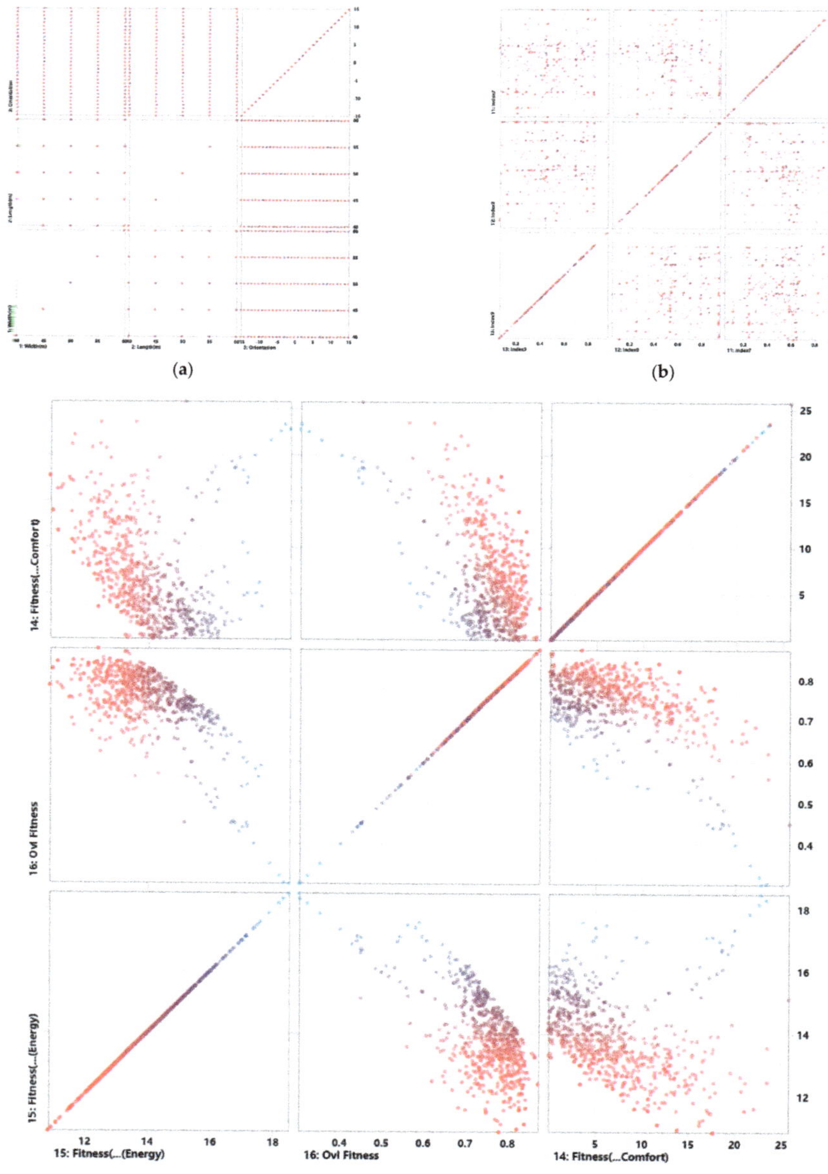

Figure 3. Distribution of optimal solutions for the optimized sample. (**a**) Performance Distribution of the Pre-Iteration Design; (**b**) Performance Distribution of Design Solutions during Iteration; (**c**) Post-Iteration Design Distribution.

The performance of the optimal solutions generated after multi-objective optimization is shown in Figure 3. The axes represent a combination of the fitness of energy consumption (Fitness of Energy), thermal comfort fitness (Fitness of Comfort), and overall fitness (Total Fitness). The dense clustering of points around the origin in the energy consumption–thermal comfort graph indicates that, under the multi-objective orientation, the iteratively generated schemes approach the performance envisioned by the objective function. The interaction of these indices is not merely additive but involves iterative optimization coordinated through the control of morphological parameters. The degree of clustering observed in the graph signifies the efficacy of multi-objective optimization, with performance enhancements realized during the iterative process.

Figure 3 depicts the iterative process of multi-objective optimization for a selected set of 650 solutions. In the initial settings of optimization, the individual geometric parameters of the building are set uniformly across gradients, as evidenced by the uniform intersection of lines in the early segments of the line graph. During the mid-stage of optimization, after evaluating multiple performance indices, a comprehensive assessment of the solutions' performance across several dimensions is conducted. Non-dominant solutions are replaced by dominant ones. In the later stages of optimization, as the number of iterations increases, the overall trend lines show a clustering tendency, and the optimal solution set on the Pareto front outperforms other non-dominated solution sets, ultimately outputting the Pareto optimal solutions.

Figure 4 illustrates the performance optimization process for 12 selected Pareto optimal solutions. The continuous lines from left to right represent each specific scheme, and the vertical axis marks indicate the standardized values of that performance standard. The distribution of fitness levels among the optimal solutions indicates that each solution exhibits distinct performance characteristics. Some schemes emphasize indices related to outdoor thermal comfort, while others focus on building energy consumption, yet all display excellent overall fitness. Thus, the multi-objective optimization not only coordinates various building performance goals but also highlights distinctive features compared to other optimal solutions.

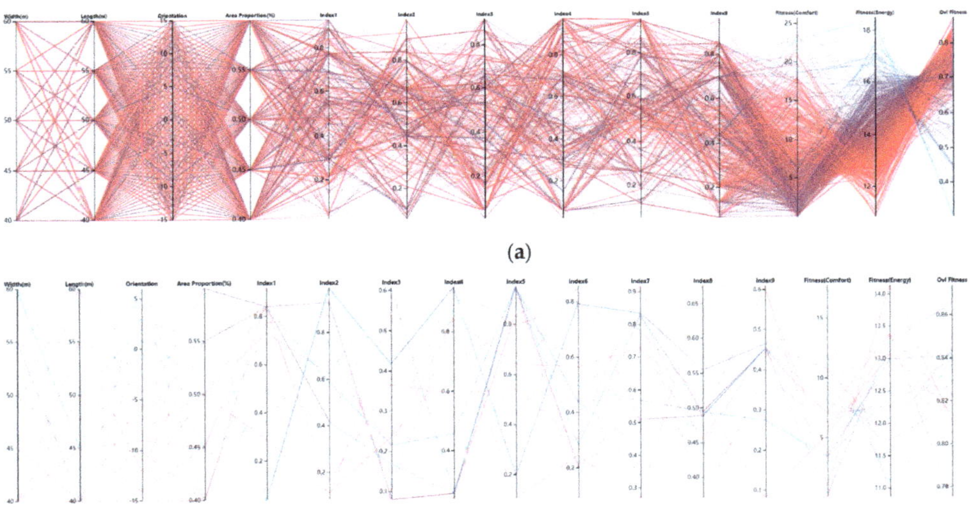

(a) (b)

Figure 4. Pareto Chart of Optimal Solutions. (**a**) Pareto Front of the Entire Solution Set; (**b**) Pareto Front of the Optimal Solutions.

Table 2 delineates the iterative process of multi-objective optimization across a set of 650 optimal solutions. Initially, individual geometric parameters of the buildings are uniformly set across gradients, as illustrated by the uniform crossing of lines in the early segment of the line graph. During the mid-phase of optimization, after evaluating various performance indices, solutions are comprehensively assessed across multiple dimensions.

Non-dominant solutions are replaced by dominant ones. In the later stages, as the number of iterations increases, the trend lines converge, indicating clustering. The optimal solution set on the Pareto front (Pareto frontier) exhibits superior performance relative to other non-dominated solution sets, culminating in the selection of the Pareto optimal solutions.

Table 2. Multi-objective optimization process.

<table>
<tr><th colspan="3">Optimization Results from Iteration 0 to Iteration 10</th></tr>
<tr><td rowspan="2">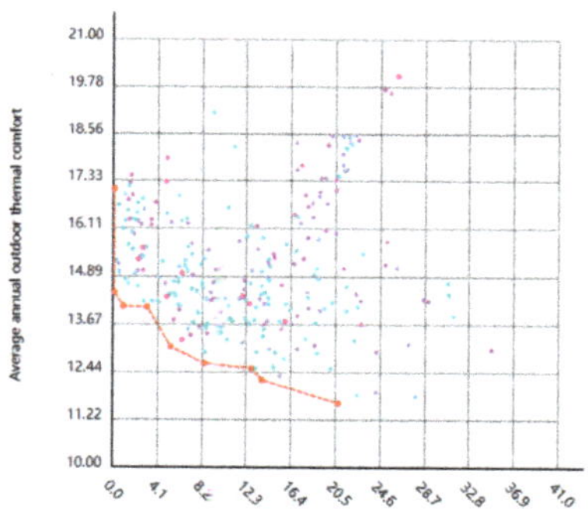

Current Mutation Rate: 0.3119
Current Mutation Radius: 0.9085</td>
<td>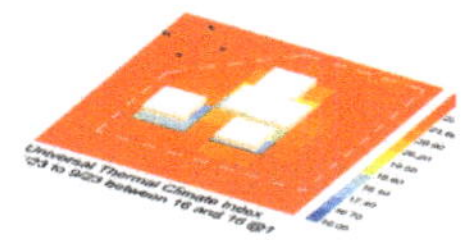
ID: 64
Fitness of load: 16.98
Fitness of comfort: 13.03</td>
<td>
ID: 128
Fitness of load: 1.03
Fitness of comfort: 16.99</td></tr>
<tr><td>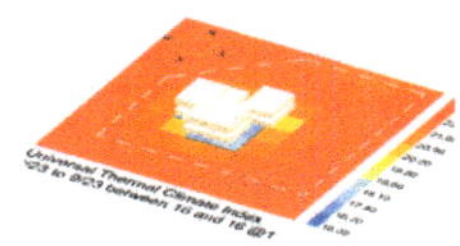
ID: 192
Fitness of load: 1.38
Fitness of comfort: 14.70</td>
<td>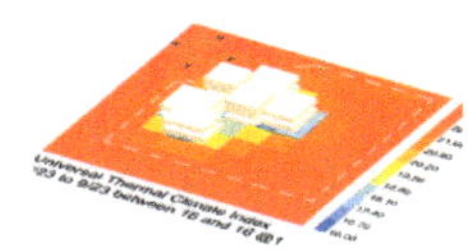
ID: 256
Fitness of load: 6.74
Fitness of comfort: 16.37</td></tr>
<tr><th colspan="3">Optimization Results from Iteration 10 to Iteration 20</th></tr>
<tr><td rowspan="2">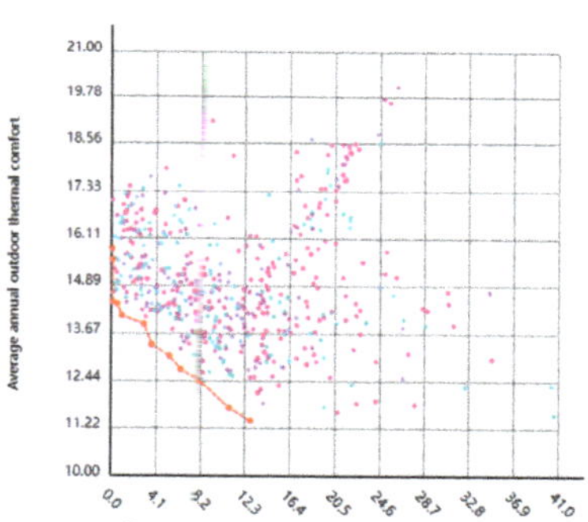

Current Mutation Rate: 0.2695
Current Mutation Radius: 0.8068</td>
<td>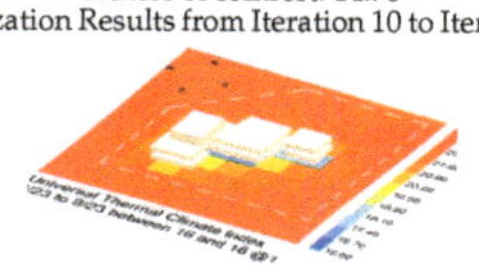
ID: 320
Fitness of load: 17.09
Fitness of comfort: 17.39</td>
<td>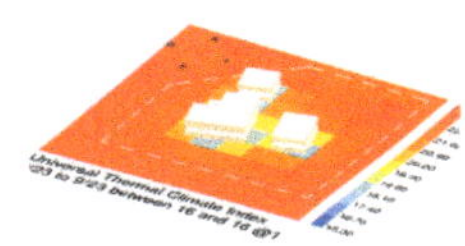
ID: 384
Fitness of load: 0.84
Fitness of comfort: 15.11</td></tr>
<tr><td>
ID: 448
Fitness of load: 19.39
Fitness of comfort: 14.18</td>
<td>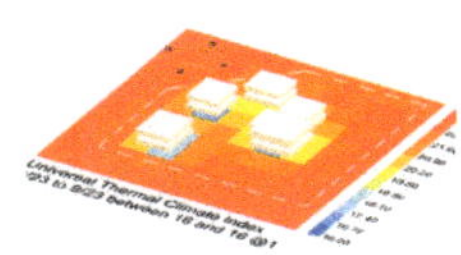
ID: 512
Fitness of load: 7.58
Fitness of comfort: 14.74</td></tr>
<tr><th colspan="3">Optimization Results from Iteration 40 to Iteration 50</th></tr>
<tr><td rowspan="2">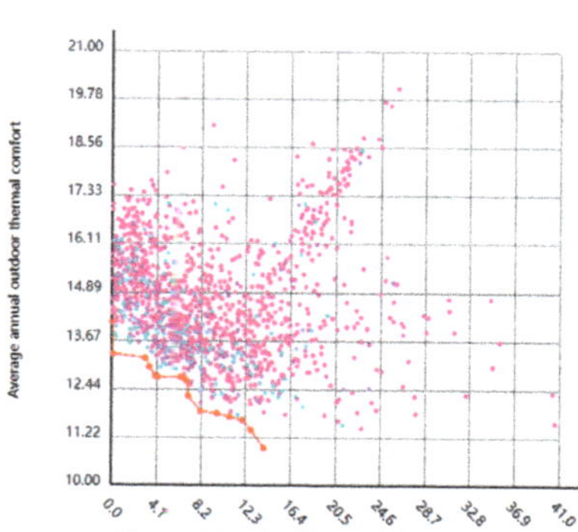

Current Mutation Rate: 0.1424
Current Mutation Radius: 0.5017</td>
<td>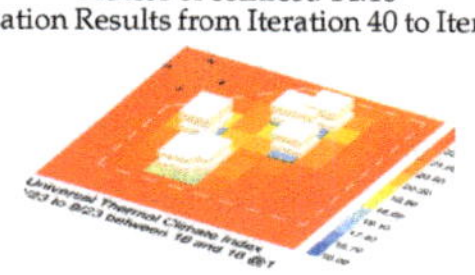
ID: 1088
Fitness of load: 1.79
Fitness of comfort: 15.43</td>
<td>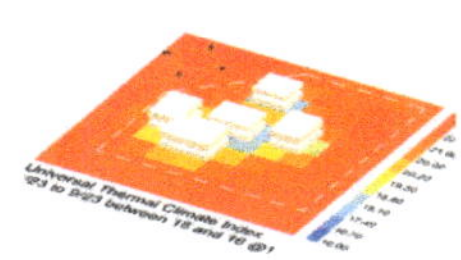
ID: 1152
Fitness of load: 4.79
Fitness of comfort: 15.67</td></tr>
<tr><td>
ID: 1216
Fitness of load: 1.59
Fitness of comfort: 13.62</td>
<td>
ID: 1280
Fitness of load: 2.89
Fitness of comfort: 15.60</td></tr>
</table>

Table 2. *Cont.*

Optimization Results from Iteration 50 to Iteration 60		
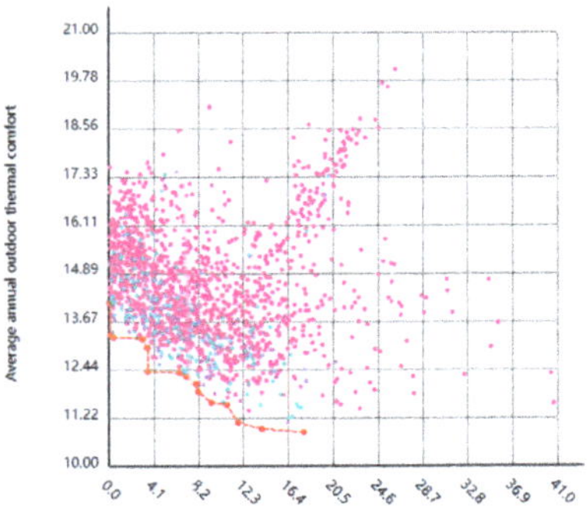 Current Mutation Rate: 0.1000 Current Mutation Radius: 0.4000	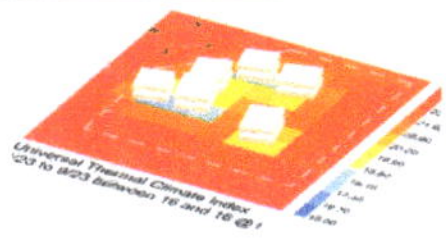 ID: 1344 Fitness of load: 8.41 Fitness of comfort: 14.90 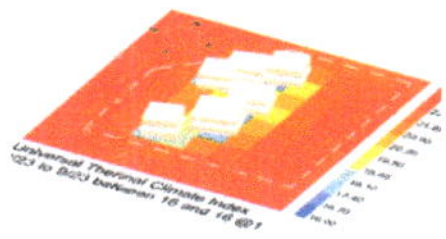 ID: 1472 Fitness of load: 3.84 Fitness of comfort: 14.01	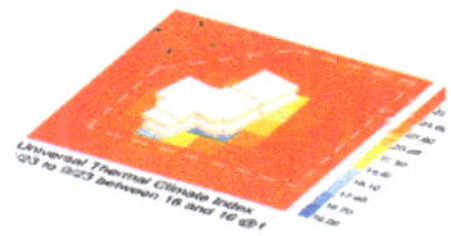 ID: 1408 Fitness of load: 13.80 Fitness of comfort: 11.54 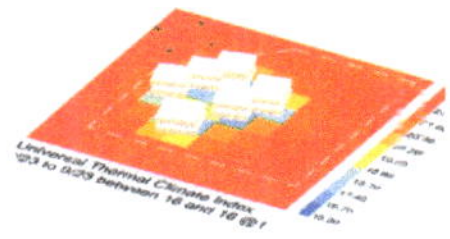 ID: 1536 Fitness of load: 1.82 Fitness of comfort: 14.12

Figure 4a,b depict the performance optimization process for 12 Pareto optimal solutions with distinct morphological features. The continuous lines from left to right in each figure represent individual schemes, with the vertical axes indicating the standardized values of the performance standards. The fitness distribution of the optimal solutions reveals significant differences in performance, with each scheme exhibiting distinct characteristics. Some schemes prioritize indices related to outdoor thermal comfort, while others focus on reducing building energy consumption, yet all demonstrate exceptional overall fitness. Therefore, the multi-objective optimization not only reconciles multiple building performance objectives but also distinguishes the optimal solutions in terms of their distinct characteristics compared to other potential solutions.

3.2. Comparison of Performance Before and After Optimization

Figure 5a,b illustrate the distribution of the Universal Thermal Climate Index (UTCI) and Predicted Mean Vote (PMV) before optimization. The original design scenario indicates an annual average daytime UTCI ranging from 22.89 °C to 27.84 °C within a 500 m × 500 m site, with an area-wide average UTCI of 10.10 °C. The minimum UTCI values are located at the northern L-shaped corner of the building. The PMV simulation identifies thermal sensations as follows: warm sensations constitute 20.04%, neutral sensations 34.69%, and cool sensations 45.26%. The zones classified as thermally neutral are concentrated around the building block but cover a small area, located 3–5 m from the building boundary and accounting for 34.69% of the total site.

Figure 5c,d present the UTCI and PMV distributions following the optimization process. Post-optimization, the UTCI values range from 9.11 °C to 10.46 °C during average daytime, with a site average UTCI of 9.79 °C. According to the UTCI thermal comfort scale, the average conditions are categorized as experiencing 'no thermal stress', showing no significant change in average UTCI compared to the original design. However, the area of thermal comfort has increased significantly from approximately 60% of the site in the original design to about 75% in the optimized design. This expansion indicates an improvement in the UTCI performance index. The PMV distribution after optimization shows thermal sensations divided as follows: warm sensations at 18.37%, neutral at 35.80%, and cool at 45.86%. The proportion of thermally neutral sensations has improved by 1.11% compared to the original scenario.

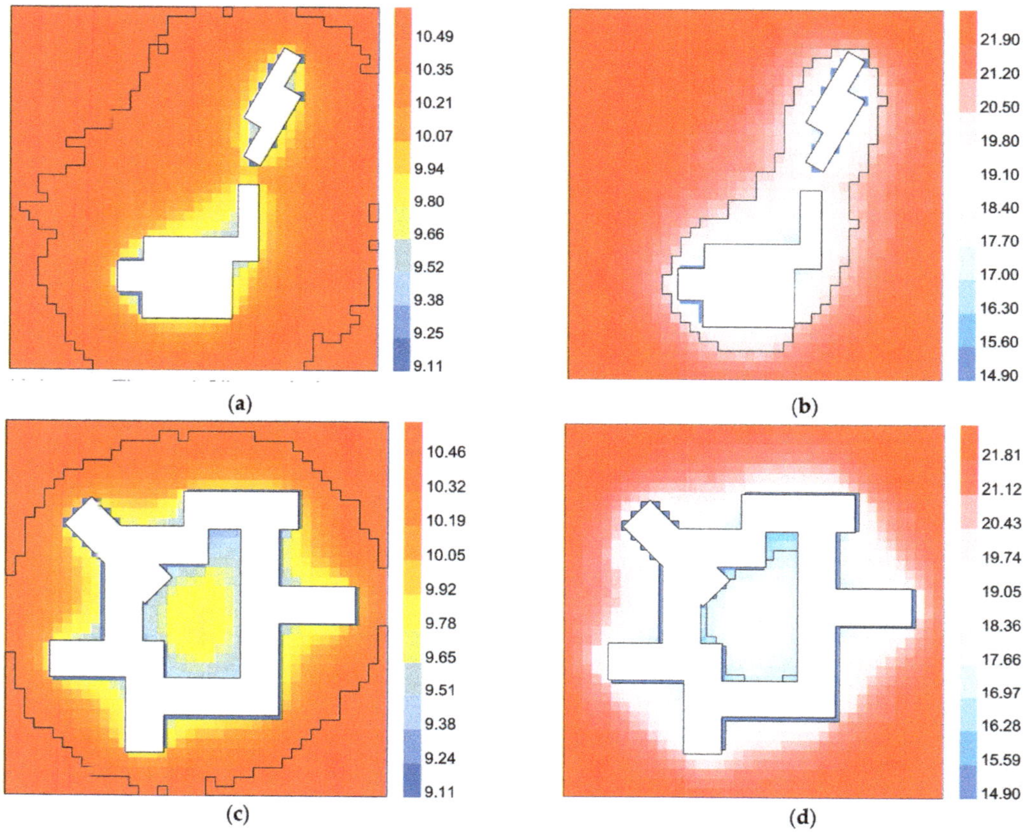

Figure 5. Distribution maps of UTCI and PMV before and after optimization. (**a**) The thermal comfort distribution heatmap before optimization: Average UTCI: 10.10 °C; (**b**) The thermal sensation vote distribution heatmap before optimization: Percentage of hot sensation: 20.04%, percentage of neutral sensation: 34.69%, percentage of cold sensation: 45.26%; (**c**) Thermal comfort distribution heatmap after optimization: Average UTCI: 9.79 °C; (**d**) The thermal sensation vote distribution heatmap after optimization: Percentage of hot sensation: 18.37%, percentage of neutral sensation: 35.80%, percentage of cold sensation: 45.86%.

The optimization of the design has enhanced thermal environment performance indicators significantly over the original scenario. The increase in the area classified as 'no thermal stress' and the improvement in the proportion of thermal neutrality highlight the effectiveness of the optimization in increasing occupant comfort and the efficient use of space with respect to thermal indices.

Table 3 presents the pre-optimization monthly building energy consumption data. According to the simulation, the total energy consumption of the original design was 75.68 kWh/m^2 per year. This included 44.43 Wh/m^2 for cooling, 1.91 Wh/m^2 for heating, 31.14 Wh/m^2 for lighting, and 31.19 Wh/m^2 for equipment usage. Table 3 illustrates the post-optimization energy consumption, where the total decreased to 62.88 kWh/m^2 per year. The breakdown is as follows: cooling energy consumption significantly reduced to 8.79 Wh/m^2, heating remained nearly unchanged at 1.87 Wh/m^2, lighting was reduced to 21.31 Wh/m^2, and equipment energy slightly increased to 31.91 Wh/m^2.

Table 3. Data Sheet of Building Energy Consumption before and after Optimization.

	January	February	March	April	May	June	July	August	September	October	November	December
Optimized Plan—Cooling	0.00	0.00	0.00	0.00	0.03	0.94	3.36	3.11	1.23	0.12	0.00	0.00
Optimized Plan—Heating	0.77	0.37	0.15	0.02	0.00	0.00	0.00	0.00	0.00	0.00	0.07	0.49
Optimized Plan—Lighting	1.80	1.64	1.87	1.67	1.87	1.79	1.74	1.87	1.74	1.80	1.79	1.74
Optimized Plan—Equipment	2.62	2.37	2.67	2.48	2.67	2.57	2.58	2.67	2.53	2.62	2.57	2.58
Original Plan—Cooling	0.00	0.00	0.00	0.00	0.03	1.18	4.45	4.12	1.53	0.13	0.00	0.00
Original Plan—Heating	0.78	0.38	0.15	0.02	0.00	0.00	0.00	0.00	0.00	0.00	0.07	0.49
Original Plan—Lighting	2.62	2.38	2.71	2.50	2.71	2.58	2.58	2.68	2.58	2.63	2.61	2.58
Original Plan—Equipment	2.67	2.37	2.68	2.52	2.69	2.61	2.59	2.67	2.54	2.67	2.58	2.62

Both schemes show similar trends in energy consumption, particularly for lighting and equipment usage, which remain consistent throughout the months, ranging between 2.4 Wh/m^2 and 2.7 Wh/m^2 for equipment and 1.6 Wh/m^2 to 1.9 Wh/m^2 for lighting. Heating energy consumption shows values from May to October, coinciding with the cooling months set according to meteorological data. Notably, the optimized design significantly reduces cooling energy consumption, especially in July and August, by approximately 1.1 Wh/m^2. This reduction is attributed to the more complex building massing in the optimized design, which creates self-shading effects that lower indoor temperatures and reduce cooling needs during summer.

Table 3 illustrates the monthly total energy consumption and energy-saving efficiency of the original and optimized schemes. The dual y-axis graph displays total energy consumption on the left and energy-saving efficiency on the right. Bar charts represent energy consumption, and line graphs depict the differences and percentage changes in energy consumption pre- and post-optimization. For instance, the original scheme consumed 6.07 Wh/m^2 in January, while the optimized scheme used 5.19 Wh/m^2, achieving an energy-saving efficiency of 14.49%. The yearly energy-saving efficiency ranges from 14.49% in January to 20.16% in July, with an average of 16.66%. The highest savings in July likely result from the self-shading provided by the optimized architectural form, which is most effective during summer, as seen in the seasonal analysis: Summer > Autumn $\geq$ Spring > Winter.

This analysis demonstrates that the multi-objective optimization set for architectural design enhances performance significantly over the original scheme, particularly in reducing energy consumption during peak cooling periods.

4. Discussion

The Pareto optimal solutions, compared with existing approaches, can significantly improve the UTCI while simultaneously reducing the EUI level. The fitness of a solution reflects its specific performance in a certain aspect. Since the optimization trend minimizes the fitness of two performance indicators, the set of solutions distributed along the Pareto front demonstrates better overall performance. The set of Pareto optimal solutions is shown in Figure 6, where the range of the sum of the fitness for outdoor thermal comfort and building energy consumption across the 16 optimal solutions is between 13.40 and 30.74. Among them, the solution with the smallest total fitness sum is ID 1444, which has an energy consumption fitness (Fitness of load) of 3.48 and an outdoor thermal comfort fitness (Fitness of comfort) of 12.99. This solution features multiple building blocks that form an enclosed central courtyard located on the east side. It can be observed that buildings with a higher enclosure degree are more adaptable to cold climates, demonstrating stronger performance in both outdoor thermal comfort and energy consumption.

However, there are still limitations in this study. (1) The HVAC system performance was simplified in the simulation process, and the simulation method has not been validated, which resulted in significant deviations in the energy consumption simulation values. (2) The weight setting for energy consumption and thermal comfort was set to w = 0.5, but in practical design, the weight coefficients for each performance aspect should be dynamically adjusted based on real-world conditions and design requirements. (3) In reality, the building design process is complex, and building forms are often the result of a combination of factors such as ideology, cultural spirit, and economic conditions. However, due to the current limitations of simulation hardware and computational power, multi-objective optimization cannot incorporate as many influencing factors into the objective functions as desired, leading to discrepancies with real-world situations. (4) The core algorithm for multi-objective optimization has been updated to NSGA-III, which has improved optimization efficiency and results compared to NSGA-II. However, this study did not use the latest multi-objective optimization technology for optimization.

Figure 6. Pareto Front Solutions of Building Forms.

5. Conclusions

Computational generative design is an important topic in the context of the era of performance enhancement. This study demonstrates the significant potential of multi-objective optimization and generative design in architectural design. The Pareto optimal solutions coordinate two primary optimization goals: "reducing building energy load" and "improving outdoor thermal comfort for occupants". These solutions can be applied to the early-stage massing design and measurement of office buildings under a green and health-oriented design approach. The results show that: (1) forms with higher enclosures (three-sided or four-sided) are more suitable for office buildings in cold regions in terms of both outdoor thermal comfort and building energy consumption; (2) compared with existing building designs, the Pareto optimal solution improves energy efficiency by 16.55% over the course of the year under 2022 meteorological conditions, and the proportion of outdoor thermal neutrality increases by 1.11%. The results indicate that the optimized design can balance building energy load and outdoor thermal comfort, achieving a multi-objective equilibrium, thus providing a specific workflow for similar engineering problems; (3) building performance is affected by seasonal changes—the impact of the parametric-generated building massing on energy consumption is most significant in the summer and least significant in the winter.

The multi-objective optimization design workflow developed in this study is of great significance for fine-tuned climate adaptation design in cold regions. This study uses real meteorological data from cold regions as input conditions, with a simulation duration for thermal comfort and building energy consumption covering an entire year, with time accuracy set at 365 days/year. Future research could further explore the coordination of different performance indicators (such as green health metrics, daylight comfort, building operational costs, CO_2 emissions, etc.) in multi-objective optimization design for different climate zones, such as the coordination of light and thermal environments in extremely cold regions. Additionally, the research could analyze sensitivity analysis of multiple performance indicators (e.g., the impact of building massing on outdoor thermal comfort) and quantify the extent to which various sub-indicators influence the optimization results. Further studies may also integrate machine learning techniques by training machine learning models with optimal results from multi-objective optimization design to address the long computation times associated with multi-objective optimization, thereby reducing the difficulty of using such designs.

This study's workflow proposes a multi-objective optimization design workflow for office buildings. To enhance the applicability of this workflow, corresponding performance evaluation indicators should be established for different building types (such as residential areas, schools, kindergartens, nursing homes, libraries, etc.) and design intentions (such as light comfort, carbon emissions throughout the life cycle, urban resilience, etc.). As the evaluation indicators are further refined, the multi-objective optimization of building forms will inevitably become more rational and objective, promoting the development of the construction industry toward the optimization of the urban built environment driven by big data.

The trend of using urban big data to assess the built environment is unstoppable, and human–machine collaboration is bound to develop further. Therefore, how to use the built environment evaluation system to improve the built environment in a high-quality manner is a direction that needs to be researched in the future. Future research could expand into thermal environment studies under extreme climates. Additionally, from the perspective of smart cities, real-time camera observation data can be utilized to conduct research on the correlation between building equipment and human behavior, with the aim of achieving energy saving and emission reduction through controlled building equipment. Finally, considering the perspective of healthy buildings, research can be conducted on the matching between medical data and built environment data. By studying the correlation and significance of the impact between medical data and the built environment, key points for improving the built environment can be identified.

Author Contributions: F.G.: conceptualization and investigation, S.M.: writing—original draft, S.X. and M.L.: Data processing and visualization, H.Z. and J.D.: writing—review and editing. All authors have read and agreed to the published version of the manuscript.

Funding: This research was jointly funded by the National Natural Science Foundation of China (52108044) for the project 'Study on Urban Thermal Environment Assessment and Planning Mechanism Based on Local Climate Zoning' and the Liaoning Province Social Science Federation's economic and social development project (2025lslybwzzkt-156) titled 'Research on Exploiting the Advantages of Liaoning's Marine Resources to Create Unique Tourism Products'.

Data Availability Statement: The original contributions presented in the study are included in the article, further inquiries can be directed to the corresponding authors.

Conflicts of Interest: We declare that we have no financial and personal relationships with other people or organizations that can inappropriately influence our work, and there is no professional or other personal interest of any nature or kind in any product, service, and/or company that could be construed as influencing the position presented in, or the review of, the manuscript.

References

1. Laino, E.; Iglesias, G. Extreme climate change hazards and impacts on European coastal cities: A review. *Renew. Sustain. Energy Rev.* **2023**, *184*, 113587. [CrossRef]
2. Javanroodi, K.; Mahdavinejad, M.; Nik, V. Impacts of urban morphology on reducing cooling load and increasing ventilation potential in hot-arid climate. *Appl. Energy* **2018**, *231*, 714–746. [CrossRef]
3. Kiss, B.; Szalay, Z. Modular approach to multi-objective environmental optimization of buildings. *Autom. Constr.* **2020**, *111*, 103044. [CrossRef]
4. Makki, M.; Showkatbakhsh, M.; Tabony, A.; Weistock, M. Evolutionary algorithms for generating urban morphology: Variations and multiple objectives. *Int. J. Archit. Comput.* **2018**, *17*, 147807711877723. [CrossRef]
5. Mukkavaara, J.; Shadram, F. An integrated optimization and sensitivity analysis approach to suppor t the life cycle energy trade-off in building design. *Energy Build.* **2021**, *253*, 111529. [CrossRef]
6. Ignatius, M.; Wong, N.H.; Martin, M.; Chen, S. Virtual Singapore integration with energy simulation and canopy modelling for climate assessment. *IOP Conf. Ser. Earth Environ. Sci.* **2019**, *294*, 012018. [CrossRef]
7. Safiri, S.; Nikoofard, A. Ladybug Beetle Optimization algorithm: Application for real-world problems. *J. Supercomput.* **2022**, *79*, 3511–3560. [CrossRef] [PubMed]
8. Zong, C.; Chen, X.; Deghim, F.; Staudt, J.; Geyer, P.; Lang, W. A holistic two-stage decision-making methodology for passive and active building design strategies under uncertainty. *Build. Environ.* **2024**, *251*, 111211. [CrossRef]
9. Wang, L.; Janssen, P.; Ji, G. Optimization-based design exploration of building massing typologiesd-EvoMass and a typology-oriented computational design optimization method for early-stage performance-based building massing design. *Front. Archit. Res.* **2024**, *13*, 1400–1422. [CrossRef]
10. Miao, Y.; Chen, Z.; Chen, Y.; Tao, Y. Sustainable Architecture for Future Climates: Optimizing a Library Building through Multi-Objective Design. *Buildings* **2024**, *14*, 1877. [CrossRef]
11. Delgarm, N.; Sajadi, B.; Delgarm, S. Multi-objective optimization of building energy performance and indoor thermal comfort: A new method using artificial bee colony(ABC). *Energy Build.* **2016**, *131*, 42–53. [CrossRef]
12. Xue, Q.; Wang, Z.; Chen, Q. Multi-objective optimization of building design for life cycle cost and CO_2 emissions: A case study of a low-energy residential building in a severe cold climate. *Build. Simul.* **2022**, *15*, 83–98. [CrossRef]
13. Li, J.; Wang, Y.; Xia, Y.; Song, Y.; Xie, H. Optimization of Urban Block Form by Adding New Volumes for Capacity Improvement and Solar Performance Using A Multi-Objective Genetic Algorithm: A Case Study of Nanjing. *Buildings* **2022**, *12*, 1710. [CrossRef]
14. Zhang, X.; Wang, J.; Zhou, Y.; Wang, H.; Xie, N.; Chen, D. A multi-objective optimization method for enclosed-space lighting design based on MOPSO. *Build. Environ.* **2024**, *250*, 1.1–1.15. [CrossRef]
15. Mirzabeigi, S.; Razkenari, M. Design optimization of urban typologies: A framework for evaluating building energy performance and outdoor thermal comfort. *Sustain. Cities Soc.* **2022**, *76*, 103515. [CrossRef]
16. Maksoud, A.; Alawneh, S.; Hussien, A.; Abdeen, A.; Abdalla, S. Computational Design for Multi-Optimized Geometry of Sustainable Flood-Resilient Urban Design Habitats in Indonesia. *Sustainability* **2024**, *16*, 2750. [CrossRef]
17. Khan, A.M.; Tariq, M.; Rehman, S.; Saeed, T.; Alqahtani, F.; Sherif, M. BIM Integration with XAI Using LIME and MOO for Automated Green Building Energy Performance Analysis. *Energies* **2024**, *17*, 3295. [CrossRef]
18. Zhao, J.; Guo, F.; Zhang, H.; Dong, J. Mechanisms of Non-Stationary Influence of Urban Form on the Diurnal Thermal Environment based on Machine Learning and MGWR Analysis. *Sustain. Cities Soc.* **2024**, *101*, 105194. [CrossRef]
19. Lin, H.; Ni, H.; Xiao, Y.; Zhu, X. Couple Simulations with CFD and Ladybug + Honeybee Tools for Green Façade Optimizing the Thermal Comfort in a Transitional Space in hot-humid climate. *J. Asian Archit. Build. Eng.* **2023**, *22*, 1317–1342. [CrossRef]

20. Li, Y.; Ding, J.; Zhang, Z.; Zhou, X.; Makvandi, M.; Yuan, P.; Xie, Y. Practical application of multi-material topology optimization to performance-based architectural design of an iconic building. *Compos. Struct.* **2023**, *325*, 117603. [CrossRef]
21. Noorollahi, Y.; Barabadi, P.; Taherahmadi, J.; Abbasizade, F. Multi-objective optimization of energy demand and net zero energy building design based on climatic conditions (Case study: Iran). *Int. J. Environ. Sci. Technol.* **2024**, 1–16. [CrossRef]
22. Bao, X.; Zhang, J. Multi-objective decision optimization design for building energy-saving retrofitting design based on improved grasshopper optimization algorithm. *Int. J. Renew. Energy Dev.* **2024**, *13*, 1058–1067. [CrossRef]
23. Deshpande, A.; Pagare, A.; Tomar, A. Assessing the efficacy of green building design strategies in minimizing energy consumption in commercial buildings of Mumbai: A building performance analysis. *Int. J. Sci. Res. Arch.* **2023**, *2024*, 31–39. [CrossRef]
24. Vandenbogaerde, L.; Verbeke, S.; Audenaert, A. Optimizing building energy consumption in office buildings: A review of building automation and control systems and factors influencing energy savings. *J. Build. Eng.* **2023**, *76*, 107233. [CrossRef]
25. Fu, H.; Baltazar, J.; Claridge, D. Review of developments in whole-building statistical energy consumption models for commercial buildings. *Renew. Sustain. Energy Rev.* **2021**, *147*, 111248. [CrossRef]
26. Bordas, A.; Le Masson, P.; Maxime, T.; Weil, B. What is generative in generative artificial intelligence? A design-based perspective. *Res. Eng. Des.* **2024**, 1–17. [CrossRef]
27. Fitriawijaya, A.; Jeng, T. Integrating Multimodal Generative AI and Blockchain for Enhancing Generative Design in the Early Phase of Architectural Design Process. *Buildings* **2024**, *14*, 2533. [CrossRef]
28. Liao, W.; Lu, X.; Fei, Y.; Gu, Y.; Huang, Y. Generative AI design for building structures. *Autom. Constr.* **2024**, *157*, 105187. [CrossRef]
29. Wang, H.-J.; Sun, J.; Chen, H.; Zhu, Y.L.; Zhang, Y.; Lang, X.-M.; Fan, K.; Yu, E.; Yang, S. Extreme Climate in China: Facts, Simulation and Projection. *Meteorol. Z.* **2012**, *21*, 279–304. [CrossRef]
30. Liu, Z.; Hu, L.; Chen, H.; Li, Z.; Jiang, L. Exploring the combined cooling effect of street canyon geometry and the surrounding built environment. *Environ. Sci. Pollut. Res.* **2024**, *31*, 1–18. [CrossRef]
31. Taheri, S.; Hosseini, P. Model predictive control of heating, ventilation, and air conditioning (HVAC) systems: A state-of-the-art review. *J. Build. Eng.* **2022**, *60*, 105067. [CrossRef]
32. Arvidsson, R.; Tillman, A.-M.; Sandén, B.; Janssen, M.; Nordelöf, A.; Kushnir, D.; Molander, S. Environmental Assessment of Emerging Technologies: Recommendations for Prospective LCA: Prospective LCA. *J. Ind. Ecol.* **2018**, *22*, 1286–1294. [CrossRef]
33. Potrc Obrecht, T.; Röck, M.; Hoxha, E.; Passer, A. BIM and LCA integration: A systematic literature review. *Sustainability* **2020**, *12*, 5534. [CrossRef]
34. Luo, X.; Du, L. Energy consumption simulations of rural residential buildings considering differences in energy use behavior among family members. *Build. Simul.* **2024**, *17*, 1335–1358. [CrossRef]
35. Ao, J.; Du, C.; Jing, M.; Li, B.; Chen, Z. A Method of Integrating Air Conditioning Usage Models to Building Simulations for Predicting Residential Cooling Energy Consumption. *Buildings* **2024**, *14*, 2026. [CrossRef]
36. Bai, C.; Liu, J. Prediction and Management of Building Energy Consumption Based on Building Environment Simulation Design Platform DeST and Meteorological Data Analysis Algorithm. *Strateg. Plan. Energy Environ.* **2024**, *43*, 357–380. [CrossRef]
37. *GB55015-2021*; General Code for Energy Efficiency and Renewable Energy Application in Buildings. Standardization Administration of China: Beijing, China, 2021.
38. Höppe, P. Different aspects of assessing indoor and outdoor thermal comfort. *Energy Build.* **2002**, *34*, 661–665. [CrossRef]
39. Lai, D.; Guo, D.; Hou, Y.; Lin, C.; Chen, Q. Studies of Outdoor Thermal Comfort in Northern China. *Build. Environ.* **2014**, *77*, 110–118. [CrossRef]
40. Coccolo, S.; Kämpf, J.; Scartezzini, J.-L.; Pearlmutter, D. Outdoor human comfort and thermal stress: A comprehensive review on models and standards. *Urban Clim.* **2016**, *18*, 33–57. [CrossRef]
41. Xi, T.; Li, Q.; Mochida, A.; Meng, Q. Study on the outdoor thermal environment and thermal comfort around campus clusters in subtropical urban areas. *Build. Environ.* **2012**, *52*, 162–170. [CrossRef]
42. Ibrahim, Y.; Kershaw, T.; Shepherd, P. A methodology For Modelling Microclimates: A Ladybug-tools and ENVI-met verification study. In Proceedings of the 35th PLEA Conference Sustainable Architecture and Urban Design: Planning Post Carbon Cities, A Coruña, Spain, 1–3 September 2020.
43. Ibrahim, Y.; Kershaw, T.; Shepherd, P. Improvement of the Ladybug-tools microclimate workflow: A verification study. In Proceedings of the IBPSA-England Building Simulation and Optimisation Conference 2020, Loughborough, UK, 21–22 September 2020.
44. Sun, R.; Chen, D.; Xu, Y.; Lai, D.; Liu, W. Improving outdoor thermal comfort of a kindergarten by optimizing its building shape with genetic algorithm. *E3S Web Conf.* **2023**, *396*, 05011. [CrossRef]
45. Sadeghipour Roudsari, M.; Pak, M. Ladybug: A parametric environmental plugin for grasshopper to help designers create an environmentally-conscious design. In Proceedings of the BS 2013: 13th Conference of the International Building Performance Simulation Association, Chambery, France, 25–28 August 2013; pp. 3128–3135.
46. *ANSI/ASHRAE Standard 55*; Thermal Environmental Conditions for Human Occupancy. ASHRAE (American Society of Heating, Refrigerating and Air-Conditioning Engineers): Atlanta, GA, USA, 2021.

47. Karl, F.; Pielok, T.; Moosbauer, J.; Pfisterer, F.; Coors, S.; Binder, M.; Schneider, L.; Thomas, J.; Richter, J.; Lang, M.; et al. Multi-Objective Hyperparameter Optimization in Machine Learning—An Overview. *ACM Trans. Evol. Learn. Optim.* **2023**, *3*, 1–50. [CrossRef]
48. Guo, W.; Dong, Y. Enhancing energy-efficient building design: A multi-agent-assisted MOEA/D approach for multi-objective optimization. *Energy Inform.* **2024**, *7*, 102. [CrossRef]
49. Wang, L.; Janssen, P.; Stouffs, R. Teaching Computational Design Optimization—An experimental course for performance-based building massing exploration. In Proceedings of the 41st Conference on Education and Research in Computer Aided Architectural Design in Europe, eCAADe 2023, Graz, Austria, 20–22 September 2023; pp. 179–188.
50. Ji, Y.; Xu, M.; Zhang, T.; He, Y. Intelligent Parametric Optimization of Building Atrium Design: A Case Study for a Sustainable and Comfortable Environment. *Sustainability* **2023**, *15*, 4362. [CrossRef]
51. Wang, L. Workflow for applying optimization-based design exploration to early-stage architectural design—Case study based on EvoMass. *Int. J. Archit. Comput.* **2022**, *20*, 41–60. [CrossRef]
52. Zhang, Y.; Teoh, B.K.; Zhang, L. Multi-objective optimization for energy-efficient building design considering urban heat island effects. *Appl. Energy* **2024**, *376*, 124117. [CrossRef]
53. Li, M.; Wang, Z.; Chang, H.; Wang, Z.; Guo, J. A novel multi-objective generative design approach for sustainable building using multi-task learning (ANN) integration. *Appl. Energy* **2024**, *376*, 124220. [CrossRef]
54. Li, L.; Qi, Z.; Qingsong, M.; Gao, W.; Wei, X. Evolving multi-objective optimization framework for early-stage building design: Improving energy efficiency, daylighting, view quality, and thermal comfort. *Build. Simul.* **2024**, *17*, 2097–2123. [CrossRef]

energies

MDPI

Review

Thermoelectric Generators Applied as a Power Source in CubeSats: State of the Art

Gabriel Brugues Soares [1], Jorge Javier Gimenez Ledesma [1,2], Eder Andrade da Silva [1,2,*] and Oswaldo Hideo Ando Junior [1,2,3,4,*]

1 Interdisciplinary Postgraduate Program in Energy & Sustainability (PPGIES), Federal University of Latin American Integration—UNILA, Paraná City 85867-000, PR, Brazil; gb.soares.2019@aluno.unila.edu.br (G.B.S.); jorge.ledesma@unila.edu.br (J.J.G L.)
2 Research Group on Energy & Energy Sustainability (GPEnSE), Academic Unit of Cabo de Santo Agostinho (UACSA), Federal Rural University of Pernambuco (UFRPE), Cabo de Santo Agostinho 54518-430, PE, Brazil
3 Smart Grid Laboratory (LabREI), Center for Alternative and Renewable Research (CEAR), Federal University of Paraiba (UFPB), João Pessoa 58051-900, PB, Brazil
4 Department of Engeneering, UFRPE, Federal Rural University of Pernambuco, R. Cento e Sessenta e Três, 300, Cabo de Santo Agostinho 55292-901, PE, Brazil
* Correspondence: ea.silva.2020@aluno.unila.edu.br (E.A.d.S.); oswaldo.ando@ufrpe.br (O.H.A.J.)

Abstract: This systematic review outlines the application of thermoelectric generators (TEGs) as energy sources in CubeSats. While CubeSats currently rely on solar cells with efficiencies between 16.8% and 32.2%, their performance diminishes with increased distance from the Sun. TEGs, although used in radioisotope thermoelectric generators (RTGs) for satellites, remain underutilized in CubeSats. A literature review revealed 33 relevant articles, with 21.2% employing simulation software to evaluate thermal behavior. Among 34 patents, only one mentioned micro-TEGs, with most focusing on structural improvements. Patent activity peaked between 2016 and 2020, emphasizing structural and thermal optimization, but no patents addressed TEGs as energy sources for CubeSats, highlighting a significant research gap. TEGs present a viable solution for harnessing residual heat in CubeSats.

Keywords: ProKnow-C; thermoelectric generators; TEG; energy source; mapping; CubeSat

1. Introduction

Since the launch of the TIROS-M satellite in 1970, satellites have played a vital role in Earth observation, meteorology, and communication, providing essential services such as weather monitoring and connectivity in remote areas [1,2]. However, the high production and launch costs of traditional satellites have historically restricted their development to major space agencies like the National Aeronautics and Space Administration (NASA) and European Space Agency (ESA) [3].

Since its inception in 1958, NASA has undertaken manned missions and launched communication satellites to enhance telecommunications and expand internet access. Similarly, ESA, founded in 1975, has contributed to global internet and telecommunications infrastructure through satellite launches and participated in scientific missions, such as the Cassini and Huygens probes, which provided valuable data on Saturn and its moons. The Soviet Union pioneered space exploration with the launch of Sputnik 1 in 1957, the world's first artificial satellite, marking the start of the space age and driving the development of subsequent communication satellites [4–6].

Advancements in Micro-Electronics and Microsystems Technologies have enabled the miniaturization of components, reducing costs and paving the way for SmallSats. These compact satellites maintain key functionalities while minimizing size and weight, making them an efficient and affordable option for LEO missions, significantly expanding access to space technology [7,8].

The CubeSat, developed in 1999 by professors Jordi Puig-Suari (California Polytechnic State University) and Bob Twiggs (Stanford University), is a standardized nanosatellite with dimensions of 10 × 10 × 10 cm, weighing up to 1.33 kg, as shown in Figure 1. One of its most notable advantages is the ability to be constructed entirely from COTS (Commercial Off-The-Shelf) components, which ensures adaptability to various mission requirements while maintaining low production costs. Additionally, its power consumption is minimal, estimated at only a few watts [7]. This standardization of size, mass, and capacity aimed to simplify satellite control and enhance payload flexibility, making it a popular choice for nanosatellite missions, with launch costs typically ranging from $50,000 to $200,000 [9,10].

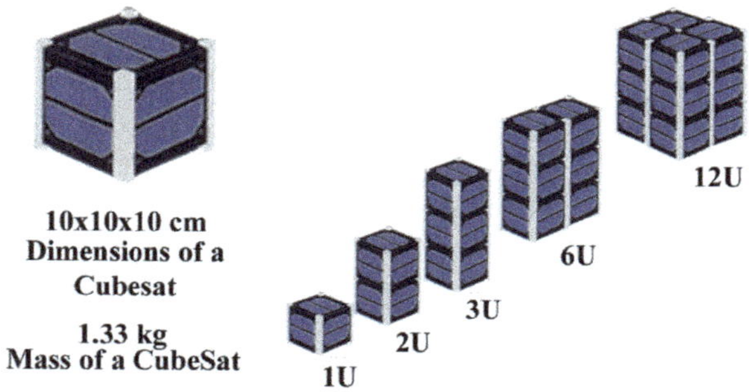

Figure 1. Standardized structure of a CubeSat. Adapted from [11].

In economic terms, the cost-per-kilogram ratio of payload has become the primary criterion for its transportation and launch, as the greater the amount of payload (in kilograms) to be launched into space, the higher its cost. This ratio has proven advantageous compared to traditional satellites, particularly for low-Earth-orbit missions, regarding construction and assembly time [10,12].

Focusing on the power supply for the systems and subsystems that make up the CubeSat, it is common to use photovoltaic solar cells as the main energy source, with efficiencies ranging from 16.8% to 32.2%, depending on their construction. Despite being widely implemented in space, these solar cells have the main disadvantage of specific power generated, measured in watts per kilogram (W/kg). This power decreases rapidly according to the $1/R^2$ relation, where R is the distance to the Sun [13,14].

Another possible unexplored solution to the use of solar cells is employing thermoelectric generators (TEGs) as a power source for CubeSats. Their application in satellites was already observed in [14]. Therefore, conducting research on the use of TEGs as a power source for CubeSats could reveal a promising application for low-Earth-orbit (LEO) space missions. Exploring this alternative could lead to significant advances in the feasibility of space missions with nanosatellites, expanding the technological and scientific capabilities of this sector.

TEGs and PVs are key energy sources for CubeSats, each with distinct advantages. TEGs efficiently convert temperature differences into electricity, leveraging waste heat from onboard systems and providing a reliable energy source in low-solar environments, such as deep-space orbits. Their compact, lightweight design suits CubeSats' weight constraints, although their conversion efficiency is relatively low (3.12–9.6%) [15,16]. PVs, on the other hand, achieve higher conversion efficiencies (16.8–32.2%) by directly harnessing solar energy, but their performance depends on consistent sunlight exposure and can be hindered by shading or orientation issues. Together, these technologies offer complementary solutions for diverse mission profiles. The comparative synthesis of the mentioned energy sources is presented in Table 1.

Table 1. Comparison between TEGs and PVs.

Characteristic	TEGs	PVs
Principle	Temperature difference conversion	Solar energy conversion
Efficiency	3.12–9.6%	16.8–32.2%
Weight	Lower in relation to generated power	Higher in relation to generated power
Dependency	Constant heat source	Solar exposure
Advantages	Operates without sunlight; utilizes residual heat; resistant to extreme conditions	High efficiency in energy conversion; greater performance in the presence of sunlight
Disadvantages	Low efficiency compared to PVs	Dependence on sunlight; reduced efficiency in shade or improper orientation

Preliminary Studies

A preliminary search identified eight scientific articles related to the research topic. Three focused on simulations using FEM software to analyze the thermal behavior of 1U CubeSats, specifically the operating temperatures of internal devices. One article explored the use of Thermoelectric Coolers (TECs) as temperature regulators for CubeSat hardware, while two investigated the application of micro-TEGs and conventional TEGs as energy sources for specific sensors in CubeSats. Additionally, two studies provided insights into advancements in thermoelectric systems: one highlighted recent progress in the modeling and simulation of thermoelectric power generation, emphasizing innovations in materials and multi-physics modeling approaches, while the other introduced a novel thermoelectric generator design incorporating stacked modules and dual heat pipes to enhance power density and efficiency.

Concerning this, the study proposed by [17] evaluated the energy generation potential of TEG modules using temperature gradients from solar panels. They conducted a numerical analysis to estimate temperature gradient variations in low Earth orbit, and the resulting data were used in an experimental test bench to map the real-time generated power. In conclusion, the experiment successfully reproduced the expected temperature gradients, but there were data discrepancies when the TEGs reached temperatures above 35 °C due to a limitation of the TEG itself.

In [18], the research addressed the development and verification of a conceptual model and an analytical model of the thermal controls of a 3U CubeSat, focusing on the study of an Orbital Satellite for Investigation of the Response of the Ionosphere to Stimulation and Space Weather (OSIRIS). The use of COMSOL Multiphysics to create FEM thermal models, along with analytical calculations, was essential to determine the worst-case temperature limits that the OSIRIS-3U might encounter during its mission. Additionally, a theoretical understanding was developed for applying different types of TEGs aimed at capturing residual energy in CubeSats. The results indicated that thin-film TEG devices provided a higher power density compared to traditional TEGs; however, due to the low expected temperature gradient in a CubeSat, the efficiency of these devices remained below 1%.

Therefore, with the mass and cost constraints of CubeSats, the current TEG technology did not seem viable for that application.

From this perspective, Ref. [19] presented a study evaluating the potential of using TEGs as an energy source to power an Assistance System, which was based on ensuring communication with a ground station or a service satellite, delivering data for maneuvers of inactive spacecraft. To achieve this, it was necessary to analyze the electrical characteristics of TEGs under space conditions to evaluate their potential for capturing residual energy in spacecraft, as commercial TEGs were developed for use on Earth. As a result, the analysis demonstrated that the proposed system was viable, with the necessary power being provided by TEGs utilizing the thermal loads present within a satellite.

In parallel, given the proposed research topic, it is extremely important to understand the thermal energy balance that occurs in a CubeSat, as the temperature distribution on the equipment's surface influences the TEGs' energy generation. Within this context, Ref. [20] focused on creating an accurate thermal model for the CubeSat MYSat-1, considering the effects of various thermal sources in orbit. A detailed model was developed using Ansys Workbench software and compared with CubeSatWizard, another MATLAB-based software The simulation results were compared with maintenance data, showing an acceptable difference of 0.45 °C in the CubeSat's average temperature between real data and the model results. However, on hot days, that difference was more significant, reaching 14 °C.

A study conducted by [21] involved the analysis of thermal simulations in a 1U CubeSat to ensure the electronic devices operated within their limits. Additionally, the study aimed to demonstrate that numerical simulations could be used to predict and optimize the temperature distribution of the space system before its construction. Ansys Icepak and Thermal Desktop software were used under LEO conditions. In conclusion, there was no overheating of the components under terrestrial hot-face conditions, as they were low-power devices. However, the study warns that in the near future, electronics were likely to continue developing while adding more power and functionality to devices. Therefore, thermal control would be necessary to ensure system reliability.

Another study proposed by [22] addressed the thermal simulation of a 1U CubeSat using COMSOL Multiphysics software Their objective was to simulate the impact of the inclination angle (beta angle) of the CubeSat on the temperature distribution of the nanosatellite, which describes the equipment's position relative to the solar vector. To investigate this parameter's effect, a set of simulations was performed for different beta angles. The simulations showed that the larger the beta angle, the greater the nanosatellite's exposure to prolonged solar heating, implying longer periods of both direct solar radiation and Albedo radiation.

In [23], the authors highlight significant progress in thermoelectric generator (TEG) systems, focusing on innovations in modeling techniques—analytical, numerical, and experimental—and advancements in materials like bismuth telluride, known for superior energy conversion properties. They explore the integration of Computational Fluid Dynamics (CFD), using COMSOL Multiphysics, with thermoelectric modeling to enhance simulation accuracy and presents practical applications, such as using TEGs to convert waste heat into electricity in automobiles, demonstrating improved energy efficiency even under variable conditions. Their research also addresses ongoing challenges, such as the need for more efficient materials and the integration of TEGs into complex systems, suggesting future investigations into novel thermoelectric materials and multi-physics models that account for phenomena like thermal conduction and electromagnetic conversion.

Lastly, Ref. [24] presents an innovative thermoelectric generator (TEG) design that utilizes stacked thermoelectric modules (TEMs) and dual heat pipes to significantly enhance power density, achieving up to 48.22 W/L in experiments, with a prototype output of 848.37 W. The use of dual heat pipes improves heat transfer efficiency between heating/cooling sources and the TEMs' hot/cold sides, enabling greater integration of TEMs in compact spaces. The study also highlights the adaptability of the design by modifying the number of layers in the stacked structure for diverse applications. Furthermore, it discusses the role of topological optimization and finite element modeling to simulate TEG performance, with experimental results validating the approach and demonstrating substantial efficiency improvements.

The data from the studies are essential for understanding the current limitations and future potential of TEGs. Advances in modeling combined with the development of new materials hold great promise for significantly enhancing the efficiency and applicability of thermoelectric systems. These improvements could make TEGs a viable option for various applications, including space exploration, where harnessing waste heat is critical for energy generation in resource-constrained environments.

This research aims to identify the application of TEGs as a power source for the Electrical Power System (EPS) of a 1U CubeSat operating in LEO. By mapping their application against existing technologies, the study seeks to address the gap in the use of TEGs in space missions and explore their potential for nanosatellite energy optimization. To achieve this, a systematic state-of-the-art review was conducted to identify the scientific literature and intellectual property (IP) related to EPS power sources, complemented by bibliometric analyses of the Bibliographic Portfolio (BP) and Patent Repository (PR). Connections between the datasets were further explored using Vosviewer software 1.6.20.

Initial research underscored the limited exploration of thermoelectric generators (TEGs) in nanosatellites, despite their widespread application in terrestrial industries such as automotive and manufacturing. This underutilization highlights a promising opportunity for LEO missions, particularly in leveraging TEGs as alternative power sources. Notably, the only prior use of TEGs in satellite applications involves radioisotope thermoelectric generators (RTGs). RTGs have been indispensable for NASA missions where solar power is impractical, such as in deep-space exploration. Often referred to as "nuclear batteries", RTGs have powered over 25 U.S. space missions since their debut in 1961, serving as a cornerstone of NASA's longstanding collaboration advancing space power systems [13].

A patent analysis identified only one direct application of TEGs for specific subsystems, neglecting geometric optimizations that could improve the watt-per-kilogram payload ratio of nanosatellites. The highlights of this study include:

- **Scientific and industrial impact**: TEGs can improve the energy efficiency of nanosatellites by increasing the W/kg payload ratio, enhancing power availability during low-Earth-orbit phases.
- **Unexplored technological approach**: the research addresses gaps in the application of TEGs as power generators for nanosatellites, utilizing temperature gradients from the space environment.
- **Utilization of waste heat**: TEGs harness waste heat generated by internal components or external space conditions, offering a novel approach to energy utilization in nanosatellites.
- **ProKnow-C methodology**: application of the ProKnow-C systematic method for mapping scientific and industrial developments related to TEGs as energy sources for CubeSats.

The study contributes to addressing this gap by investigating and analyzing key academic and industrial trends in the aerospace sector. While thermoelectric devices have primarily been used as temperature regulators for CubeSats, this research emphasizes their potential as energy sources, focusing on waste heat utilization and geometric optimization.

The results aim to enhance the performance of CubeSats, increase their power efficiency, and support future advancements in nanosatellite technology.

The article is structured into four sections. Section 1 contextualizes the topic, defines the general objective, and outlines the expected results and scientific contribution. Section 2 details the methodology, using the ProKnow-C method for a systematic review to select the Bibliographic Portfolio (BP) and Intellectual Property search, with a comprehensive bibliometric analysis of publications, patents, authors, and inventors, as well as the relevance of articles and patents. Section 3 provides a comparative analysis between the scientific portfolio and IP, utilizing Vosviewer software to examine the relationship between academic research and industrial patents. Section 4 concludes the study, summarizing the analyses and suggesting directions for future research.

2. Systematic Review

According to [25], a literature review aims to establish a line of reasoning that assists researchers in formulating hypotheses on the topic. Furthermore, a systematic review is essential for answering research questions, highlighting the main topics covered. This process aims to ensure a comprehensive and in-depth understanding of the subject.

In this context, the process of a systematic review is illustrated by the flowchart in Figure 2. Its organization involves a preliminary search for documents related to the topic to be investigated. These documents form a PDB, which should be studied to ensure alignment with the topic. Subsequently, the most common KWs should be extracted and combined, resulting in the main terms that will be used as references for creating the BP and PR. A general study of the obtained documents is conducted to ensure their relevance to the most pertinent information for investigating the proposed topic.

In this research, the scientific literature review addressing the utilization of residual heat for energy generation from TEGs as a power source for the EPS of a CubeSat was conducted using the ProKnow-C methodology, "Knowledge Development Process Constructivist".

According to [26,27], using this methodology as the primary tool to structure the BP aims to ensure its methodological foundation, based on the topic of interest, while considering the delimitations and restrictions of the articles that compose this portfolio. This establishes its scientific relevance and alignment with the research topic, based on the axes composed of combinations of KWs.

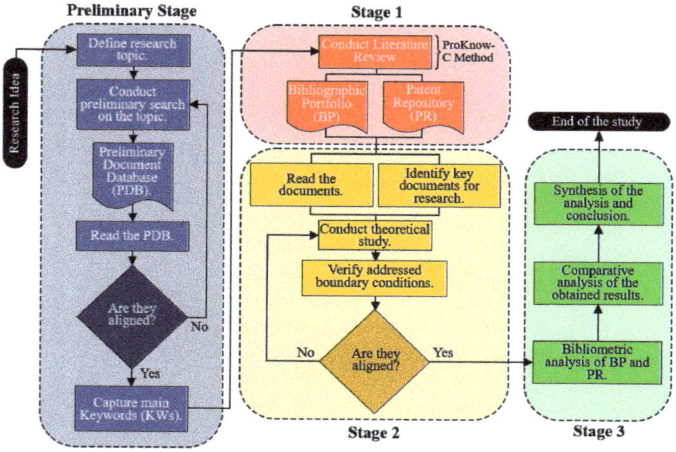

Figure 2. Basic process for conducting a systematic review.

The selected articles undergo a validation of the KWs through a verification of those present in the articles and are subsequently subjected to a bibliometric and systemic analysis. These steps form the primary stage of the ProKnow-C process, which consists of the following stages: selection of the BP; bibliometric analysis of the BP; Systemic analysis of the BP; and research question, as shown in Figure 3.

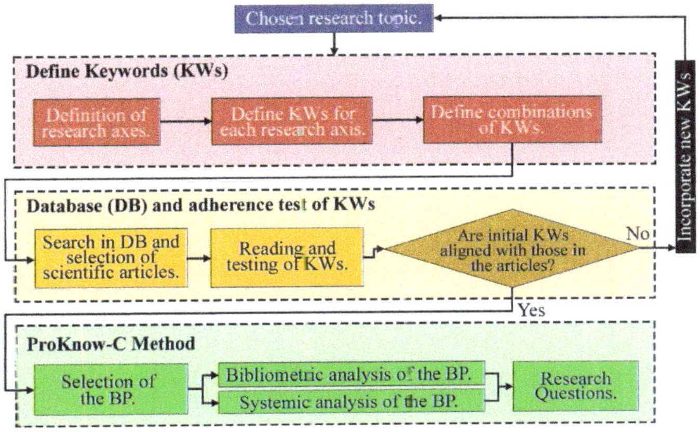

Figure 3. ProKnow-C methodology process.

2.1. Selection of the Bibliographic Portfolio

As an initial step in the execution of the ProKnow-C method, the selection of the BP is carried out through a systematic search for scientific materials, primarily in reliable databases [28]. For the present research, the chosen data sources were the WoS database, which generates the JCR and determines the impact factor of its journals, the Scopus database, which uses the SJR as its impact factor, and the SD database.

After selecting the DBs, the axes and KWs to be used for queries in these databases were defined. To clearly and comprehensively present the ProKnow-C procedures, the selection of the BP was organized into four axes within the proposed topic: Axis 1: internal and external temperature and structure of a nanosatellite; Axis 2: power sources and EPS of a nanosatellite; Axis 3: concepts of TEGs and their application in space; and Axis 4: thermal analysis in nanosatellites.

The combination of KWs for each axis was performed using logical expressions "and" and "or". Additionally, since the research area of this work focused on space conditions and the electronic systems of a nanosatellite, a category filter available in both DBs was applied, such as aerospace engineering, electrical/electronic engineering, mechanical engineering, and thermodynamics for WoS, and engineering for Scopus and SD.

The results of each query are displayed in Table 2, with a space–time delimitation over a period of 24 years. This time frame was defined based on the advancement of research that occurred from the year 2000 onward [1].

The PRAD, initially containing 1554 documents, was imported into the Mendeley bibliographic management tool [29]. After importing all the DBs into the software, it was verified whether all files from the export process were stored. This step was necessary because some titles may not have been exported due to unavailability in the desired format [30]. To address this issue, a manual search for the articles in the DBs can be performed. However, this process was not conducted at that stage.

Table 2. Results of the searches conducted in the DBs.

Database	WoS	Scopus	SD
Axis 1			
(CubeSat or Nanosatellite) and (Design or Structure)	312	157	10
(Temperature or Thermal) and (Internal or External) and (Nanosatellite or CubeSat)	27	8	80
("Finite Element analysis") and ("Thermoelectric Devices" or "Thermoelectric Generators" or "TEG") and (CubeSat)	1	0	0
Axis 2			
("Power Supply" or "Source") and (Nanosatellite)	188	156	90
Axis 3			
("Thermoelectric Generators" or "Thermoelectric Generator" or "TEG") and (Principles)	109	27	235
("Thermoelectric Generators" or "Thermoelectric Generator" or "TEG") and ("Space Applications")	26	25	50
Axis 4			
(CubeSat or Nanosatellite) and ("Thermal design" or "Thermal Modeling")	17	21	15
Preliminary Raw Articles' Database (PRAD)		**1554**	

As a result, the RAD contained 1309 titles, representing a loss of 15.76%. Subsequently, filtering actions were carried out to verify if the articles aligned with the proposed topic to obtain the K and P repositories, as described in the flowchart in Figure 4 [31]. At the end, 175 articles were obtained, whose citation numbers were verified on Google Scholar on 23 January 2024.

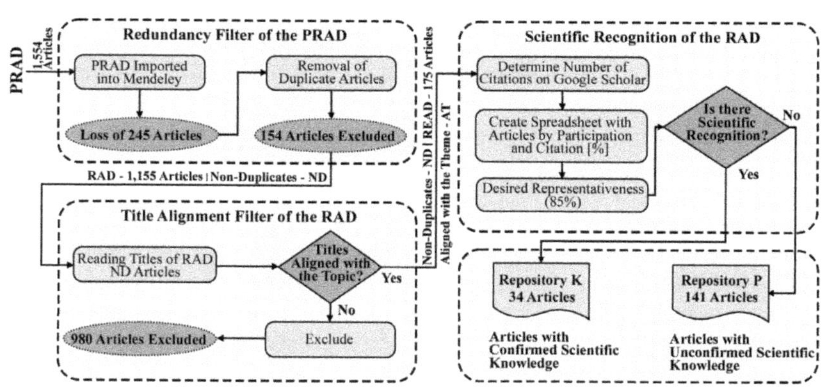

Figure 4. Initial filtering process for aligning titles with the proposed theme.

The organization of these articles was defined such that the citations were in descending order. Additionally, the minimum scientific representativeness criterion adopted followed the Pareto principle, which states that, in many phenomena, 80% of results or consequences are produced by 20% of their causes [32]. In this work, the adopted representativeness was 85%.

Using algorithms developed in Python, Repository K was composed of 34 articles with a cutoff value of 49, demonstrating proven scientific reliability. These articles represented

85% of the total citations, being cited 49 or more times. Conversely, Repository P consisted of 141 articles of lower academic relevance and without proven scientific reliability, divided between 104 journals with 1 to 48 citations and 37 without citations. Due to the number of articles in both repositories, the procedures established by [27] were automated using Python algorithms. These steps included verifying the thematic alignment of the articles, creating the AD, and validating it with articles from Repository P. Figure 5 illustrates a flowchart of the executed procedures, also presenting the final activity counts for the composition of Repository C.

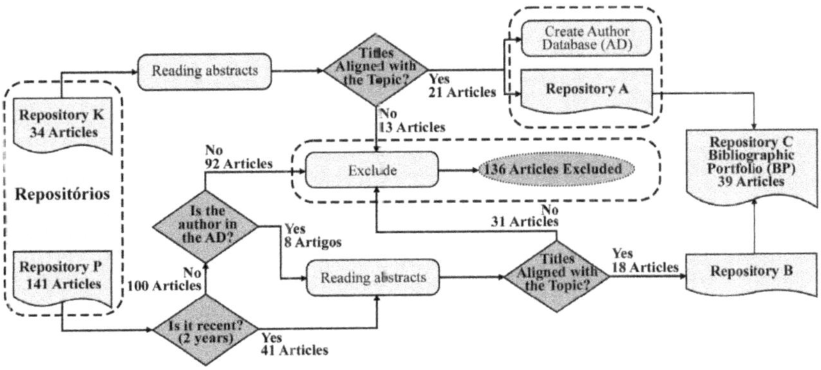

Figure 5. Process of analyzing Repositories K and P, resulting in Repository C.

In the end, Repository C was created by merging Repositories A and B, containing 39 articles that constituted the final BP, ensuring verified academic relevance and alignment with the researched topic. Each article in that portfolio was verified for availability through the CAPES portal or free access on the internet. As a result, 6 articles were excluded, totaling 33 articles in the BP. After these analysis procedures, the selection of that BP represented 2.12% of the initial 1554 articles. The 33 articles, ordered in descending order by the number of citations, are listed in Table 3.

Table 3. Presentation of the thirty-three articles that compose the Bibliographic Portfolio (BP).

Reference	BP Article	Citations
[33]	Thermoelectric generators: A review of applications. Energy Conversion and Management, 140, 167–181.	1185
[34]	Recent development and application of thermoelectric generator and cooler. Applied Energy, 143, 1–25.	762
[3]	CubeSat evolution: Analyzing CubeSat capabilities for conducting science missions. Progress in Aerospace Sciences, 88, 59–83.	534
[35]	Survey of worldwide pico- and nanosatellite missions, distributions, and subsystem technology. Acta Astronautica, 67, 854–862.	493
[36]	Safe radioisotope thermoelectric generators and heat sources for space applications. Journal of Nuclear Materials, 377, 506–521.	260
[37]	Review of wearable thermoelectric energy harvesting: From body temperature to electronic systems. Applied Energy, 258, 114069.	245

Table 3. *Cont.*

Reference	BP Article	Citations
[38]	Review of Micro Thermoelectric Generator. Journal of Microelectromechanical Systems, 27, 1–18.	219
[39]	Thermoelectric generator (TEG) technologies and applications. International Journal of Thermofluids, 9, 100063.	201
[15]	Thermoelectric Generators: A comprehensive review of characteristics and applications. Applied Thermal Engineering, 201, 117793.	151
[40]	Review of thermoelectric geometry and structure optimization for performance enhancement. Applied Energy, 268, 115075.	145
[41]	OpenOrbiter: A low-cost, educational prototype CubeSat mission architecture. Machines, 1.	109
[42]	A review of the state-of-the-art in electronic cooling. E-Prime—Advances in Electrical Engineering, Electronics and Energy, 1, 100009.	102
[7]	Small satellites and CubeSats: Survey of structures, architectures, and protocols. International Journal of Satellite Communications and Networking, 37, 343–359.	101
[43]	Thermal design and analysis of a nanosatellite in low earth orbit. Acta Astronautica, 115, 247–261.	96
[8]	CubeSat design for LEO-based Earth science missions. IEEE, 435–445.	95
[44]	Parametric study of a thermoelectric module used for both power generation and cooling. Renewable Energy, 154, 542–552.	85
[45]	Evolutionary design of a satellite thermal control system: Real experiments for a CubeSat mission. Applied Thermal Engineering, 105, 490–500.	66
[46]	Parametric study of heat-transfer design on the thermoelectric generator system. International Communications in Heat and Mass Transfer, 52, 97–105.	59
[47]	A review of battery technology in CubeSats and small satellite solutions. Energies, 13.	56
[48]	On-Orbit Thermal Design and Validation of 1U Standardized CubeSat of STEP Cube Lab. International Journal of Aerospace Engineering, 2016.	55
[49]	Analytical investigation of a nanosatellite panel surface temperatures for different altitudes and panel combinations. Applied Thermal Engineering, 75, 1076–1083.	49
[50]	Experimental Feasibility Study of Concentrating Photovoltaic Power System for CubeSat Applications. IEEE Transactions on Aerospace and Electronic Systems, 51, 1942–1949.	29
[12]	Enabling Science with CubeSats—Trends and Prospects. IEEE Journal on Miniaturization for Air and Space Systems, 3, 221–231.	8
[51]	Progress and perspectives in thermoelectric generators for waste-heat recovery and space applications. Journal of Applied Physics, 134. doi:10.1063/5.0166338	3

Table 3. *Cont.*

Reference	BP Article	Citations
[52]	Modelling CubeSat Structure for Thermal Analysis.	1
[53]	Chassis Optimization of a 1U CubeSat made in a developing Country. 2022-September.	0
[54]	Thermal modelling of a small satellite data processing unit aided by sensitivity analysis and uncertainty quantification. International Journal of Thermal Sciences, 193, 108514.	0
[55]	Numerical Simulation of Thermoelectric Based Temperature Control system for CubeSat in Space.	0
[56]	Project Irazú: Space and Ground Systems Engineering of a 1U CubeSat Store and Forward Mission for Environmental Monitoring. Transactions of the Japan Society for Aeronautical and Space Sciences, 66, 217–225.	0
[57]	Space missions in South America: Profile and evolutionary perspective of their development. Acta Astronautica, 206, 9–17.	0
[20]	Temperature Distribution of CubeSats Using Finite Element Method. 2022-September.	0
[58]	Thermal design and analysis of JZJ-5 CubeSat.	0
[59]	Morazán MRZ-SAT CubeSat: Thermal Modelling and Analysis Guide for Academic CubeSat Missions. Proceedings of the International Astronautical Congress, IAC, Vol. 2022-September.	0

Among the 33 articles that composed the BP, 7 (21.21%) were conference papers focused on the structure, designs, and missions of nanosatellites. Additionally, 18 (54.54%) aimed to present both the historical and scientific development of nanosatellites and their power systems, such as the EPS, as well as their numerical-computational modeling focused on thermal analyses at LEO altitudes, covering the range of 400 to 1500 km from Earth. Lastly, eight (24.24%) presented concepts and applicability of TEGs in terrestrial situations.

2.2. Bibliometric Analysis

The bibliometric analysis of the BP assists both in developing the theoretical framework on the topic and in accounting for the scientific recognition variables of the articles, authors, journals, and KWs [26,27]. Therefore, based on the data provided by the DBs, several combinations of the information that comprise each article are evaluated, such as the number of citations, author of each publication, year of publication, journal published, among others.

2.2.1. Citation Analysis of the Bibliographic Portfolio

The initial analysis of the 33 articles that composed the BP was structured according to the year of publication and academic relevance within the stipulated period. Figure 6a illustrates the historical distribution of the number of annual publications along with their cumulative frequency.

From this analysis, it is observed that more than half of the publications were concentrated between the years 2020 and 2023. This was confirmed by the graph presented in Figure 6b, which shows the number of annual citations that the BP contained and demonstrates the increase in the density of recent publications. Therefore, it was possible to assert that the proposed research topic was on the rise in recent years.

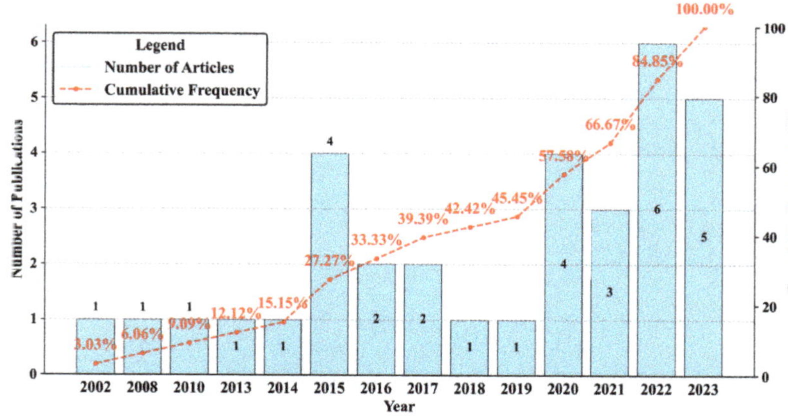

(a)

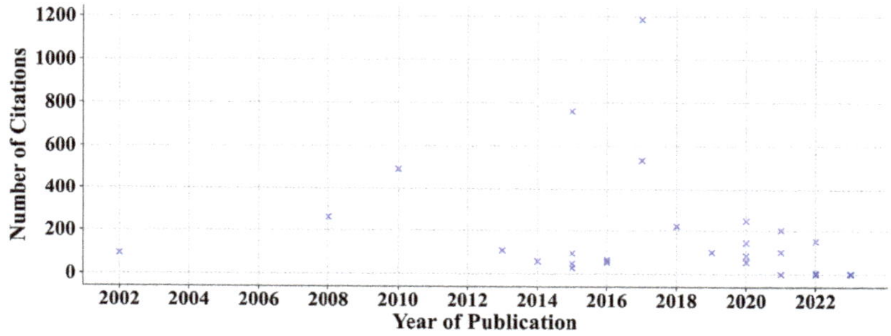

(b)

Figure 6. Distribution of (**a**) articles and (**b**) citations over the years.

2.2.2. Author Recognition

Another important analysis for determining the academic relevance of the BP, in addition to the annual publication of articles, is the recognition of the authors related to the proposed topic. In this case, a total of 138 authors were identified. Despite the number, it was found that 133 authors (96.38%) were present in only one publication, while only 5 authors (3.62%) had two publications. This distribution of the number of authors per article is illustrated in Figure 7a, which shows that even with the number of articles produced between 2020 and 2023, there were still not many branches of the proposed topic.

Furthermore, considering that the BP included both articles that studied TEGs in terrestrial conditions and in nanosatellites, Figure 7b shows the relationship of the most relevant authors to the BP research topic with 100 or more citations.

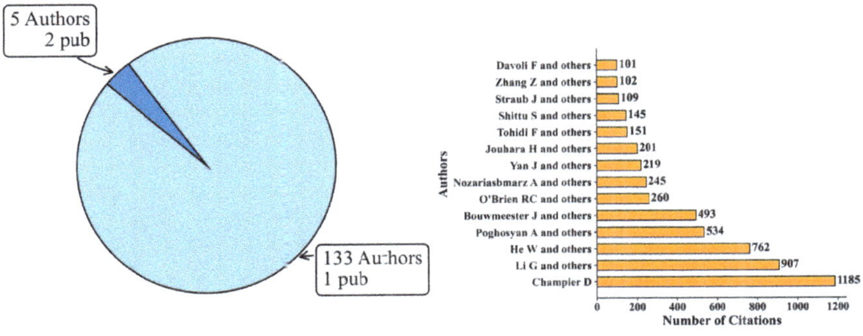

(a) (b)

Figure 7. Distribution of (**a**) the number of publications per author and (**b**) the number of citations per author.

2.2.3. Relevance of Articles and Keywords

The relevance of the articles, in terms of the publication medium and the manner in which they are presented to the academic community, whether through journals or conferences, aims to align both the articles that compose the BP according to the journal in which they were published and to present their publication frequency.

With a total of 33 articles, it was identified that 7 (21.21%) of these were conference papers presented in journals such as IEEE Journal on Miniaturization for Air and Space Systems, ResearchGate, and IOPScience. The remaining 26 (78.78%) were journal articles published in various journals, with Acta Astronautica, Applied Energy, and IEEE Journal on Miniaturization for Air and Space Systems being the most prolific, each representing 11.54% of the total articles. However, the journal Energies stood out as a promising venue for disseminating knowledge on the proposed topic, offering a suitable platform for advancing research in this field. The described results are illustrated in Figure 8.

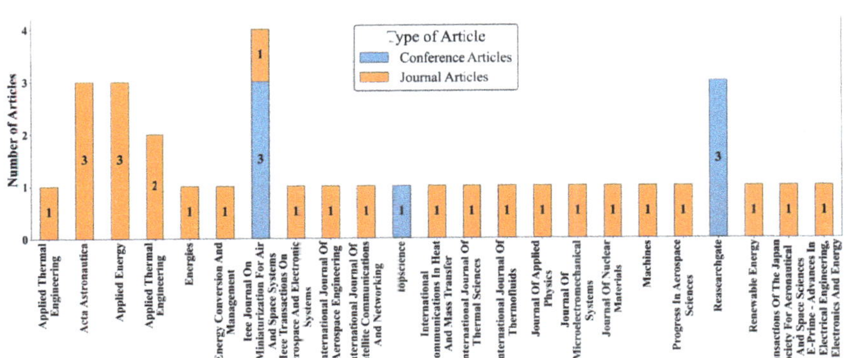

Figure 8. Scientific Relevance of the articles composing the BP and their published journals.

A total of 167 distinct KWs were identified in the academic works of the BP, with 112 (67.07%) observed only once, and 55 (32.93%) two or more times. The study of KWs helped identify which terms are most frequently used in the BP, ensuring their alignment with the proposed topic. The list of the top 30 most frequent KWs among the articles is presented in Figure 9.

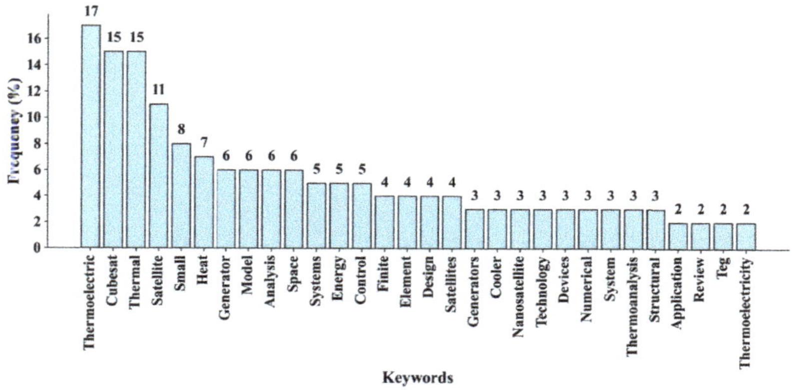

Figure 9. Representation of the 30 most frequently used KWs in the BP.

Even though the term "Thermoelectric" was the most frequently used in the BP, the terms "CubeSat" (second) and "Thermal" (third) held significant relevance in this KW analysis. Despite their individual positions, combined, they totaled 30 occurrences (17.96%), indicating their high usage in combination (or separately) for studies related to the proposed research topic. Additionally, among the 30 most cited KWs, terms like "Finite", "Element", "Numerical" and "Thermoanalysis" suggested the use of computational tools applied to the problem of thermal simulations in nanosatellites.

2.2.4. Analysis of the Bibliographic Portfolio

The BP consisting of 33 articles, it was essential to verify articles that reviewed the literature on the proposed topic, as well as those that aided in its development. These served as a basis for defining the connection between technological progress made in this area.

Therefore, to automate the process of capturing review articles, algorithms were developed in the Python programming language, along with Microsoft Excel spreadsheets, to identify the term "review" in the titles of the articles in the BP and separate them from the research articles. The distinction between research articles and review articles had to be made with the aim of identifying (i) if there had already been any review or application of TEGs within the proposed theme and (ii) if there were any research articles that had conducted an in-depth study on the application of TEGs in CubeSats as a primary energy source.

In total, seven (21.21%) review articles were identified, of which six (85.71%) presented reviews of concepts, applications, and innovations related to TEGs, while one (14.29%) presented a review of CubeSat batteries, as described in Table 4.

Table 4. Review Articles Ordered by Number of Citations.

Reference	Brief Description	Citations
[33]	Presents advancements in TEG production in terms of design and optimization, as well as classification and industrial applications. Additionally, it addresses improvements in the figure of merit (ZT), increased operating temperatures from high gradients, and the use of low-cost materials.	1185
[37]	Focuses on generally describing the use of TEGs to harness residual heat from the human body, highlighting the potential to power accessories and technological devices.	245

Table 4. *Cont.*

Reference	Brief Description	Citations
[38]	Explores the application of micro-TEGs as an economical and competitive alternative to traditional TEGs in areas such as medicine, electronic devices, and the Internet of Things (IoT).	219
[15]	Consists of a literature review on TEGs in the current energy market, aiming to present their potential and applications. Additionally, it delves into the development of TEGs using simulation software for various industrial sectors.	151
[40]	Discusses issues related to the geometry and optimization of thermoelectric device structures, focusing on enhancing their potential and mechanical performance under stress conditions.	145
[42]	Reviews current thermal management methods for electronic devices, highlighting active cooling. It covers techniques such as direct cooling, air jet, immersion, and contact. Additionally, it emphasizes the use of TEGs for contact cooling.	102
[47]	Conducts a review of battery technologies in CubeSats, illustrating the most commonly used types and their operating conditions. It also presents possible solutions to enable their use in space, considering environmental conditions, temperature, and mission requirements.	56

As identified, the set of review articles predominantly highlighted the advancements and applications of TEGs in terms of their design, optimization, and materials, as well as their widespread application in industrial and workplace environments, such as in electronic devices, IoT applications, and in the medical field, particularly for harnessing human residual heat for cardiac measurement devices, as highlighted by [33,37].

Other studies, such as those presented by [15], focused on literature reviews of the development and potential applications of TEGs, including the capture of residual heat from satellite dishes, hybrid systems with solar cells, heat sinks, and more. Notably, the text discussed the application of TEGs for deep space exploration missions, specifically describing the use of RTGs in satellites and their importance in thermal management.

There were discussions on structural challenges of thermoelectric devices under stress conditions, addressed by [40], and reviews of thermal management techniques in electronic devices, as mentioned by [42].

To conclude, Ref. [47] investigated the key technologies employed in CubeSats' batteries, identifying the main challenges associated with their use in space conditions and proposing possible solutions to optimize their efficiency given the limitations imposed by operational temperatures.

As presented, the review articles extracted from the BP indicated a gap in systematic review research regarding the application of TEGs as an energy source for CubeSats, with a primary focus on their implementation in terrestrial environments. Therefore, this work aims to fill this gap by detailing a systematic review of the scientific literature and recent industrial advancements in the aerospace sector concerning energy sources for nanosatellites.

The 26 research articles present in the BP were subdivided into three main themes: history, applications, and development of CubeSats, representing 8 (30.77%) articles; design and thermal analysis of CubeSats, comprising 13 (50.00%) articles; and finally, 5 (19.23%) articles that present studies and principles of TEGs. The articles are described in Table 5.

Table 5. Research Articles Ordered by Number of Citations.

Reference	Brief Description	Citations
[34]	Reviewed thermoelectric materials and their practical applications, highlighting existing technical barriers and proposing new research topics.	762
[3]	Reviewed CubeSats' state-of-the-art capabilities, focusing on scientific missions, and evaluated their potential for enabling high-quality, low-cost scientific missions.	534
[35]	Provided a detailed analysis of the most common technologies in nanosatellites and picosatellites, covering launch history, topologies, purposes, systems, subsystems, and applications.	493
[36]	Presented a historical review of radioisotope thermoelectric generators (RTGs) and their conversion mechanisms, providing information to enable a direct comparison with alternative isotopic systems.	260
[39]	Thoroughly analyzed the operating principles, applications, and materials used in TEGs. Additionally, it provided examples of simulations using software such as COMSOL Multiphysics and ANSYS.	201
[41]	Described the CubeSat OpenOrbiter's structural aspects and program objectives, focusing on requirements, constraints, high-level architecture, and design for small satellites.	109
[7]	Provided an overview of CubeSats, their systems and subsystems, main missions and objectives, and potential future challenges. Additionally, it classified topologies and protocols necessary for launches.	101
[43]	Presented the optimization process in the thermal analysis of a 1U CubeSat, setting boundary and environmental conditions. Developed and applied a MATLAB algorithm to compare results with ESTAN-TMS software.	96
[8]	Detailed the design and hardware/software architecture of internal systems for two 1U CubeSat missions: a combined probe/DC plasma impedance system on two satellites and two CubeSats for GPS scintillation measurements.	95
[44]	Developed a numerical model to predict the performance of a thermoelectric device as both a TEG and TEC, considering height, cross-sectional area, number of pairs, and ceramic plate's thermal conductivity.	85
[45]	Used a genetic algorithm to explore material combinations for 3U CubeSat coatings to meet operational temperature requirements, validated through FEM thermal simulation.	66
[46]	Described a high-performance platform combining heat transfer and thermoelectric conversion. Analyzed how different operating conditions affected TEG energy production using a metal pin array with forced convection.	59

Table 5. *Cont.*

Reference	Brief Description	Citations
[48]	Represented the development of the thermal design for the thermal control subsystem of the STEP Cube Lab and its validation through thermal vacuum tests, in order to estimate the thermal behavior of the components.	55
[49]	Conducted a thermal analysis of a 1U CubeSat in LEO, focusing on passive thermal control to ensure proper component operation using different surface coatings.	49
[50]	Proposed the use of a CPV system, or Concentrating Photovoltaic system, in a 1U CubeSat to improve energy generation efficiency by converting the energy provided by the Sun's light intensity and concentrating it onto the solar cells using lens arrays.	29
[12]	Provided an overview of CubeSats as scientific platforms, highlighting their potential and challenges, focusing on communication systems and efficient space utilization with integrated or shared components.	8
[51]	Discussed the current state and advancements in thermoelectric technology for use in next-generation RTGs and in waste heat recovery applications using TEGs, excluding the discussion of material property optimization strategies.	3
[52]	Analyzed different materials in transient and steady-state regimes of a 1U CubeSat to select the most suitable ones according to the specified temperatures.	1
[54]	Conducted a global sensitivity analysis to identify key elements limiting heat transfer in a 1U CubeSat's data processing unit, followed by an uncertainty analysis for the temperatures of crucial electronic components and the radiator surface.	0
[58]	Introduced the thermal control design of a 3U CubeSat, calculated the temperature field under various conditions using FEM, and compared in-orbit temperature data with thermal simulation results.	0
[20]	This study validated FEM models on the CubeSat using real data from MYSat-1, developing a thermal model in ANSYS to ensure safe CubeSat operation.	0
[57]	Proposed a comparative study of South American space programs using an adapted Wood and Weigel model, evaluating their evolution over the past three decades and providing a comparative view of ten regional programs.	0
[56]	Described the Irazú Project and its development, in a superficial manner, related to a 1U CubeSat for environmental monitoring, focusing on biomass estimation and carbon dioxide fixation in a fast-growing tree plantation.	0
[55]	Conducted a detailed simulation of a TEC-based thermal control system for a CubeSat in a 300 km circular orbit, evaluating the feasibility and performance of TECs under extreme conditions.	0

Table 5. *Cont.*

Reference	Brief Description	Citations
[59]	Explained the thermal environment in low Earth orbit for MRZ-SAT and described computational thermal modeling for this 1U CubeSat using finite element algorithms, serving as an initial guide for beginners in CubeSat thermal analysis and modeling.	0
[53]	Described the structural optimization of the Morazán Satellite (MRZ-SAT) chassis, an award-winning project for natural disaster early warning. Compared three structural patterns to achieve the lightest and most rigid design.	0

Several studies focus on the structure and development of CubeSats, covering topics from thermal analysis to system control and optimization. Works like [43], which presents optimization processes for thermal analysis in CubeSats, and [8], which details hardware and software architecture in specific missions, are fundamental for understanding key CubeSat concepts. Additionally, studies like [45] use genetic algorithms to explore different material combinations that meet the satellites' thermal needs, while [49] and [20] focused on both passive thermal control requirements and the validation of thermal models using real CubeSat data, demonstrating the practical application of theories and models in real operational contexts.

These studies, along with the analysis of specific projects like the Irazú Project detailed by [56], and the Morazán-Sat Project explored by [53,59], exemplify the structural optimization of nanosatellites intended for specific monitoring functions. These functions include verifying CO_2 emissions and implementing early warning systems for natural disasters.

For research on thermoelectric devices, notable works include those by [36,51], which offer a historical review of RTGs to compare them with alternative isotopic systems and discuss recent advances in the field, presenting new approaches to optimizing thermoelectric material properties for the next generation of RTGs. Ref. [39] delves into the operating principles of TEGs through simulations based on FEM analyses. Additionally, Ref. [44,46] develop models to predict the performance of thermoelectric devices under various conditions, highlighting the crucial role of these technologies in energy efficiency and the sustainability of space operations.

Studies by [12,57] describe the potential of space programs, focusing particularly on CubeSats and the evolution of space programs in South America, respectively. The first highlights the versatility of these nanosatellites as scientific platforms, emphasizing their challenges and potential for future space missions, with special attention to communication systems and the optimization of satellite internal space usage. The second study employs an adaptation of the Wood and Weigel model to conduct a comparative analysis of the progress of South American space programs over three decades, offering a comprehensive perspective on regional development and trends, illustrating how these initiatives have expanded and evolved in response to different technological and geopolitical challenges.

Lastly, it is important to note that, although 8 (30.77%) of the 26 research articles were not cited, they are highly relevant to this research due to their direct influence on the applicability of TEGs in CubeSats. For example, Ref. [55] present a study on the use of TECs in nanosatellites for temperature control under extreme conditions. Additionally, Ref. [20,58] focus on the modeling and thermal analysis of CubeSats to define temperature limits, directly influencing the operation of internal components. These studies are fundamental to this work, as the study of TEG application in nanosatellites depends entirely on the temperature gradient to which the satellite's face is subjected. Therefore, these studies cannot be disregarded.

In summary, the reviewed articles do not directly discuss the application of TEGs in nanosatellites, reinforcing a potential connection between thermoelectric technology and innovations in energy sources for small satellites. Future exploration in this field could open new avenues for space exploration and energy supply for nanosatellite missions.

Specifications and Boundary Conditions of CubeSats

This section aims to extract and present the main information that establishes both the specifications of CubeSats and the boundary conditions of the proposed topic, as found in the research articles of the BP. After reviewing the 26 articles, it was identified that the 13 articles focusing on the design and thermal analysis of CubeSats provided fundamental parameters for the research methodology, such as structural material, power source, number of PCBs, and others. Of these, only 11 are presented in Table 6, as they provide all the necessary information to establish the boundary conditions and appropriate structures for the application of TEGs as a power source in CubeSats.

Table 6. Structure, devices, specifications, and boundary conditions in CubeSats.

Reference	Structure and Devices	Specifications and Boundary Conditions	Software Used
[8]	1U CubeSat; structural material: AL 7075; solar cell: Tecstar Triple Junction; battery: LiPo	Circular orbit; LEO: 300 km and 60°; provides weights and consumed power	Not provided
[49]	1U CubeSat; structural material: AL6061T6; battery: LiPo; 5 PCBs	Sun-synchronous circular orbit; LEO: 600 km and 98°; transient and steady-state analysis; maximum and minimum temperatures provided	Not provided
[59]	1U CubeSat; structural material: AL6061; solar cell: GaInP2/GaAs/Ge	Circular orbit; LEO: 400 km and 51.65°; details angles and thermal analysis	GNU Octave; SINDA software
[52]	1U CubeSat; structural material: AL6061TE; 3 PCBs	LEO: 600 km and 67°; transient and steady-state analysis; operating temperatures provided	ANSYS
[55]	1U CubeSat; structural material: AL 7075; 4 PCBs; PCB material: FR4; solar cell: single crystal silicon; solar cell sizes provided; TE: CP1.031-17-06L	Circular orbit; LEO: 300 km with no inclination; period of 90.4 min; Earth represented as a sphere; satellite surface temperature: 6–60 °C; operating temperature: 0–40 °C	Not provided
[53]	1U CubeSat; structural material: 6061; solar cells and components: obtained from Endurosat 1U	Detailed component weights and consumed power provided	Autodesk Fusion 360 Education License
[50]	1U CubeSat; specifications of STEP Cube Lab; solar cells: polycrystalline single-junction silicon	LEO: 600 km, sun-synchronous; Period of 1 year	Solar Simulator
[58]	3U CubeSat; structural material: AL 7075; PCB material: FR4	LEO: 500 km, sun-synchronous; operating temperatures of components and their power provided	Not provided

Table 6. *Cont.*

Reference	Structure and Devices	Specifications and Boundary Conditions	Software Used
[43]	1U CubeSat; structural material: AL 6101T6 and AL7076T6 (plate); solar cells: gallium arsenide triple junction; reference to PiCPoT nanosatellite	LEO: 600 km and 98°; operating temperatures of components and power provided	MATLAB-ESATAN-TMS
[20]	1U CubeSat; structural material: AL 6061 and stainless steel 304; 5 PCBs; PCB material: FR4; battery: Li-Ion	LEO; operating temperatures and maximum and minimum temperatures obtained by the software provided	CubeSat Wizard with MATLAB-ANSYS
[48]	1U CubeSat; structure based on STEP Cube; structural material: AL 6061T6; battery: Li-Ion; solar cells: ITO-GaAs triple junction; 5 PCBs; PCB material: FR4	Sun-synchronous orbit; LEO: 600 km and 97.78°; period of 1 year; details components, their dissipated power, and operating temperature; provides temperatures in low and high operation; provides power generated by the solar cells	Thermal Desktop; RadCAD

Although all the presented boundary conditions are relevant for this research, the most notable ones are those that provide information about the space environment in which the nanosatellites operate. In this case, articles such as those by [20,48,53,55] provide essential data on ambient temperatures, inclination angle, weight, and internal component consumption to determine the feasibility of using TEGs as an energy source in CubeSats.

In more detail, parameters such as ambient temperature and solar incidence significantly influence the performance of TEGs in generating electrical energy. As highlighted in the reviews presented in Table 4, the operation of a TEG depends on a temperature gradient, created by a hot source on one face and a cold source on the other. Simultaneously, the weight and energy consumption of TEGs directly impact their ability to provide sufficient power for the CubeSat's operation. These factors are crucial to ensure a good watt/kg ratio of the payload, guaranteeing that the satellite operates efficiently and effectively.

2.3. *Intellectual Property*

The identification of registered products and patents within a research theme is extremely relevant for verifying existing technologies in the area, as they represent an indicator of technological advancement resulting from RD activities in universities [60]. Therefore, to verify the relationship between patent development and RD, steps similar to those applied for the selection of the BP were carried out.

2.3.1. Selection of the Patent Repository

The identification of patents was carried out using the Google Patents and Patentscope search engines. These tools are essential for a global search of patent data, including those registered in Brazil [61]. To optimize the search, combinations of KWs encompassing the investigated theme in both English and Portuguese were established, as shown in Table 7.

Table 7. Results of the patent searches conducted.

English	Portuguese	Google Patents	Patentscope
("Thermoelectric Generator" OR "TEG" OR Seebeck OR Thermoelectric) AND (Cubesat OR Nanosatellite OR Nanosat OR "Small spacecraft")	("Gerador Termoelétrico" OR "TEG" OR Seebeck OR Termoeletricidade) AND (Cubesat OR Nanossatélite OR Nanosat OR "Pequeno satélite")	93	101
("TEG") AND (Cubesat OR "Small spacecraft")	("TEG") AND ("Cubesat" OR "Pequeno satélite")	13	11
("Eletrical Power System" OR "EPS") AND (Cubesat OR Nanosatellite OR Nanosat OR "Small spacecraft")	("EPS") AND (CubeSat OR Nanosatellite OR Nanosat OR "Small spacecraft")	33	131
("Power System" OR "EPS") AND (Cubesat OR Nanosatellite OR Nanosat OR "Small spacecraft")	("Sistema de potência" OR "EPS") AND (Cubesat OR Nanossatélite OR Nanosat OR "Pequeno Satélite")	261	356
("Thermoelectric Generator" OR "TEG") AND (Cubesat OR "Small Spacecraft" OR Nanosatellite)	("Gerador Termoelétrico" OR "TEG") AND (Cubesat OR "Pequeno Satélite" OR Nanossatélite)	29	31
Raw Patent Database (RPD)		**1059**	

Subsequently, using the search data exported to Excel, data processing of the RPD containing 1059 patents was carried out, ensuring that the identified patents were aligned with the research. The process and treatment were performed using the structure shown in the flowchart in Figure 10. The initial filtering, called a correction filter, identified duplicate titles and codes within the patents in the RPB, resulting in 632 exclusions (59.68%). The patents represented in the PD were processed with an alignment filter, which involved reading all titles aligned with the research theme. At that stage, 385 patents (90.16%) that were not aligned were excluded, leading to two patent repositories: Google Patents Repository and Patentscope Repository, containing 32 and 10 patents, respectively. A check for duplicate titles between the repositories was performed through a check filter. As a result, the final PR contained 34 patents aligned with the investigated theme, representing 3.21% of the RPB.

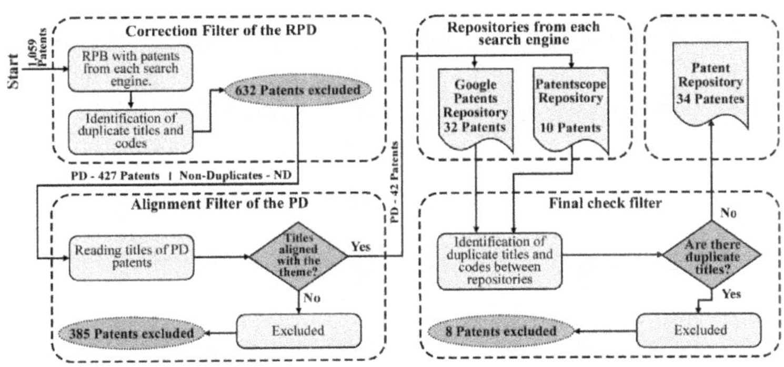

Figure 10. Flowchart representing the filtering of products and patents.

The patents from the Patent Repository (PR) are presented in Table 8, in ascending order. These patents include detailed information about the inventors, their titles and respective codes, country of origin, and a brief description of each invention. The evaluation of the relevance of these patents within the proposed theme was based on the following criteria: (i) verifying if the patented technology was directly applied to satellites or nanosatellites; (ii) assessing the impact on energy generation in the satellite or nanosatellite; and (iii) using residual heat for electricity generation.

Of the 34 patents belonging to the PR, 5 patents (14.70%) (numbers 1 to 5) met all the mentioned requirements, demonstrating a significant influence on the research. Another 26 patents (76.47%) (numbers 6 to 31) had a lesser influence, partially fulfilling the requirements. Finally, three patents (8.83%) (numbers 32 to 34) were discarded for not meeting the established criteria.

Table 8. Presentation and selection of patents that comprised the Patent Repository.

Reference (Number)	Title: Code	Brief Description
[52] (1)	A kind of device for generating electricity using satellite external surface temperature gradient: CN-106208814-B	The invention proposes a method for generating electricity by utilizing the temperature gradient on the external surface of satellites. This is achieved through a series of thermal insulation layers, a heat-conducting layer, and thermoelectric devices, which is especially useful for micro- or nanosatellites.
[62] (2)	Magneto-thermoelectric generator for energy harvesting: US-9793829-B2	The invention consists of a device called a magneto-thermoelectric generator (MTG), capable of converting residual thermal heat from a photovoltaic cell into mechanical energy and, ultimately, into electrical energy. The embodiments operate based on the principle of thermally induced switching between open and closed states of a ferromagnetic switch, aiming to generate mechanical oscillations that cause deformation in a piezoelectric material, resulting in the generation of electrical energy.
[63] (3)	Nano-engineered thin-film thermoelectric converter for photovoltaic applications: US-2022285571-A1	The patent describes systems, devices, and methods for manufacturing nano-engineered thin-film thermoelectric converters (NETT) with photovoltaic applications. These devices utilize the cold of space for satellites or improve the efficiency of terrestrial solar photovoltaic systems. Examples of methods include mounting a thin-film thermoelectric device to a photovoltaic device, followed by the assembly of a heat sink and, optionally, a radiator or heat exchanger.
[64] (4)	Nanostructured material-based thermoelectric generators: US-2009044848-A1	These are electric power generators that utilize the thermoelectric effect in arrays of nanostructured materials. A thermoelectric device is described, consisting of p-type and n-type elements made of carbon nanotubes, and it is capable of operating at high temperatures, generating considerable output power despite being lightweight. A method for manufacturing this device is also presented.
[65] (5)	Radioisotope thermoelectric generator: WO-2016138389-A1	The present invention pertains to radioisotope thermoelectric generators, focusing on miniaturized versions to provide power in space-constrained situations. These generators incorporate radiological heat sources and are designed to operate in high-temperature, high-pressure, and high-vibration environments, such as in drilling probes.
[66] (6)	A power supply: AU-2019262087-A1	The invention describes a power supply for an autonomous device with multiple power supply units, each having an input for source power, an output to a common power rail, an energy storage device, and control to manage the energy flow.

Table 8. *Cont.*

Reference (Number)	Title: Code	Brief Description
[67] (7)	Compact nuclear power system applied to space environment and working method: CN-110043338-B	A compact nuclear power system is proposed for the space environment. Using carbon dioxide as the working fluid, the system is capable of operating stably for decades or even centuries, increasing efficiency by more than 10%. It can be used as a power source in space probes for long-term missions.
[68] (8)	CubeSat system: US-11021274-B1	The invention consists of the introduction of small, ready-to-use, low-cost CubeSat systems that are more accessible, reliable, and easier to use compared to conventional CubeSats. It provides a complete solution, including a satellite, ground station, and remote field units, which can be assembled and deployed quickly.
[69] (9)	Extendable solar array for a spacecraft system: US-9856039-B2	The patent addresses a set of systems applied to a 2U CubeSat nanosatellite, focusing on generating electrical power from solar energy. These systems may include aluminum honeycomb structures or composites with glass covers, as well as rigid crystalline photovoltaic cells and rigid or flexible deployment mechanisms.
[70] (10)	Flexible thin-film solar cell: US-6410362-B1	A transparent thermal film is applied directly onto thin-film solar cells, creating a flexible solar panel. This panel can be installed on a substrate during the manufacturing process to allow illumination and heat dissipation on both sides, making it ideal for building a solar array on curved surfaces, such as spherical nanosatellites. The current inventions involve thin-film devices and processes, especially solar cells and flexible circuit boards, and their manufacturing methods.
[71] (11)	Integrated solar power module: US-6300158-B1	The invention focuses on the presentation of thin-film devices and processes, particularly thin-film solar cells and flexible circuit boards, and their manufacturing methods. An exemplary configuration involves multiple layers of transparent polyimide with metallization between them, forming a three-dimensional flexible printed circuit board. This allows for the direct deposition of thin-film solar cells, creating flexible electronic modules. These boards are ideal for electronic systems on curved surfaces, such as spherical energy nanosatellites.
[72] (12)	Integrated thin-film battery and circuit module: US-7045246-B2	The invention pertains to thin-film batteries embedded in thin-film printed circuits to provide localized power to electronic devices. These batteries are integrated into multi-layer flexible circuit boards, supporting devices such as power regulators, forming an integrated battery and circuit module.
[73] (13)	Micro–nanosatellite power system based on super capacitor: CN-106602694-A	The invention reveals a power system for micro and nano satellites based on a supercapacitor, comprising a supercapacitor and a power input module. The power input module includes an array of solar cells that charge the supercapacitor.
[74] (14)	Modular electrical power subsystem architecture: US-2022158477-A1	The present application provides an electrical power subsystem (EPS) with a dual-battery configuration that enables sufficient power supply for a spacecraft bus and a payload module carried by the spacecraft.

Table 8. *Cont.*

Reference (Number)	Title: Code	Brief Description
[75] (15)	Modular solar cell and solar cell array: US-2023299610-A1	The present invention pertains to power management in spacecraft, particularly in the context of modular solar cells and solar cell arrays that can be deployed on different types of spacecraft. Examples of the invention include modular solar cells, arrays, and energy management systems designed for use in satellite systems and constellations.
[76] (16)	Power supply module for spacecraft: WO-2017177301-A1	The present invention pertains to power converter circuits or devices, specifically power modules based on solar cells for power subsystems in spacecraft. These modules provide power to an unregulated electrical bus of the spacecraft, using a solar cell and a power converter mounted on it to convert solar energy into a DC output voltage determined by the voltage of the spacecraft's electrical bus.
[77] (17)	Satellite modular power supply: US-10536107-B1	The invention presents simplified concepts for improved power supply configurations for satellite devices. An example discussed involves power units, i.e., batteries, in the form of polygons arranged around a power control module within the satellite's chassis.
[78] (18)	Satellite storage battery system: CN-113422423-A	The invention illustrates a battery system for satellites, enabling the conversion of solar energy into electricity, the storage of that energy, and its use when solar energy is unavailable. It includes a solar energy production device, a storage device, and a set of batteries that work together to provide power as needed.
[79] (19)	Solar array remote acoustic sensing (SARAS): US-10078328-B1	The invention relates to sensing technologies, specifically systems and devices that utilize solar arrays, photovoltaic cells, or photodetectors to obtain or provide acousto-optic signatures. These devices act as remote acoustic sensors, detecting fluctuations in the intensity of optically modulated light at audio frequencies.
[80] (20)	Solar battery and the integrated device of slot antenna: CN-106711576-B	The invention presents a device that combines a slot antenna and a solar battery into a single system. This resolves space competition issues in micro- and nanosatellites, reducing the satellite's volume and weight while increasing the light capture area of the solar battery.
[81] (21)	Solar energy conversion and transmission system and method: US-9815573-B2	The present invention relates to systems and methods for converting solar energy into microwave electrical energy and transmitting it to Earth.
[82] (22)	Solar panel and flexible radiator for a spacecraft: EP-3239057-B1	The invention relates to a combination of a solar array and a flexible radiator for a spacecraft.
[83] (23)	Solar panel array: US-9882330-B2	The invention relates to methods and devices for interconnecting a set of solar panels composed of elongated PCBs, each with solar cells mounted linearly and connected in series to form a panel. There are also hinge assemblies consisting of hinge pieces, support plates, a spring, and a hinge pin, which are mounted between the panels to allow them to be folded in an accordion arrangement or deployed up to 180 degrees.
[84] (24)	Space vehicles including multijunction metamorphic solar cells: US-2021202777-A1	The document addresses spacecraft with multijunction solar cell panels and their manufacturing, focusing on the design and specification of bandgap ranges in four- or five-junction solar cells to optimize efficiency in orbit and ensure performance over an operational lifespan of five to fifteen years.

Table 8. *Cont.*

Reference (Number)	Title: Code	Brief Description
[85] (25)	Space-based radioisotope production and methods of use: US-20220367077-A1	The text describes a system for producing radioisotopes in space and its methods of use. It highlights the acceleration of a propellant by decay energy and the possibility of recharging radioisotope rocket thrusters at an orbital charging station. Additionally, it mentions that activated isotopes can be used to generate electricity and irradiate other items in space for various purposes.
[86] (26)	Spacecraft solar cell monitoring system: US-20091190600-A1	The invention deals with monitoring the characteristics of a solar cell array in small spacecraft. A method is used to determine the current and voltage of the solar cells, allowing the determination of the cells' orientation relative to the sun. This is useful for monitoring the degradation of the solar cells over time in picosatellites.
[87] (27)	Thermoelectric rocket propellant tank pressurization system: US-10495028-B1	The text describes advanced rocket engine systems, where one system uses TEGs to produce electrical power for the pump, based on the temperature difference between the liquid propellant and the heat produced in the combustion chamber. This pressurizes the propellant tanks to feed the engine.
[88] (28)	CubeSat system, method, and apparatus: US-9248924-B2	It addresses a satellite system consisting of a chassis with an avionics package in the upper portion. This package includes various boards, such as the main board, payload interface board, daughter board, and battery board, all arranged in parallel planes. They are connected through stackable connectors. The method of operating the satellite is also described.
[89] (29)	Power distribution system: US-6396167-B1	The present invention relates to power distribution systems that manage the distribution of electrical power between power sources and energy storage elements on a regulated bus in micro- and nanosatellites. The system allows the coupling of multiple power sources and energy storage devices to the bus, with individual regulators and chargers for each device.
[90] (30)	Power supply and distribution module applied to CubeSat: CN-106059266-A	The invention relates to a power and distribution module designed for CubeSats, offering 3.3 V and 5 V converters connected in parallel to a power distribution unit. This compact and highly integrated module is developed to meet the high-reliability, high-performance, and high-integration power requirements of CubeSat equipment.
[91] (31)	Satellite-constructor, training-demonstration model: RU-2693722-C2	The invention relates to the design and equipment of small modular CubeSats, including models for training. The CubeSats feature a basic modular platform, internal network, central processor, control and communication module, and service systems for selecting and replacing elements. A service connector allows access to the internal network, the debugging of programmable devices, and external power supply.
[92] (32)	Satellite thermal control WO-2024003013-A1	The invention pertains to a satellite with a body containing a radiative surface to emit heat into space, and an attitude control system to orient the satellite in Earth's orbit. This system can adjust the satellite's orientation relative to the sun, switching between a position where the radiative surface points away from the Sun and one where it is exposed to the Sun. The goal is to control the satellite's heat transfer.

Table 8. *Cont.*

Reference (Number)	Title: Code	Brief Description
[93] (33)	Scenario-based method for testing software: US-11138100-B2	The present invention pertains to the field of software testing, specifically a method for testing software, such as that used in satellites, using scenario-based programming. The method involves testing the operational software of a system with various internal subsystems, operating in an environment controlled by a hardware board using electronic signals and operational software.
[94] (34)	Spherical mobility system: US-11420777-B1	The text describes a lunar exploration rover designed to detect subsurface ice and traverse varied terrains in extraterrestrial environments. The rover features a spherical structure that houses an avionics core. The avionics core contains components for data acquisition and motor assemblies that enable the rover to rotate the spherical shell for locomotion.

Description:
US—United States; CN—China; AU—Australia; RU—Russia; EP—European Patents Official; WO—World Intellectual Property Organization.

2.3.2. Analysis of the Patent Repository

The initial analysis of the PR focused on identifying the countries with the most patents related to the theme. As a result, Figure 11 shows the number of patents from each country, indicating that 64.71% of the total patents in the PR originated from the United States, followed by China with 17.65%, patents administered by WIPO with 8.82%, and countries like Russia, Australia, and the EP institution each representing 2.94%.

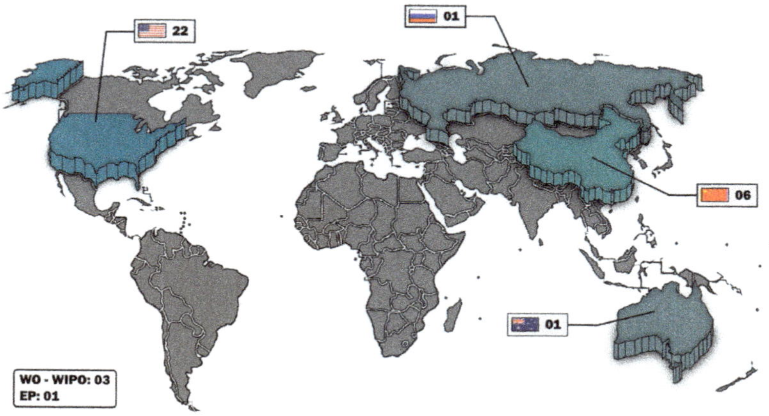

Figure 11. Number of patents registered by country.

Using The Lens Patent Search mechanism provided by The Lens site allowed us to capture legally characterizing information about the patents in the PR. Data such as annual patent publications, most cited patents, discontinued, inactive, expired, or active patents, main inventors and institutions, among others, are made available for free by the platform. Thus, analyses of the patents and the inventors/institutions were carried out to verify their industrial relevance within the proposed theme.

By adding the codes of each patent in the PR to the platform, a "Dynamic Collection" was formed. This guide allowed access to the associated legal records. Of the patents analyzed, 29 patents (85.29%) referenced other inventions, totaling 335 citations, highlighting their importance for conceptual and innovative development.

Additionally, 12 patents (35.29%) included citations of non-patent literature documents, providing additional context and aiding in defining the state of the art in the relevant field. Concurrently, 21 patents (61.76%) served as a foundation for developing new inventions.

Figure 12 illustrates the evolution in the number of patents created over the past 20 years, according to their year of publication, classified into two categories: granted patents, referring to those approved by a licensed office, and patent applications, referring to inventions that have entered the patenting process [95]. In this case, a significant increase in the production of patents related to the theme was observed starting from 2016.

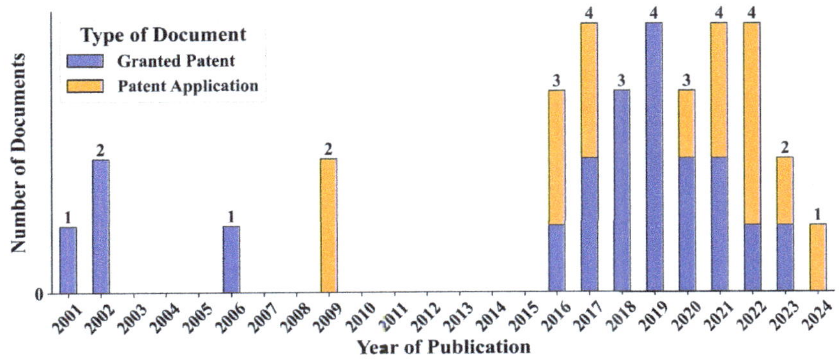

Figure 12. Number of patents registered per year.

In parallel, it was essential to identify the legal conditions associated with each patent, considering their year of registration. This process involved determining the current legal status of the patent, which can vary between active, inactive, expired, discontinued, pending, patented, or unknown. Additionally, it was important to quantify the number of citations each patent had received to highlight the relevance of the previously selected ones in the field of study.

Thus, Figure 13 shows the number of citations per patent for each year of registration, accompanied by Table 9, which presents the number of documents by legal status. This analysis indicated a significant growth in the production of new patents starting from 2016, showing increased interest from companies and inventors in the aerospace sector. Furthermore, 52.9% of the total patents remained active, while 26.4% represented patents that had been discontinued, expired, or were inactive, and 20.6% were pending approval by an accredited office.

Table 9. Number of patents by legal status.

Legal Status	Quantity	
Active/patented	18	(52.9%)
Pending	7	(20.6%)
Expired	3	(8.8%)
Inactive	3	(8.8%)
Discontinued	3	(8.8%)

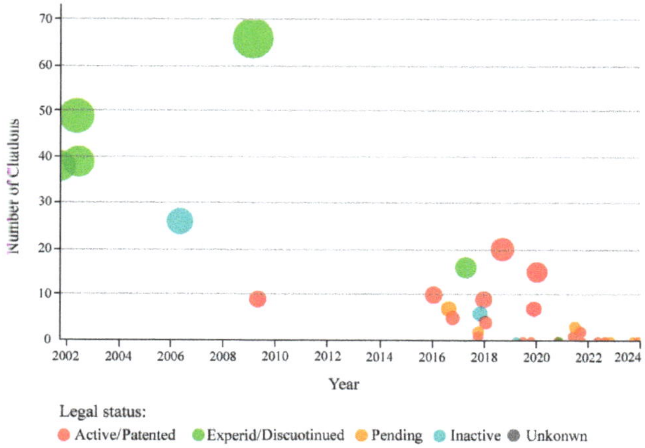

Figure 13. Number of citations per patent by year of publication. Adapted from [96].

In view of this, Table 10 describes the legal information of the patents that met all the requirements influencing the research, briefly described in Table 8.

Table 10. Description of the legal status of the filtered patents.

Reference	Code	Legal Status	Citations per Patent
[64]	US-2009044848-A1	Discontinued	66
[65]	WO-2016138389-A1	Pending	7
[62]	US-9793829-B2	Active	1
[97]	CN-106208814-B	Inactive	0
[63]	US-2022285571-A1	Pending	0

The analysis of the filtered patents revealed the legal status and citation impact of each. Among the patents, one was active (US-9793829-B2), two were pending (WO-2016138389-A1 and US-2022285571-A1), one was inactive (CN-106208814-B), and one was discontinued (US-2009044848-A1). Notably, the discontinued patent had the highest number of citations (66), while others had significantly fewer or none.

With this analysis, it was possible to identify the inventors with the highest number of registered patents in their names, as well as their involvement in other patents. Figure 14 presents the top 20 inventors in the PR, organized by the number of patents produced. It shows that inventor Edward Simburger had the highest number of produced patents, followed by Paul Gierow, Frank Jeffrey, and James Matsumoto.

Similar to the analysis of patent production by inventor, it is advisable to identify which institutions focus on the research theme. Additionally, it is important to determine the purposes for which the patents were designated to ensure their relevance to the present work.

Using the same procedures as before, the number of patents registered by company was obtained through a heat map, illustrated in Figure 15a, and the top 20 codes related to CPC are presented in Figure 15b. The CPC is a classification system developed by the USPTO and the EPO. This system has a designation letter for each study area, detailed in [98].

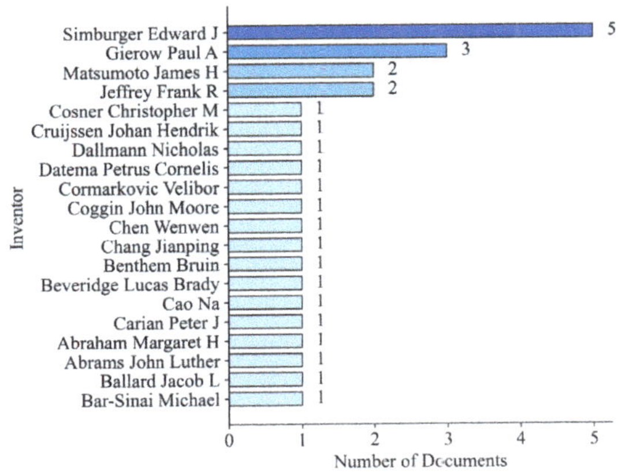

Figure 14. Number of patents produced by inventor. Adapted from [96].

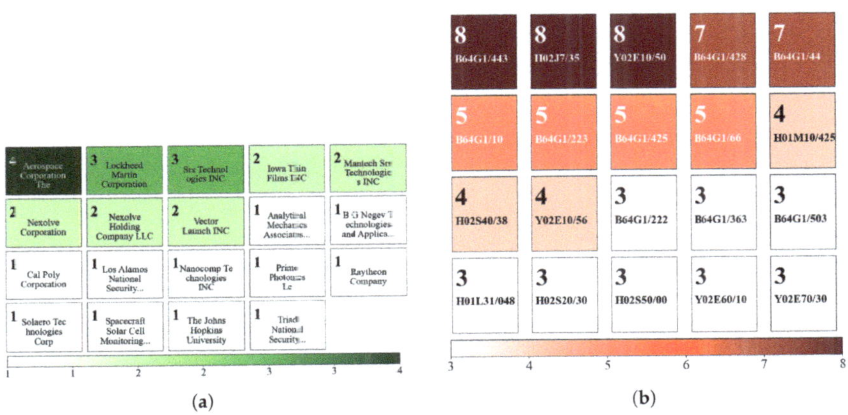

Figure 15. Number of documents by (**a**) institutions and (**b**) according to CPC classification.

In some cases, a patent may have more than one CPC classification. This occurs because the invention can cover multiple technical areas, making it relevant to different aspects of classification. For the PR, which contained 34 patents, some documents may also have had multiple categories.

Thus, out of a total of 31 documents registered by companies, it was noted that Aerospace Corporation had the highest number of patents, with four patents (12.90%), followed by Lockheed Martin Corporation and Srs Technologies INC with three patents each (9.68%), Iowa Thin Films, Mantech Srs Technologies INC, Nexolve Corporation, Nexolve Holding Company LLC, and Vector Launch INC with two patents each (6.45%), and other institutions with one patent each (3.23%).

Regarding the CPC classification, it was identified that a large portion of the patents were related to the code B64G1, which is classified under Performing Operations Transporting. More specifically, they focused significantly on energy and communication systems for space applications. Patents classified under codes H02J7, H01M10, H01L31, H02S50, and H02S20 were directed towards the development of structural components, energy storage systems, and photovoltaic system testing. Lastly, those with codes Y02E10, Y02E60, and Y02E70 were related to the development of batteries and photovoltaic systems.

2.3.3. Most Relevant Patents

In the context of the research, Ref. [97] describes the use of micro-thermoelectric devices, composed of thermoelectric materials with high ZT values at low temperatures and high-thermal-conductivity ceramic plates, to generate electricity using the temperature gradient on the external surface of the nanosatellite. Positioned centrally on the six faces of a CubeSat (Figure 16a), the structure of the invention is shown in Figure 16b. Other details are presented in [97].

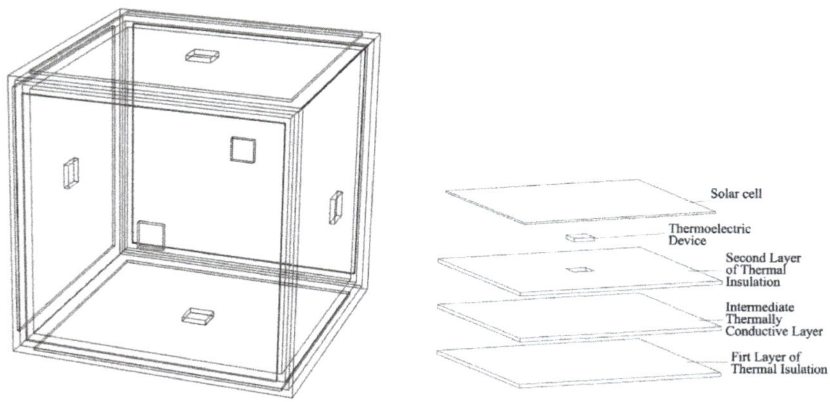

(a) (b)

Figure 16. Drawing of the (**a**) three-dimensional schematic and (**b**) layer structure of the invention [97].

The approach presented in [62] describes a device capable of converting residual thermal heat from a photovoltaic cell into mechanical energy and, ultimately, into electrical energy. Called the MTG, this invention operates based on the thermally induced change in magnetic moment within a magnetic material due to the thermal transition between a heat source and a heat sink. This requires the use of a medium to force the nonlinear restoration of motion, exemplified by the use of a spring coupled to a ferromagnetic material. Its construction is observed in Figure 17, from top to bottom, where (10) details the MTG device.

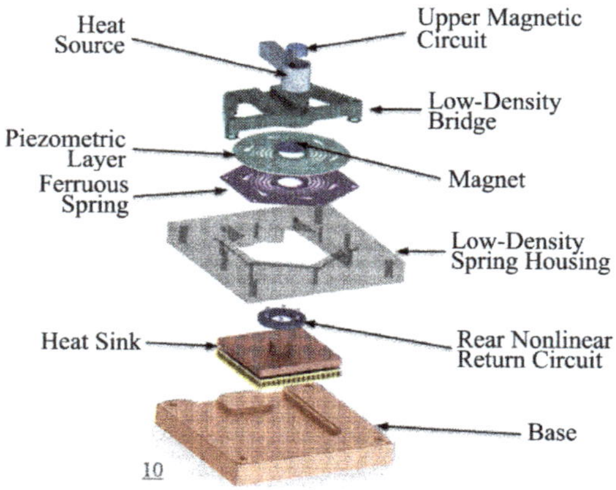

Figure 17. MTG device. Adapted from [62].

In the cold state, the magnetic material is in the closed position, i.e., in thermal contact with the hard magnet on the hot side. As heat from the source is conducted to the magnetic material, it approaches its transition temperature and experiences a change in the magnitude or direction of its magnetization. With the opposing magnetic force weakened, the means for the nonlinear restoring force return the magnetic material to the cold side of the device. After sufficient heat transfer to the cold side, the magnetic moment returns to its original state, moving the material back to the hot side. The movement of the magnetic material between the hot and cold sides causes mechanical oscillations, which are a form of nonlinear restoring force. Electrical energy is generated by coupling the means for the nonlinear restoring force with a piezoelectric material, resulting in the electromechanical production of electrical energy. Its operation and methodology are presented in [62].

Meanwhile, Ref. [63] describes methods for the fabrication of NETT devices for MJPV applications. Focusing on space employment, their use in satellites is appreciated in terms of harnessing residual energy, optimizing size, weight, and power. In certain cases, NETT devices can be used between the thermal blankets used by satellites, providing distributed energy to various sensors without reliance on the solar panel array. This means that the residual thermal energy output can be considered as the sum of the total incident solar power minus the energy output of the photovoltaic solar panels and minus the energy output of the NETT. The structure of the MJPV-NETT device is shown in Figure 18, where (190) shows the fabrication of the MJPV-NETT device. Additional details are available in [63].

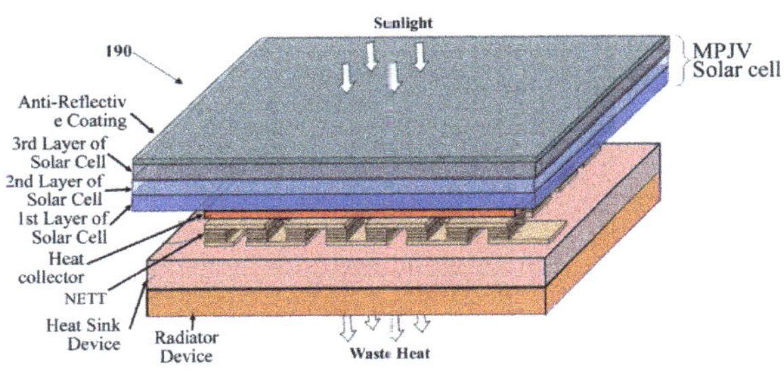

Figure 18. MJPV-NETT device. Adapted from [63].

In parallel, Ref. [64] conducted studies on the changes in geometry, material, and weight of TEGs for space applications. As a result, they patented a thermoelectric device consisting of (i) a first element designed to collect heat from a heat source, (ii) a second element spaced from the first element to dissipate heat from the first element, and (iii) a core positioned between the first and second elements to convert heat into useful energy. The core was composed of a carbon nanotube that exhibited a relatively high Seebeck coefficient, which increased as the temperature rose. Additionally, it had a relatively high transition temperature, allowing the coupled elements to operate in a substantially high-temperature range. Its construction is shown in Figure 19a, and detailed in Figure 19b, where (130) shows the device.

In summary, Table 11 addresses the main characteristics identified in the PR. It highlights that the only patents that closely aligned with the proposal of using TEGs as an energy source for CubeSats were those by [62,97]. However, these patents differed mainly because they integrated a hybrid TEG system with solar cells (or another type) and did not provide power for the entire nanosatellite. Additionally, [62] discussed thermoelectric devices to regulate the temperature of specific sensors and subsystems, rather than providing power for the entire system.

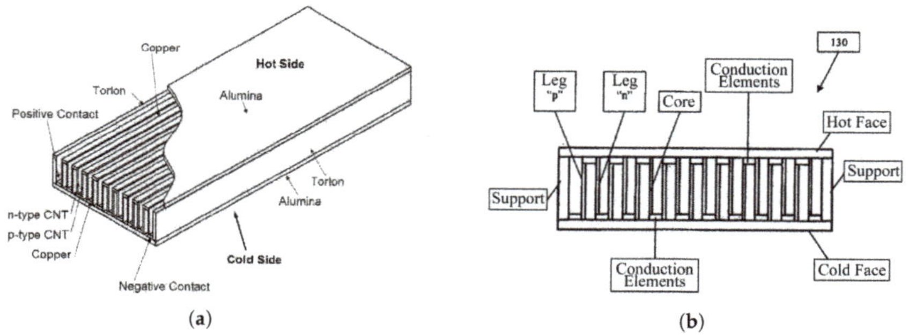

Figure 19. Schematic drawing of the thermoelectric device composed of carbon nanotube cores: (**a**) three-dimensional and (**b**) two-dimensional [64].

Table 11. Mapping of the main characteristics addressed in the topic.

Analyzed Characteristics	IP Publications [97]	[62]	[63]	[64]	[65]	**Proposal**
Study of thermoelectric devices as electrical power generators	✓	✗	✗	✓	✓	✓
Presents modification in the standard structure of TEGs	✗	✓	✗	✓	✓	✓
Has space applications	✓	✓	✓	✓	✓	✓
Utilizes only residual heat from ambient conditions	✓	✓	✓	✓	✗	✓
Forms a hybrid system with solar cells or another type	✓	✓	✓	?	✓	✗
Application in CubeSats	✓	✓	✓	?	✓	✓
Impact on energy supply to CubeSat subsystems	✓	✓	?	?	?	✓
Uses only TECs for the integral power supply of CubeSats	✗	✗	?	?	?	✓

Complies: ✓ **Does Not Comply:** ✗ **Not informed/Not identified:** ?

The proposed research distinguishes itself from existing studies by exclusively focusing on the use of TEGs as the primary and integral power source for the EPS of a CubeSat, in contrast to prior works that predominantly explore hybrid systems or specific subsystem applications. Furthermore, the proposal introduces structural modifications to TEGs to optimize their functionality in space environments, an aspect not addressed in previous references. While many existing studies provide limited or unclear assessments of the impact of TEGs on nanosatellite subsystems, this research offers a comprehensive evaluation of their integration into the entire power system, thereby addressing significant gaps in the current literature.

3. Comparative Analysis of the Scientific and Patent Portfolio

This section aims to establish the relationships between the BP identified through the ProKnow-C methodology and the IP search conducted, resulting in the PR. It evaluates the connections between the authors and inventors of each repository, as well as the progress of RD in relation to the advancement of industrial innovations on the researched topic.

3.1. Relationship Between Scientific and Industrial Advancement

The importance of the comparative analysis between the BP, which comprised the most academically relevant articles, and the PR, constructed from the most relevant patents in the industrial sector, lay in establishing the main relationships regarding the production and development of new technologies applied to the energy supply for a CubeSat's EPS. This analysis structured the evaluation of scientific and industrial advances derived from the bibliometric analysis and compared the connections between authors and inventors, keywords, and the annual production history of works.

Thus, the evaluation of the annual production of works based on the existing relationship between scientific articles and patents related to the proposed research topic was conducted, aiding in the evolutionary diagnosis of documents and their possible historical links. The result is illustrated by Figure 20, where, for a better behavior analysis, rectangular regions with performance criteria are defined. These are red rectangles indicating the intervals between years in which there were no publications for both documents and green rectangles emphasizing the years in which the total number of both documents was greater than one.

The analysis indicated an increase in the production of works related to the topic studied in this research starting from the year 2014, with at least a total of two publications per year. Additionally, periods without publications were observed in the years 2002 to 2005, 2006 to 2008, and 2010 to 2013. This pattern suggested growing interest in the topic investigated by articles linked to both RD and the industry over the past 10 years.

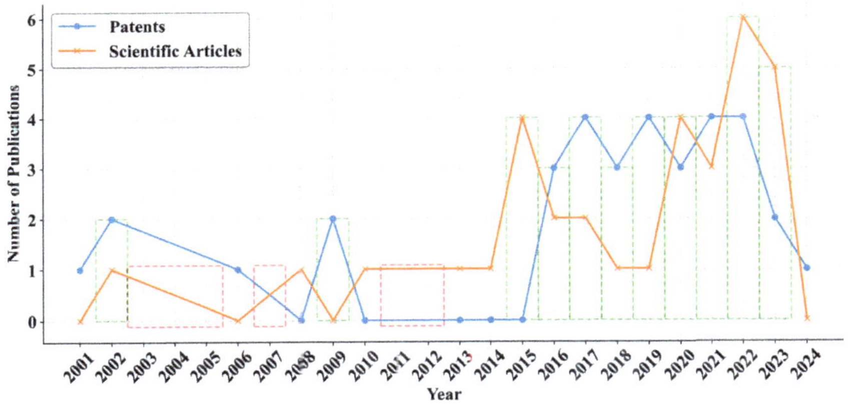

Figure 20. Number of scientific articles and patents over the years.

3.2. Interrelations of the Bibliographic Portfolio and Patent Repository

In order to explore the interrelationships among both the authors and KWs of the BP articles and the inventors of the PR patents, a robust analytical approach was adopted using the open-source software Vosviewer. This analysis focused on mapping the main connections between the authors and inventors, aiming to identify the key influencers in the scientific and industrial fields and contribute to a deeper understanding of RD trends. Additionally, it sought to highlight collaboration networks that may influence future academic work.

For the patents, detailed mappings of authors and co-authorships were obtained, illustrated in Figure 21. A total of 32 isolated and unrelated groups of inventors were identified, with a particular highlight on inventor Edward Simburger. He showed the highest density of interconnections among inventors, which supports the analysis conducted in the previous section (Figure 14) and confirms his relevance in inventions within this industrial sector. Additionally, the publication years of the related patents were highlighted and linked to their inventors, revealing a trend of emerging new inventors such as Stephen David, Jacob Ballard, and others after 2016.

An additional analysis revealed the lack of connection between the 32 groups of inventors. This suggested that despite the progress in patent inventions related to the research, there was no direct correlation between the inventions of each subgroup. This indicates possible confidentiality and restrictions when it comes to sharing information among inventors from different institutions. As result, Table 12 presents patents related to the main connections of each inventor, subdivided into low, medium, and high density.

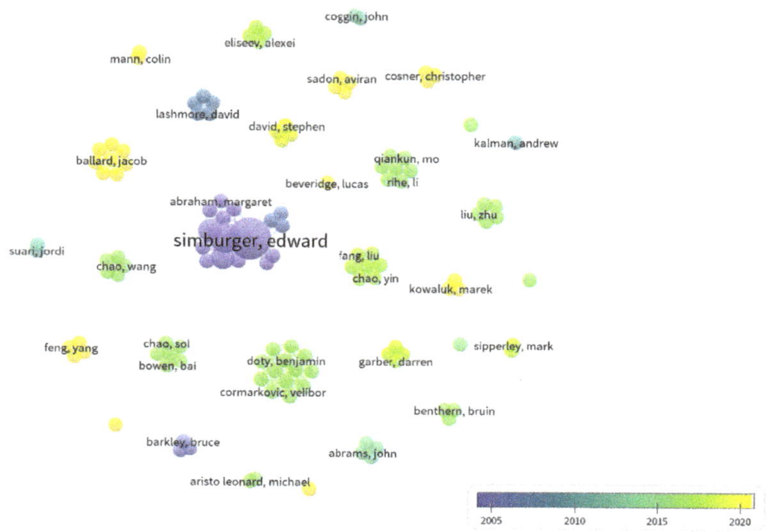

Figure 21. Mapping the existing connections between the inventors.

High-density patents indicate studies related to solar cells and energy storage systems, such as flexible thin-film solar cells and the integrated thin-film battery and circuit module. Medium-density patents present variations in innovations such as thin-film thermoelectric generators and the application of RTG. Low-density patents address both storage and energy supply systems for CubeSats, utilizing batteries, solar cells, and RTG modifications, as well as communication systems. Examples include power distribution modules applied to CubeSats and compact nuclear energy systems for space environments.

Table 12. Relationship between patents and the degree of connection density.

Density	Title
High density	Flexible thin-film solar cell. Integrated solar power module. Integrated thin-film battery and circuit module. Power distribution system. Spacecraft solar cell monitoring system.

Table 12. *Cont.*

Density	Title
Medium Density	Nano-engineered thin-film thermoelectric converter for photovoltaic applications. Radioisotope thermoelectric generator.
Low Density	Micro–nanosatellite power system based on supercapacitor. A power supply. Power supply and distribution module applied to CubeSat. Compact nuclear power system applied to space environment and working method. Solar battery and the integrated device of slot antenna. Satellite-constructor, training-demonstration model.

Regarding the scientific articles, similar to the previously adopted criteria, the density map is illustrated in Figure 22a, presenting a total of 31 isolated and unrelated groups of authors, with existing links between authors for specific articles. By evaluating the density of each group, it was possible to determine its intensity degree according to the number of citations per author presented in Figure 7b.

As a result, Figure 22b was obtained, demonstrating that, in this study, the citation relationship was not directly proportional to the connection density. For example, Champier D, who had 1185 citations, exhibited a low density, whereas Sratub et al. and Shittu et al., with 109 and 154 citations, respectively, exhibited a high density. Similarly, the same pattern of isolation observed with the patents was evident, highlighting the isolation of each scientific article concerning the academic development related to the research topic.

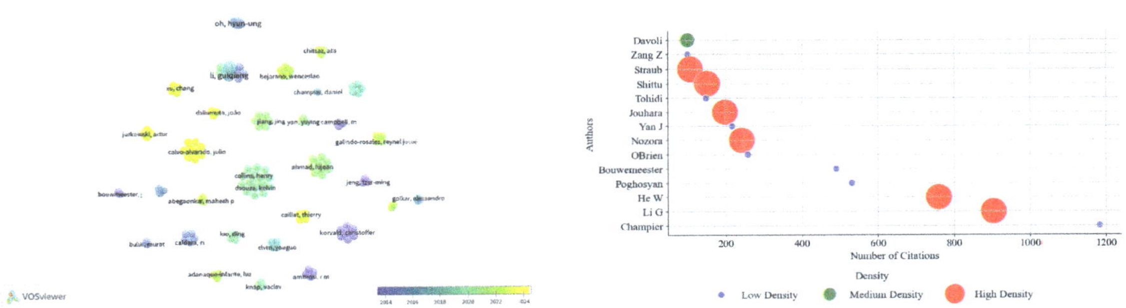

(a) (b)

Figure 22. Representation of (**a**) density mapping and (**b**) the relationship between the number of citations per author and density.

At the end, Table 13 presents the density levels of scientific articles based on the existing connections between the authors, categorized into low, medium, and high densities. Based on the distribution of the titles of scientific articles within these density categories, articles classified as "High Density" focus on technologies related to CubeSats, suggesting a detailed technical discussion on the development and capabilities of these nanosatellites. Additionally, they highlight the optimization of geometry and applications of TEGs, reflecting a deep exploration of how to maximize the efficiency of these devices.

The article classified as "Medium Density" presents a slightly broader scope, covering an overview of CubeSats structures, architectures, and protocols. Articles addressed as "Low Density" discuss both the various applications of TEGs more generally and studies dealing with the global analysis of nano- and picosatellite missions and the use of RTG in space missions.

Table 13. Relationship between scientific articles and the degree of connection density.

Density	Title
High density	OpenOrbiter: A low-cost, educational prototype CubeSat mission architecture. Review of thermoelectric geometry and structure optimization for performance enhancement. Thermoelectric generator (TEG) technologies and applications. Review of wearable thermoelectric energy harvesting: from body temperature to electronic systems. CubeSat evolution: analyzing CubeSat capabilities for conducting science missions. Review of thermoelectric geometry and structure optimization for performance enhancement.
Medium density	Small satellites and CubeSats: survey of structures, architectures, and protocols.
Low density	A review of the state of the art in electronic cooling. Thermoelectric generators: a comprehensive review of characteristics and applications. Review of micro thermoelectric generator. Safe radioisotope thermoelectric generators and heat sources for space applications. Survey of worldwide pico- and nanosatellite missions, distributions and subsystem technology. CubeSat evolution: analyzing CubeSat capabilities for conducting science missions. Thermoelectric generators: a review of applications.

5.3. *Keyword Analysis*

This section aims to establish the relationships between the BP, identified through the ProKnow-C methodology, and the search conducted by the IP, resulting in the PR, through the existing relationships between the KWs of each article. The objective is to identify the most relevant terms associated with CubeSats, focusing on their subsystems and energy sources. Additionally, it seeks to diagnose the connection between the application of TEGs and thermal analysis in the structures of this category of nanosatellite.

As a consequence, Figure 23 illustrates that the term "CubeSat" is prominently associated with branches in thermal analysis, boundary conditions, communication and distribution technologies, and energy storage and consumption.

In conclusion, no direct relationships were identified with the application of TEGs as the main energy source, but rather as temperature regulators for these subsystems. Nevertheless, this analysis corroborated the combination of KWs initially used and presented both in Table 7 and in the graph shown in Figure 9, demonstrating that they were coherently aligned with the proposed research theme.

Another important highlight is the absence of the "Keywords" topic in patent documents, indicating the recommendation to synthesize the documents to identify the main words that describe them. This synthesis would enable a comparison with the most frequent words in the BP, to verify the relationship between the axes established by ProKnow-C and those identified in the IP search.

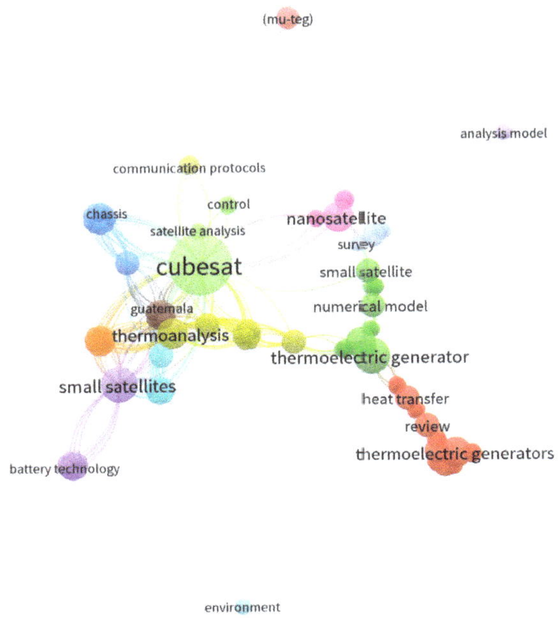

Figure 23. Mapping the existing connections between the inventors.

4. Final Considerations

This review aimed to recognize the application of TEGs as the main energy source in nanosatellites, particularly CubeSat models, identifying the impacted structures and systems under space conditions. The study highlighted the absence of TEG application, prompting the use of the ProKnow-C methodology for a state-of-the-art search. This search identified 33 relevant scientific articles and, after filtering, defined the necessary conditions to evaluate TEGs in CubeSats.

A similar patent search resulted in 34 patents, with only 5 meeting the criteria and 2 closely aligning with the research proposal. Initial assessments showed extensive use of simulation software like ANSYS and Matlab in 63.63% of the reviewed articles to evaluate CubeSat thermal behavior, considering construction specifications and enabling TEG application evaluation based on a CubeSat's face temperatures. However, there was a notable absence of studies directly addressing the use of TEGs as energy sources in nanosatellites, with existing research predominantly focused on terrestrial applications. Both internationally and in Brazil, the scientific literature and patent production on this topic remain limited, with the field being notably dominated by North American contributions.

The patent analysis revealed a predominance of U.S.-origin inventions, indicating a lack of incentives for this topic in Brazil and internationally. Only one patent addressed micro-TEGs, while others suggested structural modifications without direct applications in nanosatellites. The analysis of patents and scientific articles highlighted a peak in patent creation between 2016 and 2020, with a decline in scientific article production during the same period, but an increase since 2019, suggesting industry influence on academia.

Since 2020, the focus shifted to CubeSat structural and thermal optimization, crucial for potential TEG applications. Patents were mostly classified under "Execution of Transport

Operations", focusing on space application energy and communication systems, structural development of electronic components, energy storage, and photovoltaic system testing. No patents were specifically for TEG application in CubeSats, highlighting an industrial innovation gap in nanosatellite energy supply systems.

The methodology used facilitated a detailed bibliometric analysis of scientific articles and patents, leading to key conclusions. TEGs were identified more as thermoelectric temperature regulators rather than energy sources for CubeSats, highlighting the importance of comprehensive state-of-the-art reviews. Among identified review articles, none directly addressed TEGs in CubeSats, indicating pioneering research. The 33 scientific articles did not show TEG applications as energy sources in CubeSats, instead focusing on temperature regulation. The analysis of the relationship between academic research and industrial sector patents revealed increased patent production since 2015, with a decline in the scientific literature until 2019, when it began to correlate more with patent development, indicating a connection between research and industrial progress.

The use of TEGs in CubeSats presents both opportunities and challenges. On the one hand, TEGs offer unique advantages, such as their ability to harvest waste heat from internal components and external space conditions, providing a supplementary power source that complements solar cells. Their compact and lightweight design aligns well with the constraints of nanosatellites, and their lack of moving parts ensures high reliability in the harsh space environment. In hybrid systems, TEGs could enhance the overall performance of the EPS, particularly during orbital eclipses or periods of reduced solar exposure.

On the other hand, significant limitations remain in the use of TEGs for CubeSats, particularly their low power output and reliance on thermal gradients, which are challenging to achieve on small satellite surfaces. A potential solution lies in modifying the geometry of TEGs to better adapt to the 10×10 cm dimensions of a 1U CubeSat's faces. By optimizing their design to conform to specific face orientations—such as those exposed to higher thermal flux—TEGs could improve heat absorption and dissipation, enhancing power generation. This geometric optimization would enable better integration into the CubeSat structure without compromising payload space or mass constraints, making TEGs a more viable option for hybrid energy systems or even as standalone power sources in the future.

Future work should consider modifying procedures to include lost relevant articles, using new criteria like journal impact factors, and reading abstracts of excluded articles for valuable information. Additionally, numerical-computational simulations to describe TEG properties when optimally applied to CubeSats could reveal innovative methods for using thermoelectric devices in space missions, enhancing both temperature distribution analysis and power systems.

Author Contributions: Conceptualization, O.H.A.J.; methodology, G.B.S.; software, G.B.S.; validation, O.H.A.J. and G.B.S.; formal analysis, G.B.S.; investigation, G.B.S.; resources, O.H.A.J.; data curation, G.B.S.; writing—original draft preparation, G.B.S.; writing—review and editing, J.J.G.L. and O.H.A.J.; visualization, E.A.d.S.; supervision, O.H.A.J.; project administration, O.H.A.J. and J.J.G.L.; funding acquisition, O.H.A.J. All authors have read and agreed to the published version of the manuscript.

Funding: This research was partially supported by the FACEPE agency throughout the project with references APQ-0616-9.25/21 and APQ-0642-9.25/22. O.H.A.J. was funded by the CNPq, grant numbers 407531/2018-1, 303293/2020-9, 405385/2022-6, 405350/2022-8 and 40666/2022-3, as well as the Program in Energy Systems Engineering (PPGESE) Academic Unit of Cabo de Santo Agostinho (UACSA), Federal Rural University of Pernambuco (UFRPE). G.B.S were funded by the Federal University of Latin American Integration (UNILA).

Data Availability Statement: Data collection related to the articles studied was entirely carried out through the CAPES Cafe governmental portal, a platform widely recognized for the integration of scientific journals and academic databases, guaranteeing access to highly relevant publications. Regarding patents, data was obtained via Google Patents, which offers an accessible and comprehensive interface for searching technological and innovative records. Additional information about patents was explored in Patentscope, a tool made available by the World Intellectual Property Organization (WIPO), which allows free access to patent documents from various international offices, ensuring a broad and detailed analysis of registrations.

Conflicts of Interest: The authors declare no conflicts of interest.

Abbreviations

The following abbreviations are used in this manuscript:

NASA	National Aeronautics and Space Administration
ESA	European Space Agency
LEO	Low Earth orbit
COTS	Commercial Off-The-Shelf
TEGs	Thermoelectric generators
RTG	Radioisotope thermoelectric generator
FEM	Finite Element Method
TECs	Thermoelectric Coolers
KWs	Keywords
BP	Bibliographic Portfolio
PR	Patent Repository
EPS	Electrical Power System
IP	Intellectual Property
RD	Research and Development
PDB	Preliminary Database
WoS	Web of Science
JCR	Journal Citation Report
SJR	SCImago Journal Rank
SD	Science Direct
DBs	Databases
PRAD	Preliminary Raw Articles' Database
RAD	Raw Articles' Database
AD	Author Database
PCBs	Printed circuit boards
RPD	Raw Patent Database
PD	Patent Database
CPC	Cooperative Patent Classification
USPTO	United States Patent and Trademark Office
EPO	European Patent Office
MTG	Magneto-thermoelectric generator
NETT	Nano-engineered thin-film thermoelectric
MJPV	Multiple Junction Photovoltaic
IoT	Internet of Things
CNPq	Brazilian National Council for Scientific and Technological Development
FACEPE	Fundação de Amparo a Pesquisa de Pernambuco
PPGIES	Programa de Pós-Graduação Interdisciplinar em Energia e Sustentabilidade

References

1. Evans, B.G.; Thompson, P.T.; Corazza, G.E.; Vanelli-Coralli, A.; Candreva, E.A. 1945–2010: 65 years of satellite history from early visions to latest missions. *Proc. IEEE* **2011**, *99*, 1840–1857. [CrossRef]
2. NASA. *TIROS M Spacecraft (ITOS 1) Final Engineering Report*; NASA: Washington, DC, USA, 1970.
3. Poghosyan, A.; Golkar, A. CubeSat evolution: Analyzing CubeSat capabilities for conducting science missions. *Prog. Aerosp. Sci.* **2017**, *88*, 59–83. [CrossRef]
4. NASA. *Powering Cassini*; NASA: Washington, DC, USA, 2024.
5. ESA. *Mission Navigator*; ESA: Paris, France, 2023.
6. Nobrega, L. *Capítulo 10: Histórico da Exploração Espacial*; UFRGS: Farroupilha, Brazil, 2021.
7. Davoli, F.; Kourogiorgas, C.; Marchese, M.; Panagopoulos, A.; Patrone, F. Small satellites and CubeSats: Survey of structures, architectures, and protocols. *Int. J. Satell. Commun. Netw.* **2019**, *37*, 343–359. [CrossRef]
8. Waydo, S.; Henry, D.; Campbell, M. CubeSat design for LEO-based Earth science missions. In Proceedings of the IEEE Aerospace Conference, Big Sky, MT, USA, 9–16 March 2002; pp. 435–445.
9. Heidt, H.; Puig-Suari, J.; Moore, A.S.; Nakasuka, S.; Twiggs, R.J. CubeSat: A new Generation of Picosatellite for Education and Industry Low-Cost Space Experimentation.In Proceedings of the 2000 Small Satellite Conference, Logan, UT, USA, 21–24 August 2000.
10. Selva, D.; Krejci, D. A survey and assessment of the capabilities of Cubesats for Earth observation. *Acta Astronaut.* **2012**, *74*, 50–68. [CrossRef]
11. Alén Space. *A Basic Guide to Nanosatellites*; Alén Space: Nigrán, Spain, 2022.
12. De, R.; Abegaonkar, M.P.; Basu, A. Enabling Science with CubeSats—Trends and Prospects. *IEEE J. Miniaturization Air Space Syst.* **2022**, *3*, 221–231. [CrossRef]
13. NASA. *State-of-the-Art: Small Spacecraft Techonology*; NASA: Washington, DC, USA, 2023.
14. Rowe, D.M. Applications of Nuclear-Powered Thermoelectric Generators in Space. *Appl. Energy* **1991**, *40*, 241–271. [CrossRef]
15. Tohidi, F.; Holagh, S.G.; Chitsaz, A. Thermoelectric Generators: A comprehensive review of characteristics and applications. *Appl. Therm. Eng.* **2022**, *201*, 117793. [CrossRef]
16. Doraghi, Q.; Jouhara, H. Thermoelectric generator efficiency: An experimental and computational approach to analysing thermoelectric generator performance. *Therm. Sci. Eng. Prog.* **2024**, *55*, 102884. [CrossRef]
17. Ostrufka, A.L.A.; Filho, E.M.; Borba, A.C.; Spengler, A.W.; Possamai, T.S.; Paiva, K.V. Experimental evaluation of thermoelectric generators for nanosatellites application. *Acta Astronaut.* **2019**, *162*, 32–40. [CrossRef]
18. Maria, G. Development and Analysis of the Thermal Design for the OSIRIS-3U. Master's Thesis, The Pennsylvania State University, University Park, PA, USA, 2012.
19. von Lukowicz, M.; Schmiel, T.; Rosenfeld, M.; Heisig, J.; Tajmar, M. Characterisation of TEGs Under Extreme Environments and Integration Efforts Onto Satellites. *J. Electron. Mater.* **2015**, *44*, 362–370. [CrossRef]
20. Khater, A.; Jafrar, F. Temperature Distribution of CubeSats Using Finite Element Method. In Proceedings of the 73rd International Astronautical Congress, IAC 2022, Paris, France, 18–22 September 2022.
21. Dinh, D. Thermal Modeling of Nanosat. Master's Thesis, San Jose State University, San Jose, CA, USA, 2012. [CrossRef]
22. Tribak, F.; Bendaou, O.; Nejma, F.B. Impact of orbit inclination on heat transfer in a 1U LEO CubeSat. *MATEC Web Conf.* **2022**, *371*, 02001. [CrossRef]
23. Luo, D.; Liu, Z.; Yan, Y.; Li, Y.; Wang, R.; Zhang, L.; Yang, X. Recent advances in modeling and simulation of thermoelectric power generation. *Energy Convers. Manag.* **2022**, *273*, 116389. [CrossRef]
24. Luo, D.; Yu, Y.; Yan, Y.; Chen, W.H.; Cao, B. Increasing power densities in a thermoelectric generator by stacking and incorporating dual heat pipes. *Device* **2024**, *2*, 100435. [CrossRef]
25. Dorsa, A. O papel da revisão da literatura na escrita de artigos científicos. *Interações* **2020**, *21*, 681–683. [CrossRef]
26. Lacerda, R.; Ensslin, L.; Ensslin, S.R. Uma análise bibliométrica da literatura sobre estratégia e avaliação de desempenho. *Gestão Produção* **2012**, *19*, 59–78. [CrossRef]
27. Ensslin, L.; Ensslin, S.R.; Lacerda, R.T.O.; Tasca, J.E. ProKnow-C, Knowledge Develpment Process-Constructivist. *Processo Para Mapeamento Conhecimento Tema* **2010**, *10*, 2015.
28. de Souza, T.S.; de Carvalho, F.L.S.; Gomes, T.C.; Junior, M.W.J.S. Aplicação do ProKnow-C para Seleção e Análise de um Portfólio Bibliográfico Sobre Aplicação dos Recusos da Tecnologia 4.0. *Gestão Desenvolv. Rev.* **2021**, *8*, 68–89. [CrossRef]
29. Elsevier. *Mendeley*; Elsevier: Amsterdam, The Netherlands, 2024.
30. Clarivate. *Web of Science: Option to Save to BibTeX Format Not Available in Every Database*; Clarivate: Philadelphia, PA, USA, 2024.
31. Junior, O.H.A.; Ledesma, J.J.G.; Maciel, J.N. Forecasting Solar Power Outpit Generation—A Systematic Revier with Proknow-C. *IEEE Lat. Am. Trans.* **2021**, *29*, 612–624.
32. Dunford, R.; Su, Q.; Tamang, E. The Pareto Principle. *Plymouth Stud. Sci.* **2014**, *7*, 140–148.
33. Champier, D. Thermoelectric generators: A review of applications. *Energy Convers. Manag.* **2017**, *140*, 167–181. [CrossRef]

34. He, W.; Zhang, G.; Zhang, X.; Ji, J.; Li, G.; Zhao, X. Recent development and application of thermoelectric generator and cooler. *Appl. Energy* **2015**, *143*, 1–25. [CrossRef]
35. Bouwmeester, J.; Guo, J. Survey of worldwide pico- and nanosatellite missions, distributions and subsystem technology. *Acta Astronaut.* **2010**, *67*, 854–862. [CrossRef]
36. O'Brien, R.C.; Ambrosi, R.M.; Bannister, N.P.; Howe, S.D.; Atkinson, H.V. Safe radioisotope thermoelectric generators and heat sources for space applications. *J. Nucl. Mater.* **2008**, *377*, 506–521. [CrossRef]
37. Nozariasbmarz, A.; Collins, H.; Dsouza, K.; Polash, M.H.; Hosseini, M.; Hyland, M.; Liu, J.; Malhotra, A.; Ortiz, F.M.; Mohaddes, F.; et al. Review of wearable thermoelectric energy harvesting: From body temperature to electronic systems. *Appl. Energy* **2020**, *258*, 114069. [CrossRef]
38. Yan, J.; Liao, X.; Yan, D.; Chen, Y. Review of Micro Thermoelectric Generator. *J. Microelectromechanical Syst.* **2018**, *27*, 1–18. [CrossRef]
39. Jouhara, H.; Żabnieńska Góra, A.; Khordehgah, N.; Doraghi, Q.; Ahmad, L.; Norman, L.; Axcell, B.; Wrobel, L.; Dai, S. Thermoelectric generator (TEG) technologies and applications. *Int. J. Thermofluids* **2021**, *9*, 100063. [CrossRef]
40. Shittu, S.; Li, G.; Zhao, X.; Ma, X. Review of thermoelectric geometry and structure optimization for performance enhancement. *Appl. Energy* **2020**, *268*, 115075. [CrossRef]
41. Straub, J.; Korvald, C.; Nervold, A.; Mohammad, A.; Root, N.; Long, N.; Torgerson, D. OpenOrbiter: A low-cost, educational prototype CubeSat mission architecture. *Machines* **2013**, *1*, 1–32. [CrossRef]
42. Zhang, Z.; Wang, X.; Yan, Y. A review of the state-of-the-art in electronic cooling. *e-Prime Adv. Electr. Eng. Electron. Energy* **2021**, *1*, 100009. [CrossRef]
43. Corpino, S.; Caldera, M.; Nichele, F.; Masoero, M.; Viola, N. Thermal design and analysis of a nanosatellite in low earth orbit. *Acta Astronaut.* **2015**, *115*, 247–261. [CrossRef]
44. Luo, D.; Wang, R.; Yu, W.; Zhou, W. Parametric study of a thermoelectric module used for both power generation and cooling. *Renew. Energy* **2020**, *154*, 542–552. [CrossRef]
45. Escobar, E.; Diaz, M.; Zagal, J.C. Evolutionary design of a satellite thermal control system: Real experiments for a CubeSat mission. *Appl. Therm. Eng.* **2016**, *105*, 490–500. [CrossRef]
46. Tzeng, S.C.; Jeng, T.M.; Lin, Y.L. Parametric study of heat-transfer design on the thermoelectric generator system. *Int. Commun. Heat Mass Transf.* **2014**, *52*, 97–105. [CrossRef]
47. Knap, V.; Vestergaard, L.K.; Stroe, D.I. A review of battery technology in cubesats and small satellite solutions. *Energies* **2020**, *13*, 4097. [CrossRef]
48. Kang, S.J.; Oh, H.U. On-Orbit Thermal Design and Validation of 1 U Standardized CubeSat of STEP Cube Lab. *Int. J. Aerosp. Eng.* **2016**, *2016*, 4213189. [CrossRef]
49. Bulut, M.; Sozbir, N. Analytical investigation of a nanosatellite panel surface temperatures for different altitudes and panel combinations. *Appl. Therm. Eng.* **2015**, *75*, 1076–1083. [CrossRef]
50. Oh, H.U.; Park, T. Experimental Feasibility Study of Concentrating Photovoltaic Power System for CubeSat Applications. *IEEE Trans. Aerosp. Electron. Syst.* **2015**, *51*, 1942–1949. [CrossRef]
51. Candolfi, C.; Oualid, S.E.; Lenoir, B.; Caillat, T. Progress and perspectives in thermoelectric generators for waste-heat recovery and space applications. *J. Appl. Phys.* **2023**, *134*, 100901. [CrossRef]
52. Chacon, S.; Gutierrez, J.; Adanaque-Infante, L. Modelling cubesat Structure for Thermal Analysis. In Proceedings of the 29th IEEE International Conference on Electronics, Electrical Engineering and Computing, INTERCON 2022, Lima, Peru, 11–13 August 2022. [CrossRef]
53. Morales, F.O.; Bejarano, W.; Rosales, D.; Aguilera, F.Z.; Mejuto, J. Chassis Optimization of a 1U CubeSat made in a developing Country. In Proceedings of the 73rd International Astronautical Congress, IAC 2022, Paris, France, 18–22 September 2022.
54. Jurkowski, A.; Paluch, R.; Wójcik, M.; Klimanek, A. Thermal modelling of a small satellite data processing unit aided by sensitivity analysis and uncertainty quantification. *Int. J. Therm. Sci.* **2023**, *193*, 108514. [CrossRef]
55. Yu, P.; Zhang, C.; Zhang, Z.; Yang, Y.; Niu, Y.; Jiang, J.; Wang, C. Numerical Simulation of Thermoelectric Based Temperature Control system for CubeSat in Space. In Proceedings of the IEEE International Conference on Imaging Systems and Techniques (IST), ELECTR NETWORK, Kaohsiung, Taiwan, 24–26 August 2021. [CrossRef]
56. Gomez-Jenkins, M.; Calvo-Alvarado, J.; Chaves-Jimenez, A.; Carvajal-Godinez, J.; Martinez, E.; Yeine, A.J.A.; Calvo-Obando, A.J.; Jimenez, V.; Rojas, J.J.; Valverde-Salazar, A.; et al. Project Irazú: Space and Ground Systems Engineering of a 1U CubeSat Store and Forward Mission for Environmental Monitoring. *Trans. Jpn. Soc. Aeronaut. Space Sci.* **2023**, *66*, 217–225. [CrossRef]
57. Dallamuta, J.; Perondi, L.F.; de Oliveira, M.E.R. Space missions in South America: Profile and evolutionary perspective of their development. *Acta Astronaut.* **2023**, *206*, 9–17. [CrossRef]
58. Zhou, Y.; Zhang, X.; Xu, C.; Yang, X. Thermal design and analysis of JZJ-5 CubeSat. In Proceedings of the 13th Asia Conference on Mechanical and Aerospace Engineering (ACMAE), Chengdu, China, 22–24 December 2022; Volume 2472. [CrossRef]

59. Hipp-Méndez, J.R.; Sánchez-Varela, V.; Galindo-Rosales, R.J.; Zorto-Aguilera, F.J. Morazán MRZ-SAT CubeSat: Thermal Modelling and Analysis Guide for Academic CubeSat Missions. In Proceedings of the 73rd International Astronautical Congress IAC 2022, Paris, France, 18–22 September 2022.
60. Kartus, R.; Kukrus, A. Innovation, product development and patents at universities. *Est. J. Eng.* **2013**, *19*, 477–482. [CrossRef]
61. Wipo. PATENTSCOPE. Available online: https://www.wipo.int/patentscope/en/ (accessed on 11 April 2024).
62. Gray, D.; Coggin, J. Magneto-Thermoelectric Generator for Energy Harvesting. U.S. Patent 9,793,829, 17 October 2017.
63. Venkatasubramanian, R.; Himmtann, M.; Gajendiran, P.; Pierce, J.; Fairbanks, N.; Ung, R.; Ballard, J.; Maranchi, J. Nano-Engineered Thin-Film Thermoelectric Converter for Photovoltaic Applications. U.S. Patent Application 17/686,788, 8 September 2022.
64. Lashmore, D.; White, M.; White, B.; Degtiarov, D.; Mann, J. Nanostructured Material-Based Thermoelectric Generators. U.S Patent Application 12/191,765, 19 February 2009.
65. Ewell, R.; Fleurial, J.P.; Patel, J.; Nesmith, B.; Li, B.; Smith, K.; Doty, B.; Molina, G.; Cormarkovic, V.; Keyawa, N.; et al. Radioisotope Thermoelectric Generator. WIPO Patent WO 2016/138389 A1, 1 September 2016.
66. Sysouphat, A.J.; Kennedy, S.D. A Power Supply. WIPO Patent WO 2019/210358 A1, 7 November 2019.
67. Wang, M.; Ma, W.; Zhang, Z.; Wang, Y.; Zhu, Q.; Liu, X. Compact Nuclear Power System Applied to Space Environment and Working Method. Chinese Patent CN 110043338 B, 28 November 2023.
68. Leonard, M.A.; Patel, S.S. CubeSat System. U.S. Patent 11,021,274 B1, 1 June 2021.
69. Abrams, J.; Duchek, M.; Paz, A.; Harbach, R. Extendable Solar Array for a Spacecraft System. U.S. Patent 9,856,039, 2 January 2018.
70. Simburger, E.; Meshishnek, M.; Gilmore, D.; Smith, D.; Abraham, M.; Jeffrey, F.; Gierow, P. Flexible Thin Film Solar Cell. U.S. Patent 6,410,362 B1, 25 June 2002.
71. Simburger, E.; Matsumoto, J.; Giants, T.; Garcia, A.; Jeffrey, F.; Gierow, P. Integrated Solar Power Module. U.S. Patent 6,300,158 B1, 9 October 2001.
72. Simburger, E.; Matsumoto, J.; Gierow, P.; Hepp, A. Integrated Thin Film Battery and Circuit Module. U.S. Patent 7,045,246, 16 May 2006.
73. Wenwen, C.; Ning, S.; Hao, W.; Jianping, C.; Na, C.; Chao, Y.; Fang, L. Micro-Nano Satellite Power System Based on Super Capacitor. Chinese Patent CN106602694A, 26 April 2017.
74. Cosner, C.; Stickelmaier, J.; Wootan, J. Modular Electrical Power Subsystem Architecture. U.S. Patent 11,750,021, 5 September 2022.
75. Mann, C.; Walker, D. Modular Solar Cell and Solar Cell Array. U.S. Patent 2023/0299610 A1, 21 September 2023.
76. Payrebrune, M. Power Supply Module for Spacecraft. WIPO Patent WO 2017/177301 A1, 19 October 2017.
77. Ning, A.; Teel, G.; Riskas, B.; Garber, D. Satellite Modular Power Supply. U.S. Patent 10,536,107 B1, 14 January 2020.
78. Xiaoxia, X.; Feng, Y.; Weijia, R.; Jian, D. Satellite Storage Battery System. Chinese Patent CN113422423B, 30 November 2021.
79. Slater, D. Solar Array Remote Acoustic Sensing (SARAS). U.S. Patent US-10078328-B1, 18 September 2018.
80. Bowen, B.; Xiaoping, L.; Yanchao, Z.; Yanming, L.; Chao, S.; Lei, S. Solar Battery and the Integrated Device of Slot Antenna. Chinese Patent CN 106711576 B, 25 October 2019.
81. Woods, J. Solar Energy Conversion and Transmission System and Method. U.S. Patent 9,815,573, 14 November 2017.
82. Cruijssen, J.; Datema, P.; Benthern, B. Solar Panel and Flexible Radiator for a Spacecraft. U.S. Patent 10,737,808, 11 August 2020.
83. Kalman, A. Solar Panel Array. U.S. Patent 9,882,330 B2, 30 January 2018.
84. Derkacs, D. Space Vehicles Including Multijunction Metamorphic Solar Cells. U.S. Patent Application 17/180,210, 1 July 2021.
85. Beveridge, L. Space-Based Radioisotope Production and Methods of Use. U.S. Patent Application 17/736,083, 17 November 2022.
86. Simburger, E.; Rumsey, D.; Liu, S.; Halpine, J. Spacecraft Solar Cell Monitoring System. U.S. Patent 2009/0119060 A1, 7 May 2009.
87. Sipperley, M.; Smith, C. Thermoelectric Rocket Propellant Tank Pressurization System. U.S. Patent 10,495,028, 3 December 2019.
88. Suari, J.; Williams, A. CubeSat System, Method and Apparatus. U.S. Patent US20140039729A1, 6 October 2015.
89. Simburger, E.; Prater, A.; Carian, P. Power Distribution System. U.S. Patent 6,396,167 B1, 28 May 2002.
90. Haining, M.; Qiankun, M.; Wenhe, L.; Yongjun, Y.; Xiang, Z.; Rihe, L.; Yuchen, H. Power Supply and Distribution Module Applied to CubeSat. Chinese Patent CN 106059266 A, 26 October 2016.
91. Eliseev, A.; Zharenov, I.; Zharkikh, R.; Purikov, A. Satellite-Constructor, Training-Demonstration Model. Russian Patent RU 2 693 722 C2, 4 July 2019.
92. Poirot, P.; Smith, C.; Kowaluk, M. Satellite Thermal Control. WIPO Patent WO 2024/003013 A1, 4 January 2024.
93. Weiss, G.; Sadon, A.; Elyasaf, A.; Sinai, M. Scenario Based Method for Testing Software. U.S. Patent US11138100, 5 October 2021.
94. Ehrlich, J. Spherical Mobility System. U.S. Patent 11,420,777 B1, 23 August 2022.
95. Patents, B. What's the Difference Between a Pending and a Granted Patent? Available online: https://boldip.com/whats-the-difference-between-a-pending-and-a-granted-patent/ (accessed on 23 April 2024).
96. Lens, T. Explore o conhecimento global em ciência e tecnologia. Available online: https://www.lens.org/?locale=pt (accessed on 25 April 2024).

97. Chao, W.; Chenggu, Z.; Lin, L.; Chunhong, L.; Zezhan, Z. A Kind of Device to be Generated Electricity Using Satellite External Surface Temperature Gradient. Chinese Patent CN 106208814 B, 5 April 2019.
98. USPTO. *Cooperative Patent Classification*; USPTO: Alexandria, VA, USA, 2024.

energies

MDPI

Article

Energy-Saving Design Strategies for Industrial Heritage in Northeast China Under the Concept of Ultra-Low Energy Consumption

Shiqi Yang [1], Hui Ma [1,*], Na Li [2], Sheng Xu [1] and Fei Guo [1]

[1] School of Architecture and Fine Art, Dalian University of Technology, Dalian 116024, China; augyang@mail.dlut.edu.cn (S.Y.); xusheng@mail.dlut.edu.cn (S.X.); guofei@dlut.edu.cn (F.G.)
[2] Design Institute of Civil Engineering & Architecture of Dalian University of Technology Co., Ltd., Dalian 116023, China
* Correspondence: huima@dlut.edu.cn

Abstract: Countries around the world have developed standards for ultra-low energy consumption building design and future plans. Unfortunately, these standards lack specific requirements for industrial heritage. As an important carrier of urban context, history, and the transmission of residents' memories, industrial heritage cannot be overlooked in urban development. This study uses DesignBuilder energy simulation software to model industrial heritage (taking the Changchun Tractor Factory as an example) and compares the energy consumption before and after renovation strategies. The results show that in the Case 4 plan, after implementing the renovation strategy, heating energy consumption can be reduced by about 11,648 (kWh/m^2) over the heating season, the total primary energy was reduced by about 4 million (kgce/tce), and total energy consumption decreases by approximately 95%. This demonstrates the effectiveness of the industrial heritage reuse design strategy proposed in this paper. It provides a new direction for reuse design under ultra-low energy consumption requirements in related case studies.

Keywords: industrial heritage; ultra-low energy consumption; DesignBuilder; building reuse

1. Introduction

The global building roadmap released by the International Energy Agency (IEA) aims to achieve zero emissions for new and existing buildings gradually from 2020 to 2050 [1]. Governments worldwide have prioritized the building sector's focus on low carbon and ultra-low energy consumption. In March 2024, the Central Committee of the Communist Party of China and the State Council issued the Opinions on Accelerating the Comprehensive Green Transformation of Economic and Social Development, which stated that ultra-low-energy buildings in China are expected to achieve large-scale development by 2027. It emphasized that advancing green and low carbon development in urban and rural areas should begin with a strong focus on the development of ultra-low-energy buildings [2]. Globally, 39% of carbon emissions originate from the construction industry and buildings themselves. The issues of energy consumption and sustainability in buildings have become focal points for countries worldwide [3,4]. Various nations have established standards and regulations for energy-efficient, low-energy, and green buildings. Unfortunately, these standards lack specific considerations for industrial heritage [5].

The first project for the conservation and adaptive reuse of industrial heritage was the Ironbridge Gorge in the UK, repurposed in the 1960s [6]. In fact, the process of protecting industrial heritage evolved from architectural heritage. As such, like architectural heritage, it is essential to the development of a city, as it symbolizes and represents the historical period of the past [7]. However, extending the lifespan of industrial heritage and sustaining its historical and cultural value requires consideration of not only the impacts and limitations of traditional codes [5], the influences of sustainability requirements [8], and regional differences [7,9], but also energy consumption [10] and the balance between conservation and reuse [11]. Among these influencing factors, aligning with current global energy development requirements, the greatest and most pressing issue is achieving low energy consumption and ultra-low energy consumption.

However, a review of existing research literature reveals a significant lack of studies and practices related to low energy consumption and ultra-low energy consumption in the context of architectural heritage, particularly in the field of industrial heritage, which undoubtedly hinders the conservation and sustainable development of industrial heritage severely [12]. In addition, a disconnect exists between China's industrial heritage reuse cases and the critical international and domestic strategic objectives, such as carbon peaking and sustainable development. It is a significant risk to the full lifecycle utilization of industrial heritage. The research on the energy consumption aspects of industrial heritage still has large gaps. On one hand, since most early industrial buildings were constructed to meet the needs of industrial production at minimal cost, their envelope structures were designed with the lowest standards for insulation and heating. Therefore, such buildings are inherently vulnerable in terms of energy-saving. On the other hand, industrial heritage is an important and continuous branch of the architectural field, which cannot be overlooked.

In addition, building simulation and computational technologies have been widely applied in architectural design for decades [13]. Building energy simulation tools have consistently assisted architects, designers, engineers, and researchers in identifying the most suitable and the most cost-effective energy-saving measures to reduce and optimize building energy consumption during the design, construction, and operation phases [14]. Currently, these software tools have become critical in the design phase of energy-efficient buildings [15]. Among them, DesignBuilder, as one of the most comprehensive energy simulation software tools, ranks among the top three in terms of usage frequency [16].

In summary, given the research gaps regarding industrial heritage under the concept of ultra-low energy consumption, the urgent need for renewal, and the alignment with sustainable development goals, this study proposes a novel reuse strategy based on the current status and needs of representative cases. DesignBuilder energy simulation software is utilized to conduct comparative studies of reused plans under different strategies. This research employs a combination of the Case Study Method, Simulation Analysis Method, and Comparative Analysis Method. By integrating theory and case studies, as well as qualitative analysis and quantitative simulations, this study develops instructive reuse strategies. Furthermore, the effectiveness and applicability of these strategies are demonstrated through scientific simulations and data validation methods. The main goals of this study are as follows:

1. To investigate the current energy consumption status and energy-saving potential of industrial heritage buildings in Northeast China using DesignBuilder energy simulation software.
2. To propose the "Two Enhancements and Two Reductions" design strategy, which balances ultra-low energy consumption design principles with the imperative to conserve and adaptively reuse industrial heritage structures.
3. To verify the effectiveness of the "Two Enhancements and Two Reductions" strategy and demonstrate its potential for achieving ultra-low energy consumption in practice.

2. State of the Art

2.1. Ultra-Low-Energy Buildings

Ultra-low-energy buildings are a subset of green buildings, a concept that can be traced back to the energy crisis following World War II [17]. In 1969, Italian architect Paolo Soleri first introduced the concept of "arcology", or ecological architecture [18]. In 1976, Torben V. Esbensen conducted theoretical and experimental research on using solar energy for building heating, proposing the concept of "zero-energy buildings" for the first time [19]. In 1990, the British Building Research Establishment introduced the world's first green buildings assessment method, Building Research Establishment's Environmental Assessment Method (BREEAM), marking a significant step toward the formalization of green buildings practice [20]. In 1991, German physicist Wolfgang Feist constructed the first "Passive House", initiating extensive research into green buildings development worldwide [21]. In 2006, the UK proposed the concept of low-carbon buildings and introduced the goal of "zero-carbon homes", propelling the advancement of low-carbon and zero-carbon buildings [22]. By 2010, Germany raised the energy-saving standards for buildings, combining low-energy building standards with passive house standards and introducing the concept of ultra-low-energy buildings. These architectural concepts originated from global energy crises, climate change, and the need for sustainable development. As a result, building standards in different countries have evolved into distinct systems and regulations worldwide [17,23,24]. In China, the development of green buildings began with energy-saving. Since the standard of its first design standard in 1986, the central and local governments have issued hundreds of building energy-saving codes, revising many standards regularly [25–27]. The evolution and derivation of these related concepts have been complex and lengthy, but they remain interconnected and share commonalities. The relationships between some of these concepts are illustrated in Figure 1.

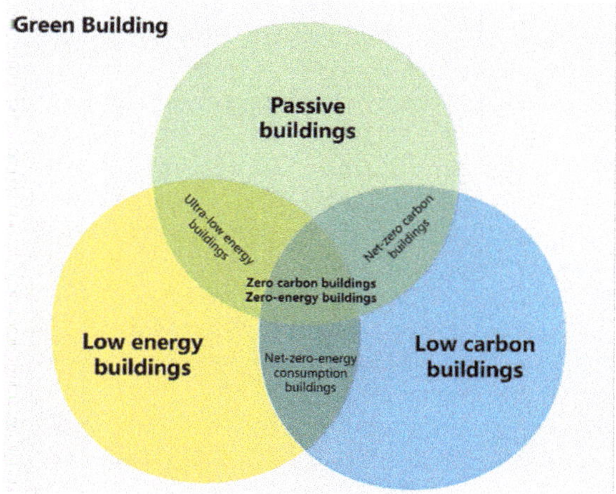

Figure 1. Relationship diagram of concepts related to ultra-low-energy buildings.

2.2. Reuse of Industrial Heritage

As early as 1965, UNESCO proposed the establishment of the International Council on Monuments and Sites (ICOMOS) [28], initiating global attention to the conservation and adaptive reuse of architectural heritage. Currently, 168 countries are signatories to ICOMOS, with the ranking of countries by the number of World Heritage Sites shown in Figure 2 (Data from ICOMOS official website) [29]. Among them, Italy, China, Germany, France, and Spain occupy the top five positions. Notably, among these top five countries, all except China include industrial heritage within their World Heritage Sites. Examples include Italy's Rhaetian Railway and the 20th-century industrial city of Ivrea; Germany's Völklingen Ironworks; France's Nord-Pas de Calais Mining Basin; and Spain's Almaden mercury mining area. Additional examples include Belgium's Major Mining Sites of Wallonia, Japan's Tomioka Silk Mill, and the United Kingdom's famous Ironbridge Gorge. This highlights the insufficient conservation and utilization of industrial heritage in China. However, this does not imply a severe lack of industrial heritage in China. On the contrary, China possesses a significant amount of industrial heritage.

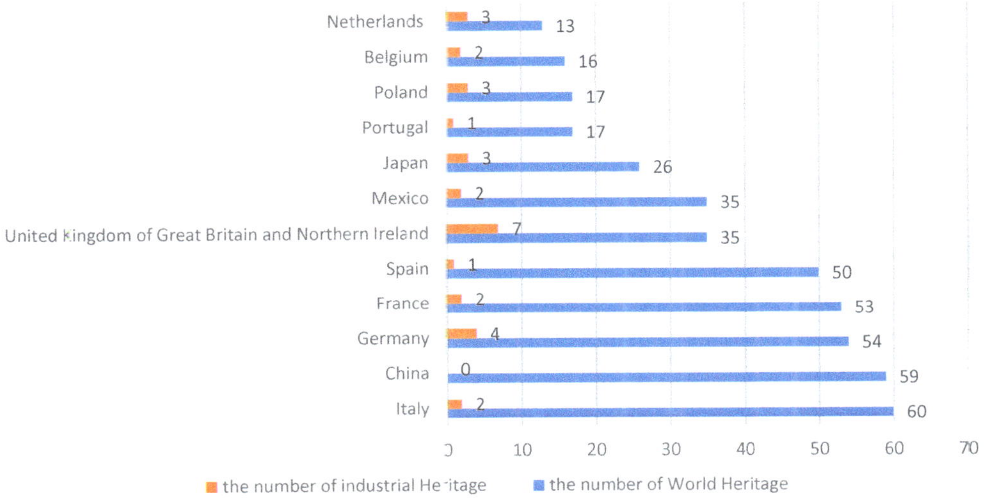

Figure 2. Total number of World Heritage properties and industrial heritage sites in some of the ICOMOS member countries.

China's industrial development started relatively late. However, for a significant period after 1949, the country's economy was primarily driven by heavy industry, light industry, and handicrafts [30]. With the continuous advancement of technology and urban development, the transformation of former industrial cities has become inevitable in the face of urban and architectural renewal. Within these cities, a large amount of industrial heritage has been preserved, serving as a new type of historical and cultural reserve resource for urban development, which is significant and cannot be overlooked [31–33].

The research on ecological architecture, ecological conservation, and architectural site preservation originated in the 1960s. Subsequently, studies on ecological architecture in various countries gradually diversified, giving rise to fields such as building carbon emissions and energy-saving technologies. Around the same time, research on industrial heritage began to emerge as an independent discipline from the broader category of architectural heritage, with architects and researchers extensively debating the issues of preservation versus renewal. However, it was not until the 1980s that energy-saving technologies were first applied to architectural heritage, while B. Jankovich et al. retrofitted passive and active solar energy systems and cooling systems suitable for architectural heritage [34]. From that point onward, the conservation and reuse of architectural and industrial heritage started to address energy consumption issues, as shown in Figure 3.

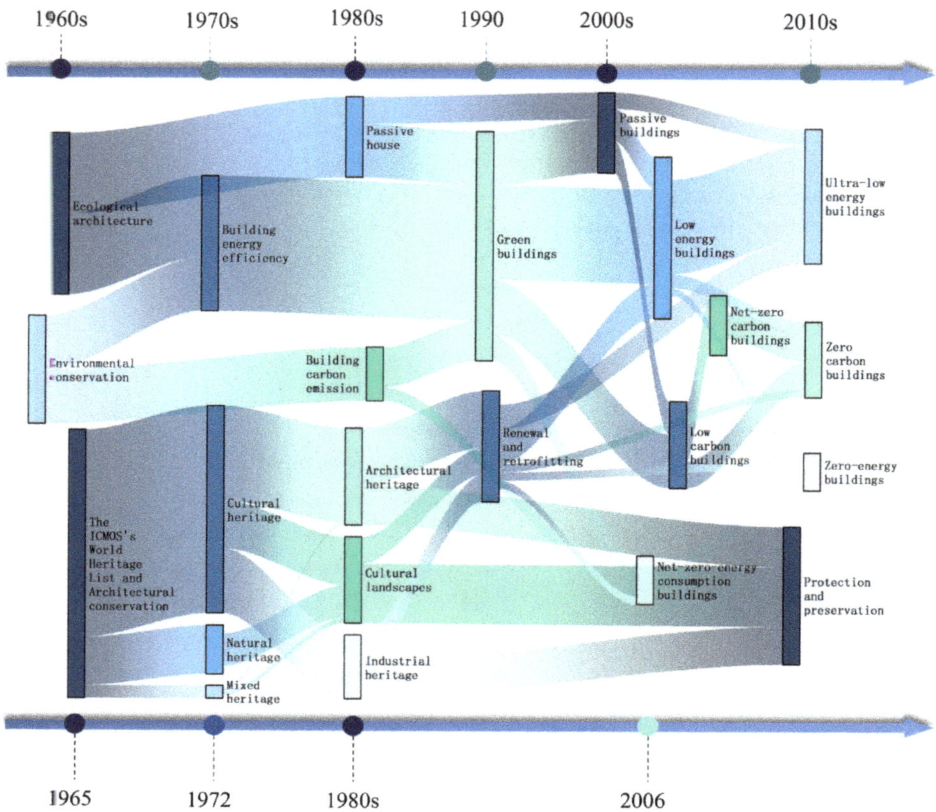

Figure 3. Development and correlation chart of research related to ultra-low energy consumption and conservation and reuse of built heritage.

2.3. Selection of Energy Simulation Software

In the 1970s, the United States took the lead in developing energy simulation software, the BLAST and DOE-2. Subsequently, the United Kingdom developed BREDEM, China created DeST, and Japan produced HASP. By the end of the 20th century, many European countries had also developed their own building energy simulation software, as shown in Figure 4 [35]. Although there are more than 30 types of energy simulation software globally, after 50 years of updates and eliminations, a relatively mainstream selection has emerged. Choosing software suitable for the specific project has become an essential step in the research process. A comparison of these tools is shown in Table 1.

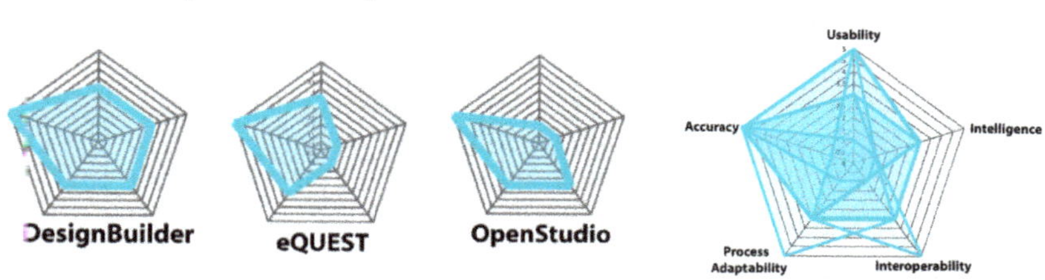

Figure 4. Some of the results from the analysis of Attia, Shady [35].

Table 1. Comparison of major international building energy simulation softwares.

	TRNSYS	PHPP	EnergyPlus	eQUEST	DeST	DesignBuilder
Country	United States	Germany	United States	United States of America	China	United Kingdom
Development Company	University of Wisconsin	Passive House Institute	U.S. Department of Energy (DOE)	JB Engineering	Tsinghua University	DesignBuilder Software Ltd.
Development	1975	1996	1999	2000	2004	2006
Load Simulation Methods	Modular Thermodynamic Modeling	Heat Balance Method	Heat Balance and Fluid Dynamics	Simplified Models	Heat Balance Method	Heat Balance and Fluid Dynamics
Model Complexity	high	medium	high	low	medium	medium
System Restructuring	highly flexible	limited	flexible	limited	limited	flexible
Accurate Temperature Calculations	accurate	accurate	accurate	accurate	accurate	accurate
Applicable Structures	all types of complex systems	primarily for passive houses	all types of buildings	residential and small businesses	Chinese buildings	all types of buildings
Photovoltaic Module	support	partial support	support	partial support	partial support	support
Data Input Method	manual or script	manual	manual or batch input	simple import	manual	graphical input
Output of Results	reports and data output	reports	graphs and data tables	graphs and reports	reports	graphs and reports

Attia, Shady conducted a comparative analysis of mainstream building energy simulation tools in the early 21st century, evaluating them from five aspects: usability, intelligence, accuracy, process suitability, and interoperability. Among these, DesignBuilder demonstrated the best overall performance [35]. Sağdıçoğlu, M. S., using Vosviewer and Bibliometrix tools, performed bibliometric analyses on the usage of energy simulation software, revealing that DesignBuilder ranked first in terms of usage frequency [36]. Haidong Wang et al. found that DesignBuilder integrates environmental simulation with building energy modeling, constructing an integrated simulation program that significantly enhances the accuracy of predictions for thermal and flow behaviors [37]. Subsequently, more scholars began utilizing DesignBuilder to simulate and calculate various building components. Piero Bevilacqua et al. used DesignBuilder to evaluate the performance of new envelope materials [38]. M.M. Rahman et al. compared energy-saving under various HVAC systems and high COP chiller investment strategies using the software [39]. Yang Liu et al. explored the relationship between the window/wall ratio and energy consumption [40]. Meanwhile, A. Goenaga-Pérez et al. conducted simulations to verify whether Spain's nZEB standards are applicable to the various regional climate conditions within the country [41]. Moreover, DesignBuilder has demonstrated extensive applicability in energy simulation and is widely employed in studies analyzing overall building energy performance [42–44]. Based on previous research, DesignBuilder simulation software is applicable to almost all types of buildings. In summary, given DesignBuilder's applicability and precision in simulating building energy consumption, this study selects DesignBuilder as the software for representative case simulation and analysis.

3. Methods and Strategies

3.1. Experiment Overview

This study focuses on ultra-low energy consumption and energy-saving to address the practical conditions of industrial heritage. We comprehensively consider sustainable development, urbanization, technological progress, adaptive reuse, social needs, and ecological balance to explore more suitable reuse strategies, providing reference and direction for the development and reuse of industrial heritage. The flowchart of this study is illustrated in Figure 5: Firstly, based on a comprehensive analysis of geographical location, specific conditions, and typical characteristics, the Changchun Tractor Factory in Changchun, Jilin Province, is selected as the research subject. Subsequently, the original case is simulated and analyzed to evaluate its current features, surrounding context, and urban development demands, forming a "Two Enhancements and Two Reductions" design strategy tailored to industrial heritage reuse. Finally, using this strategy, four schemes are developed for the case. These schemes are then subjected to comparative analysis and validation calculations using DesignBuilder (v 7.0.2) software to determine the potential of the strategy for achieving ultra-low energy consumption.

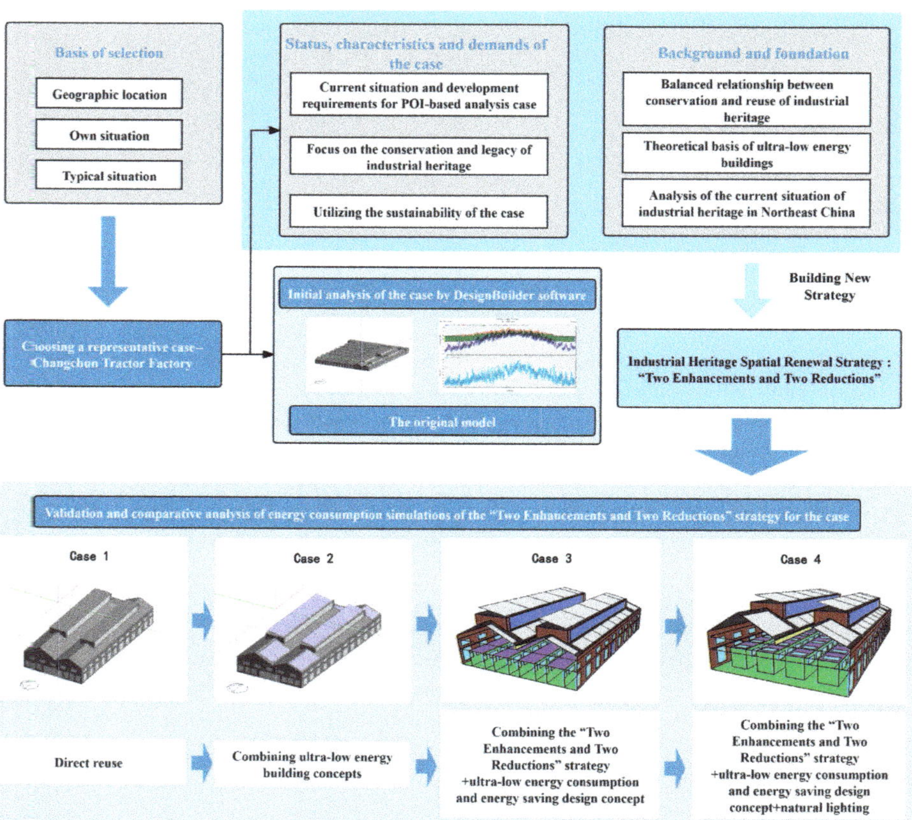

Figure 5. Experimental flowchart diagram.

3.2. The Case Study

Among the list of old industrial bases released in China, there are 95 prefecture-level cities, 23 of which are located in Northeast China (Including Heilongjiang, Jilin, and Liaoning Provinces), accounting for 24.2% of the total. Within the "156 Projects" of the "First Five-Year Plan" implemented in the early years of the People's Republic of China, Northeast China hosted 56 projects, accounting for 35.9% of the total, as shown in Figure 6. Additionally, the three provincial capitals in Northeast China are all key cities for revitalizing old industrial bases [45]. As a critical area for the old industrial base revitalization, Northeast China, compared to other regions of the country, has undergone more profound industrial transformations, shifts in dominant sectors, adjustments in industrial structures, urban development demands, and significant policy changes. These factors have resulted in a larger number of abandoned or underutilized industrial heritage resources in need of renewal and reuse. Furthermore, Northeast China belongs to the severe cold B and C zones in China's building thermal design classification, making it the coldest region in the country. Its unique geographical and climatic conditions are also among the critical factors influencing energy consumption.

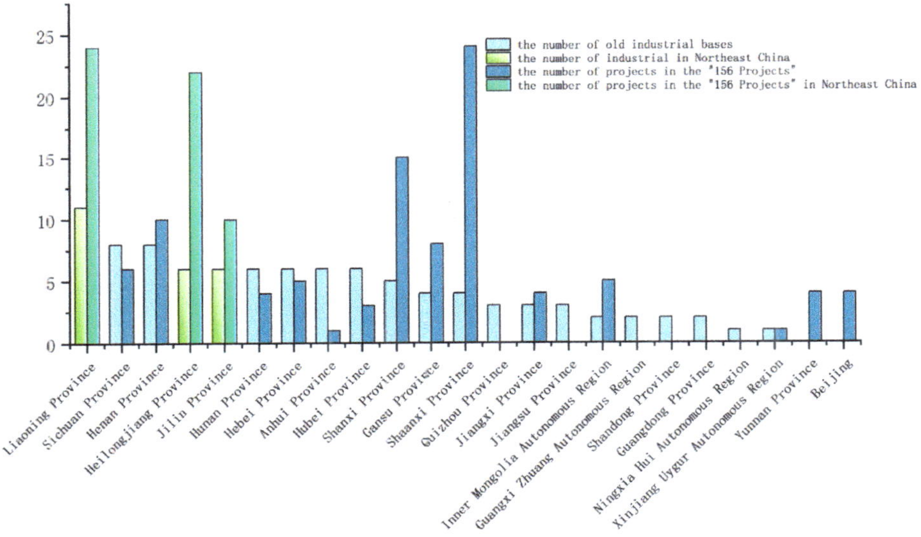

Figure 6. Distribution of the number of old industrial cities in China by province (the number of provinces not appearing is 0).

China's Ministry of Housing and Urban-Rural Development has divided thermal design zones into primary levels based on the coldest average temperatures and secondary levels based on heating degree days, as shown in Table 2 [46]. Most cities in the northeastern region fall under the severe cold zone, with cities in the severe cold zones B and C accounting for 72.5% of the total cities in the northeast, as indicated in Table 3. The northeastern region, located in the northern part of China, is one of the coldest areas in the country. Moreover, the cold climatic conditions are a critical factor in building energy consumption, directly resulting in heating energy use being the highest throughout a building's lifecycle. Therefore, selecting a representative case from the cold northeastern region, which also has the highest concentration of industrial heritage in China, adds significant value to the development of energy consumption research for industrial heritage. Among the three provincial capitals in the northeast, Changchun is centrally located, and its thermal design zone is classified as severe cold zone C, representing a midpoint within the thermal zoning range of all northeastern cities. Consequently, this study selects industrial heritage in Changchun as a representative case.

Table 2. Some thermal design zoning in China.

Name of the First-Level Zone	Main Indexes of the Classification	Name of the Second-Level Zone	Main Indexes of the Classification
Severe cold zone	$t_{min \cdot m} \leq -10$ °C	Severe cold zone A (1A) Severe cold zone B (1B) Severe cold zone C (1C)	$6000 \leq$ HDD18 $5000 \leq$ HDD18 < 6000 $3800 \leq$ HDD18 < 5000
Cold zone	-10 °C < $t_{min \cdot m} \leq 0$ °C	Cold zone A (2A)	$2000 \leq$ HDD18 < 3800 CDD26 ≤ 90
		Cold zone B (2B)	$2000 \leq$ HDD18 < 3800 CDD26 > 90

Table 3. Number of thermal design zones for cities in Northeast China.

Provinces in Northeast China	Number and Percentage of 1A	Number and Percentage of 1B	Number and Percentage of 1C	Number and Percentage of 2A
Heilongjiang Province	6; 30%	14; 70%	0	0
Jilin Province	0	3; 30%	7; 70%	0
Liaoning Province	0	0	5; 50%	5; 50%
Total number	6	17	12	5
Total Percentage	15%	42.5%	30%	12.5%

Through field research on the present situation of industrial heritage in the region, the following was found: First, most industrial buildings were constructed with assistance from the former Soviet Union, showing specific characteristics of that era. Second, northeast China primarily had developed heavy industry, resulting in most industrial sites containing similar large-span factory buildings. Third, due to the rapid development of heavy industrial areas in Northeast China during the early years of the People's Republic of China, many of the industrial heritage sites in this region are located in relatively favorable geographical positions. And some projects are even located in city centers, meaning that their abandonment or demolition would impose a burden on surrounding residents and the environment [30].

Based on the background conditions, a comparative analysis of the existing industrial heritage in the Changchun region was conducted. Ultimately, a tractor factory located in the Erdao District (within the Second Ring Road) of Changchun was selected as the representative industrial heritage case. This selection was made because the site is situated in the city center, making it urgently in need of renovation. Additionally, most industrial heritage sites in the area are large-space factory buildings, which are the most prevalent and representative building type among industrial heritage sites in Northeast China. Established in 1958, the factory covers an area of 26.9 km^2 and was once the largest production base for wheeled tractors in China [47]. Over more than half a century of development, the surrounding area of the case study has transitioned to residential and educational land use. However, the site itself is still in a vacant situation, representing not only a waste of land resources but also a disconnection from the functional integration of the surrounding areas. This situation urgently calls for improvement.

It is worth noting that partial demolition and renovation have been carried out on the No. 3 Production Workshop within the factory site, while the other workshops remain vacant. Notably, the No. 2 Machining Workshop, located at the core of the factory, holds a superior geographical position and sunlight intensity compared to other workshops in the area. At present, it is being used as a temporary parking lot. We focus on conducting simulation calculations and reuse design for the overall structure of the No. 2 Machining Workshop and its A space. Based on the status analysis and development needs outlined in

Section 3.3, the renovation is primarily focused on commercial use. Thermal zone setup is detailed in Section 3.5. The historical site layout is shown in Figure 7 (left), while the current site layout is depicted in Figure 7 (right). The current condition of the No. 2 Processing Workshop is illustrated in Figure 8.

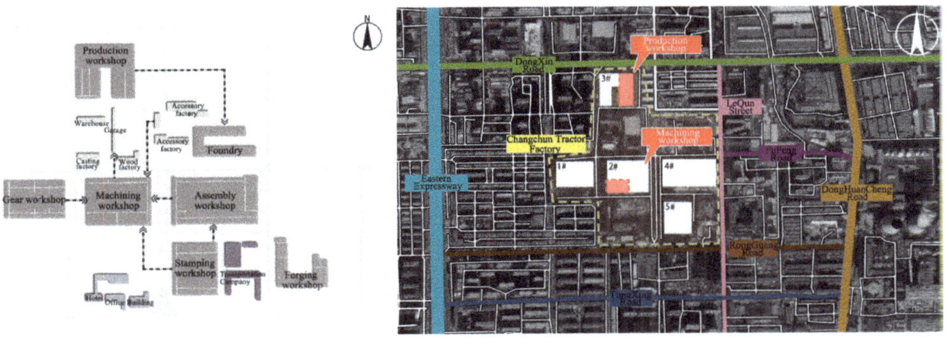

Figure 7. Layout of Changchun Tractor Factory (transferred by the author).

Figure 8. Current condition of the No. 2 Processing Workshop of Changchun Tractor Factory.

3.3. *Model Construction and Simulation Settings*

Based on the analysis in Section 2.2, the selected representative building—Factory No. 2—was originally a machining workshop located at the center of the Changchun Tractor Factory area. Surrounded by internal factory roads, it has a building area of 18,300 m^2, a total building length of 154 m, a total building width of 119.2 m, and an average interior clear height of 8 m. For various reasons, it is still unused and temporarily utilized as a parking space. Its structural form combines reinforced concrete columns with a prefabricated steel roof truss, categorized as a row of reinforced concrete column structures. Through field surveys and investigations, we redrew the floor plan and section of Factory No. 2, as shown in Figure 9. Subsequently, a model was constructed in DesignBuilder based on the measurement results.

- Construction and Glazing Template: Factory No. 2 exhibits distinct characteristics of an industrial factory building in terms of structural form and internal space. According to China's Code for Thermal Design of Civil Buildings (GB50176-2016) [46], the external opaque envelope consists entirely of 490 mm-thick red brick walls, is lightweight mortar masonry clay brick with a thermal conductivity of 0.76 (W/m·K), and exposed both internally and externally. The transparent envelope consists of double-layer ordinary glass, with a thermal conductivity of 0.76 (W/m·K). The roof is composed of prefabricated concrete slabs combined with composite steel plates. The thermal conductivity of the concrete slab is 1.74 (W/m·K), while that of the composite steel plate is approximately 0.04 (W/m·K). And the roof skylights are made of non-operable single-layer ordinary glass. Based on these features, the corresponding original structural model was established as shown in Figure 10.

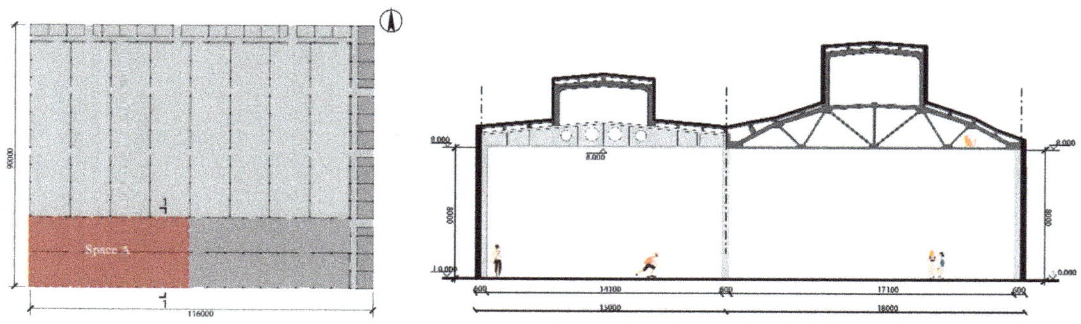

Figure 9. The redraw of floor plan and 1-1 section of Factory No. 2.

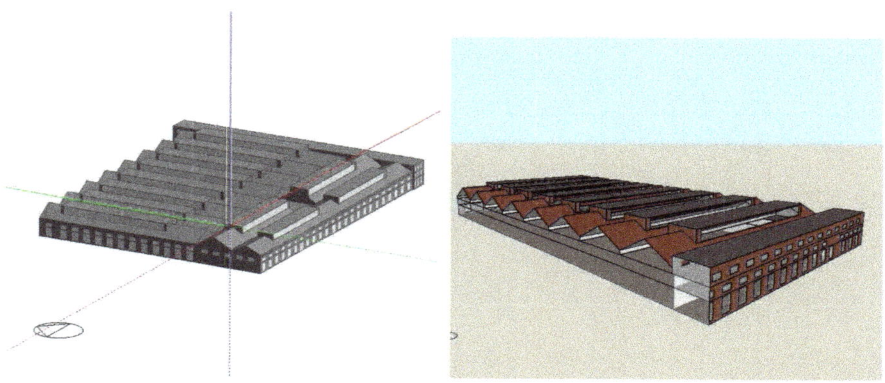

Figure 10. Model of Factory No. 2 in DesignBuilder software.

- Climatic Conditions: The project is located in Changchun, the capital city of Northeast China, classified as a Severe Cold Zone C in China's building thermal design zoning. The annual total solar radiation on a horizontal plane is 5000 MJ/m^2, with an average annual outdoor temperature of 4.9 °C. The climate here is short and warm in summers, while cold and long in the winters, and there are significant daily temperature variations in spring and autumn. DesignBuilder software supports the CSWD's (Chinese Standard Weather Data) weather database, which was developed by Dr. Jiang Yi, Department of Building Science and Technology at Tsinghua University and China Meteorological Bureau, and calculates heating and cooling loads using the heat balance method [48,49]. Therefore, the meteorological data of the Changchun area in the CSWD database selected for this project, detailed information, and outdoor temperature changes are shown in Table 4.

Table 4. Weather information and daily frequency of site outdoor air dewpoint temperature and dry bulb temperature in Changchun.

Location Templates		Daily Frequency of Site Outdoor Air Dewpoint Temperature and Dry Bulb Temperature in Changchun
Location		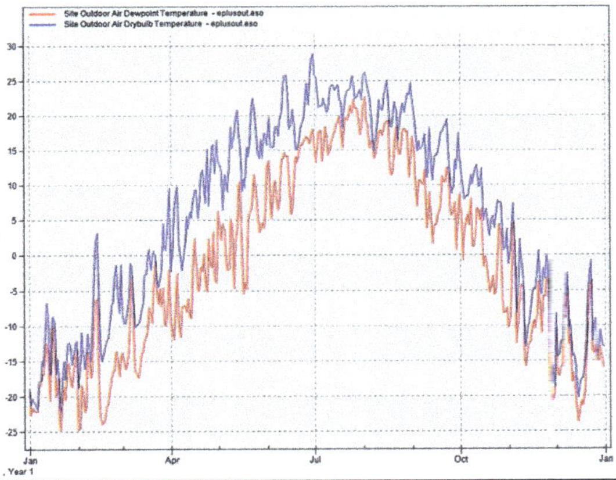
Source	ASHRAE/CSWD	
WMO	541,610	
ASHRAE climate zone	6A	
Latitude	43.90	
Longitude	125.22	
Elevation (m)	238.0	
Standard pressure (kPa)	98.5	
Time zone	(GMT+8:00) Beijing	
Winter design weather		
Outside design temperature	−25 °C	
Wind speed	2.0	
Wind direction	220.0	

- Activity Template: This study combines the Design Standard for Energy Efficiency of Public Buildings by Jilin Province (energy efficiency 72%) (DB22/T 5160-2024) [50], the Design standard for Green Building by Jilin Province (DB22/T 5055-2021) [51], and other standards. Additionally, the surrounding Points of Interest (POIs) were considered and analyzed. Based on this, the project is preliminarily defined with a focus on a cultural and creative business and educational template, supported by flexible structural forms, to expand the building's adaptability and enhance its sustainability, thereby meeting more diverse user needs. The time-by-time parameterization for various types of room use requirements are shown in Figure 11.

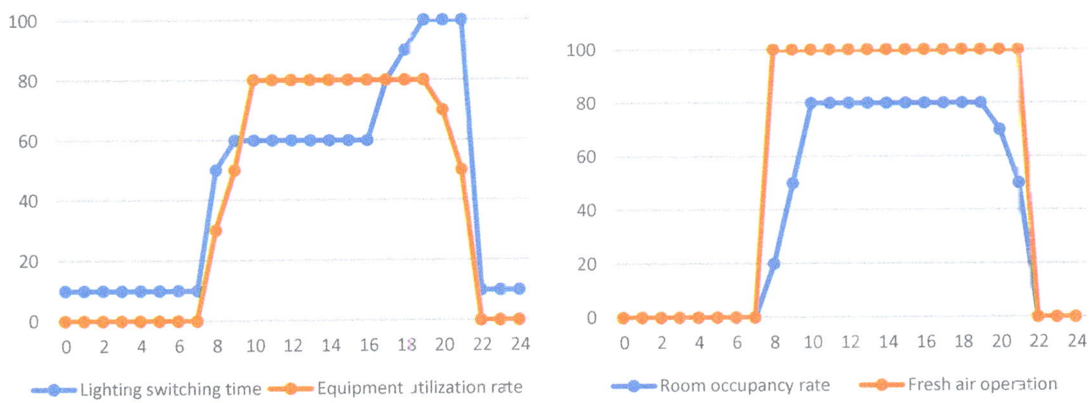

Figure 11. Time-by-time parameterization for room.

- Other Templates: During the previous use of the factory building, the indoor temperature was maintained at the minimum standard to prevent freezing, primarily due to the heat generated by machinery, technical limitations, and cost-saving considerations. Human comfort limits and other requirements were not taken into account. Field investigations revealed that the building was originally equipped with a boiler heating system but lacked an air conditioning system. Therefore, the original model parameters exclude mechanical ventilation and cooling systems. Based on the average temperature during the heating period in Changchun, the heating temperature was set to 24 °C. The heating time was set to the annual heating cycle schedule of Changchun from October 20 of each year to April 6 of the following year.

The DesignBuilder software calculates heating and cooling loads based on the ASHRAE-approved heat balance method in EnergyPlus. Below are some commonly used ASHRAE heating and cooling load calculation formulas, which are used to determine the heating or cooling capacity required for a building [13].

The overall cooling load calculation formula is as follows:

$$Q = U \times A \times CLTD \tag{1}$$

Here, Q is the cooling load (W); U is the conduction heat flow of the building shell (W/m^2 K); A is the surface area (m^2); and CLTD is the color decorated temperature difference (K).

The overall heat load calculation formula is as follows:

$$Q = u \times A \times (\Delta T) \tag{2}$$

Here, Q is the heat load (W); U is the conduction heat flow of the building shell (W/m^2 K); A is the surface area (m^2); and ΔT is the temperature difference (K).

3.4. *Building a Reuse Strategy*

3.4.1. Analysis of Current Situation

This study explores the reuse of industrial heritage under ultra-low energy consumption theories, emphasizing the conservation of the historical value of the building envelope while integrating new eco-friendly materials and low-energy technologies. The analysis is conducted from the following three views:

1. Current status and development needs of the Changchun Tractor Factory surrounding: A data analysis was performed on six categories of POIs (Points of Interest) within a 5 km radius of the Changchun Tractor Factory, using data sourced from Gaode Map [52]. The project samples were visualized and organized into a dataset using Kepler.gl software, as illustrated in Figure 12. It can be observed that education categories in this area are distributed very densely, followed by sports categories and mall categories. Additionally, the data obtained from POIs after data analysis are shown in Table 5. To verify the accuracy of the data, we used the triple standard deviation method, and the guidelines were shown as follows:

Data Di (Deviation Average) that satisfy Di < 2σ are considered as normal; 2. Data Di that satisfy 2σ < Di < 3σ are considered early warning data, requiring expert assessment to determine their adherence to expectations; 3. Data that satisfy Di > 3σ are regarded as abnormal and necessitate re-collection and recalculation (σ is standard deviation) [53].

The results of the calculations are: $\sigma = \pm 0.0309$, $2\sigma = \pm 0.0619$, and $3\sigma = \pm 0.0928$. Based on the data computed in Table 5, it is evident that all of the data fall within two times the standard deviation range, and the validation passes. In order to make the data more readily available for statistical purposes, the results were normalized in this study. From the data in Table 5, it can be concluded that exhibition category and mall category have the highest share in the entire city, followed closely by education category, with only a small difference. Currently, after renovation of the Changchun Tractor Factory's Building #3, it is primarily focused on shopping mall use, with the addition of multiple fitness and children's entertainment facilities. In the update design for Building #2, this study takes into account the city's evolution and development. The reuse of industrial heritage should prioritize long-term applicability and sustainability. By integrating the density and regional proportion data from POIs, this study ultimately focuses on the demand for education category, exhibition category, and some mall spaces.

Table 5. The POI data of Changchun Tractor Factory.

Categories Name	Area	Number	Percentage P_i	Deviation Average (Di)	Normalized Percentage P_i'
sports	Surrounding Citywide	197 3912	5.04%	−0.0317	10.23%
malls	Surrounding Citywide	94 783	12.00%	−0.0380	24.37%
parks	Surrounding Citywide	15 223	6.73%	−0.0148	13.66%
tourism	Surrounding Citywide	26 619	4.20%	−0.0401	8.53%
education	Surrounding Citywide	435 4610	9.44%	0.0123	19.17%
exhibition hall	Surrounding Citywide	9 76	11.84%	0.0363	24.03%

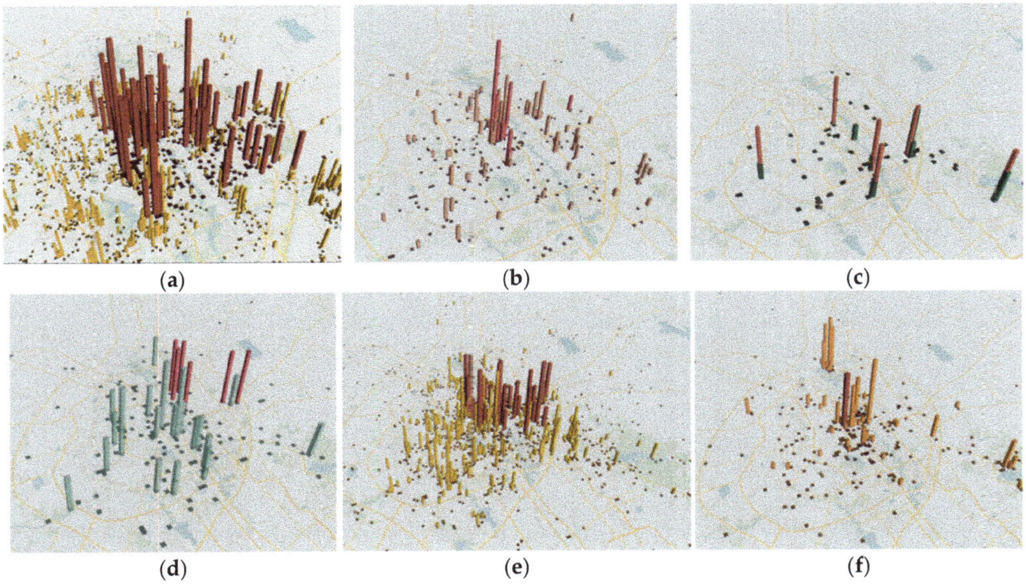

Figure 12. POI data visualization and analysis map for 5 km around Changchun Tractor Factory: (**a**) Kernel density analysis of education category; (**b**) Kernel density analysis map for mall category; (**c**) Kernel density analysis of exhibition hall category; (**d**) Kernel density analysis of park category; (**e**) Kernel density analysis chart of sport category; (**f**) Kernel density analysis of tourism category. (Red in these figures represents data within 5 kilometers of this project).

2. Key concerns for the conservation and inheritance of industrial heritage: The U.S. Department of the Interior, through discussions with experts, identified windows as a consistently significant element of historical buildings. This point has also been recognized in European countries, where windows are considered a unique and integral part of historical architecture, warranting maximum restoration and preservation efforts [54–57]. Consequently, this project focuses on protecting the external envelope and overall form, specifically preserving historically valuable windows, red brick walls, and building façades. Meanwhile, reuse strategies are applied to enhance the adaptability and flexibility of internal spaces [48].
3. Sustainability of reuse projects: This study employs ultra-low energy concepts to reduce building energy consumption and extend the service life of the structure. However, due to the large spatial volume of factory buildings and the principle of preserving external enclosures, calculations using DesignBuilder indicate that achieving ultra-low energy performance while maintaining the existing enclosure is nearly impossible with current technologies. Consequently, this study incorporates the "loggias" retrofitting concept proposed by Reina Oki et al. from Waseda University in Japan [54,58] and combines it with the design of interior walking spaces, resulting in a sustainable, low-energy, and adaptable prefabricated design solution.

In addition, we learned from the design requirements outlined in China's Code for Design of Underground Air Defense Shelters, which mandates a one-time conversion between peacetime and wartime functions [59]. This adaptable form addresses spatial limitations, technological evolution, and changing user needs. In summary, this study innovatively proposes the "Two Enhancements and Two Reductions" strategy for industrial heritage reuse design.

3.4.2. "Two Enhancements": Increasing Functionality of Space and Diversity of Energy-Saving Materials

In terms of spatial functionality, after analyzing the spatial, structural, and regional characteristics of various industrial factory buildings, it is concluded that the reuse of industrial heritage can accommodate multiple service-oriented space functions: Under regular circumstances, it serves as a multi-school joint practice base integrating teaching, experimentation, and practical application, with additional spaces for cultural innovation business spaces; during summer and winter breaks and public holidays, it hosts exhibitions, sales events, or similar projects with commercial objectives to increase economic income, while under special requirements, it provides full dismantling capability to offer larger usable spaces when necessary. Accordingly, a variable structural design is adopted for the spatial layout, aiming to meet the requirements of different spatial volumes through flexibility and thereby enhancing utilization over time. Similarly, for other design parameters mandated by the codes, a full-coverage design is employed.

In terms of materials, new energy-saving products available on the market are selected. Research by Dorota Chwieduk has found that recent energy-saving materials—especially envelope materials—exhibit total heat transfer coefficients approximately 0.15–0.45 [W/(m^2·K)] lower than those of traditional envelope materials, undoubtedly providing a better foundation for low-energy buildings [60]. The physical properties of building materials determine a building's energy consumption [61]. In recent years, building materials have transitioned from traditional, singular options such as reinforced concrete and masonry to composite materials [62], laminated materials [63], and phase change materials [64]. Moreover, these new materials share the common characteristic of effectively reducing building energy consumption and carbon emissions. Therefore, in the structural system constructed under this strategy, the load-bearing frame utilizes prefabricated H-section steel beams and columns, offering excellent tensile and compressive strength while ensuring ease of construction and dismantling. For short-term non-removable walls, dual-cavity ultra-low energy light steel composite panels are employed, with a tested thermal conductivity value of 0.25 (W/m^2 K), providing exceptional insulation, soundproofing, and measures to reduce thermal bridging. Movable walls are constructed using rock wool composite panels paired with perimeter steel tracks, achieving a thermal conductivity value of 0.41 (W/m^2 K). These panels are lightweight, soundproof, and easily repositioned, enabling flexible spatial configurations. The glazing components feature triple-glass, two-cavity windows filled with inert gas, designed for adjustable top-and-bottom tilting, ensuring excellent insulation and soundproofing while maximizing natural light penetration. The selection of new energy-saving materials and their performance characteristics are derived from the "China Passive Ultra-Low Energy Building Annual Development Research Report 2022".

3.4.3. "Two Reductions": Reducing Profile Sizes and Carbon Emissions

The term "modular architecture" originated in the late 19th century and has evolved with continuous technological advancements, offering more diverse forms. Research indicates that if a building features repetitive spaces, the advantages of modularity are significantly enhanced [65]. Therefore, to meet diverse service demands, the strategy in this study minimizes the variety of structural profiles and establishes standardized modular dimensions to shorten construction periods, facilitate transportation, storage, and assembly, and enable easy disassembly and reconfiguration. Based on the factory building's floor height, vertical structural components are standardized with a 4m module to meet the height requirements of various functional spaces and reserve enough space for natural ventilation. Horizontal structural components are modularized with dimensions of 4.5 m × 3 Nm (where N represents a multiple of the extended module), enabling seamless transitions between different functional uses and faster, simpler implementation in expansion projects.

Compared to traditional house construction, the use of steel structures and new environmentally friendly materials can effectively reduce carbon dioxide emissions. The use of numerous prefabricated steel structural components can effectively reduce carbon emissions. According to the China Annual Development Report on Building Energy Efficiency 2022 (Public Building Edition), the carbon emissions during the production of building materials primarily originate from steel and cement production, which together account for over 70% of the total energy consumption in the construction industry [66]. Furthermore, steel structures produce approximately 27% lower carbon emissions during production and transportation compared to concrete structures, with minimal differences observed during construction [67]. Therefore, this project extensively employs steel structures, prefabricated components, and modular systems, significantly reducing resource consumption, environmental pollution, construction-related pollution, and carbon emissions. This strategy primarily adopts steel structures, prefabricated components, and precast construction to significantly reduce carbon dioxide emissions. Additionally, when selecting material suppliers, we prioritize those located closer to the project site to minimize transportation energy consumption and carbon emissions. Furthermore, by implementing a standardized modular design strategy, the reusability of components is enhanced, thereby reducing resource consumption, environmental pollution, construction-related pollution, and carbon emissions. The overall technical pathway for the "Two Enhancements and Two Reductions" reuse design strategy is presented in Table 6.

Table 6. The overall technical pathway for the "Two Enhancements and Two Reductions" reuse design strategy.

Reuse Strategy Deconstruction	Technology Pathway	Technology Support
Increase the functionality of the space	Constructing Variable Composite Structure System	Movable wall panel track system Foldable three-glass, two-cavity inert gas-filled windows Removable dual-cavity ultra-low-energy light steel composite panels
Increase the diversity of energy-saving materials	Reducing the energy consumption of renewed buildings	Reduction of heating (cooling) loss through the high performance of new energy-saving materials themselves
reducing profile sizes	Modular architecture	Reduced profile sizes of modularized buildings 4.5 m × 3 Nm as the standard module size
reducing carbon emissions	New energy-saving materials and standardization	Carbon reduction advantages and reuse of energy-saving materials

3.5. Model Construction and Simulation Settings Under the "Two Enhancements and Two Reductions" Strategy

Given the vast scale and high spatial repetition of the factory, this study selects the A space of Changchun Tractor Factory's Factory No. 2 as a representative case. The total building area of the original Factory No. 2 is approximately 18,300 m^2, while the A space covers about 2376 m^2, with a total length of 72 m and a width of 33 m, accounting for roughly 12% of the total area.

To achieve an enhanced ultra-low-energy design, this study constructs a progressive technical verification model based on the sustainable development principles of industrial heritage and the "Two Enhancements and Two Reductions" design strategy, comprising four cases as presented in Table 7. Case 1, the baseline model, simulates the energy consumption of reuse without altering the original conditions. Case 2 integrates the "Two Enhancements and Two Reductions" structural system while accounting for the advantageous aspects of photovoltaics. Case 3 not only incorporates the "Two Enhancements and Two Reductions" structural system but also simulates zonal control and the addition of photovoltaic solar panels. Building on Case 3, Case 4 addresses the lighting shortcomings caused by the factory's 8 m clear height and internal partitioning by introducing internal structural skylights. It should be noted that while Case 1 serves as the baseline, Cases 2–4 represent progressive scheme models developed by combining the "Two Enhancements and Two Reductions" design strategy with various technical measures. The basic information for Cases 1–4 is provided in Tables 8 and 9, and the model construction is illustrated in Figure 13.

These usage spaces can be categorized into commercial, educational, and exhibition types, which achieve more efficient utilization by staggering their use over time. Consequently, during the construction of the energy model in DesignBuilder, the division of thermal zones becomes more complex. Considering that actual usage scenarios are complex and variable—with the time proportions allocated to different functions over the year being unpredictable and subject to uncontrollable changes—it is impossible to accurately determine the operating time and proportional duration of each function. Therefore, a full-coverage design is adopted in the simulation, while in practice, some equipment can be manually adjusted according to varying usage demands. In the model constructed using DesignBuilder, all spaces, except for corridors, are designated as usage spaces (i.e., spaces that meet the requirements for commercial, educational, and exhibition uses). Their full-coverage design parameters are presented in Table 10, and the corridor parameters for Cases 2–4 are provided in Table 11. These data are sourced from the Chinese building code: Standard for Green Performance Calculation of Civil Buildings (JGJ/T 449-2018) [68]. In addition, there are auxiliary usage spaces such as restrooms and storage rooms; however, since this study simulates the same case using different techniques, the impact of these constant-parameter usage spaces on the simulation results is minimal. To more intuitively analyze the effectiveness of the "Two Enhancements and Two Reductions" design strategy, detailed simulations of auxiliary spaces are not conducted. Furthermore, based on the requirements of China's General Code for Energy Efficiency and Renewable Energy Application in Buildings (GB55015-2021) [69], the nominal efficiency of each piece of equipment is determined, as detailed in Table 12.

Table 7. Explanation of the technical program level for the application of the "Two Enhancements and Two Reductions" strategy to the case.

	Program Level	Characterization	Validation Objectives
Case 1	L0—Baseline model	Partial area of the original plant	Establishing a Baseline for building energy consumption
Case 2	L1—Energy Complement Model	+The structural system of the "Two Enhancements and Two Reductions" strategy + Solar panel	Effectiveness of spatial reconfiguration and renewable energy
Case 3	L2—Thermal control model with spatial reconfiguration zoning	+The structural system of the "Two Enhancements and Two Reductions" strategy + Solar panel + Zone temperature control	Effectiveness of spatial reconfiguration, renewable energy, and zonal temperature control
Case 4	L3—Comfort Optimization Model	+The structural system of the "Two Enhancements and Two Reductions" strategy + Solar panel + Zone temperature control + Dynamic natural lighting	Dynamic optimization of energy consumption changes and comfort

Table 8. Basic information for Cases 1–4.

	Cooling System	Solar Panels	Zone Temperature Control	Internal Space	
				Internal Enclosure Dividing	Interior Structure Roof Lighting
Case 1	Applicable	-	-	-	-
Case 2	Applicable	16 groups	-	-	-
Case 3	Applicable	30 groups	Applicable	Applicable	-
Case 4	Applicable	30 groups	Applicable	Applicable	Applicable

Table 9. Material information for Cases 1–4.

	Roof Material	Envelope (Restoration Only)	Internal Space			
			Removable Dividing Wall	Interior Side Windows	Interior Roof	Internal Roof Windows
Case 1	250 mm prefabricated concrete panel – 80 mm laminated steel panel	490 mm red brick repair	-	-	-	-
Case 2			-	-	-	-
Case 3			Rockwool composite panel	Triple-glazed, double-cavity inert gas-filled windows	Double-cavity ultra-low energy consumption light steel composite panel	-
Case 4						Double-cavity inert gas-filled windows

Table 10. Full-coverage design parameter information for use of space for Cases 1–4.

Spaces	Lighting Power Density for Different Functional Spaces (W/m^2)	Lighting Power Density with Full-Coverage Design (W/m^2)	Equipment Power Density for Different Functional Spaces (W/m^2)	Equipment Power Density with Full-Coverage Design (W/m^2)	Fresh Air Volume for Different Functional Spaces (m^3/h·person)	Fresh Air Volume with Full-Coverage Design (m^3/h·Person)
Usage spaces of Case 1–4	General commercial = 10 Upscale commercial = 16 General classroom = 9 Art classroom = 15 Exhibition = 10	16	General commercial = 13 Upscale commercial = 13 General classroom = 5 Art classroom = 5 Exhibition = 10	13	General commercial = 19 Upscale commercial = 19 General classroom = 24 Art classroom = 20 Exhibition = 20	24
Spaces	**Personnel Density for Different Functional Spaces (m^2/h·Person)**	**Personnel Density with Full-Coverage Design (m^2/h·Person)**	**Room Summer Set Temperature for Different Functional Spaces (°C)**	**Room Summer Set Temperature with Full-Coverage Design (°C)**	**Room Winter Set Temperature for Different Functional Spaces (°C)**	**Room Winter Set Temperature with Full-Coverage Design (°C)**
Usage spaces of Case 1–4	General commercial = 4 Upscale commercial = 4 General classroom = 1.39 Art classroom = 4 Exhibition = 4	4	General commercial = 26 Upscale commercial = 26 General classroom = 26 Art classroom = 26 Exhibition = 20	26	General commercial = 20 Upscale commercial = 20 General classroom = 18 Art classroom = 18 Exhibition = 20	20

Table 11. Parametric information for the hallway for Cases 1–4.

Name	Lighting Power Density (W/m^2)	Equipment Power Density (W/m^2)	Personnel Density (m^2/Person)	Fresh Air	Lighting Power Density (W/m^2)	Equipment Power Density (W/m^2)
Case 2 hallway	7	13	4	19	26	20
Cases 3–4 hallway	7	13	4	19	26	10

Table 12. Nominal efficiency of individual equipment.

Type	Nominal Efficiency
Heating system: coal-fired boiler	η = 85%
LED lighting	η = 95%
Cooling system: inverter air conditioner (GB 21455-2019 standard) [70]	Annual Performance Factor: APF = 4.8
Domestic Hot Water systems (DHW)	CoP = 0.85
Solar Panel	PV Constant Efficiency = 20%

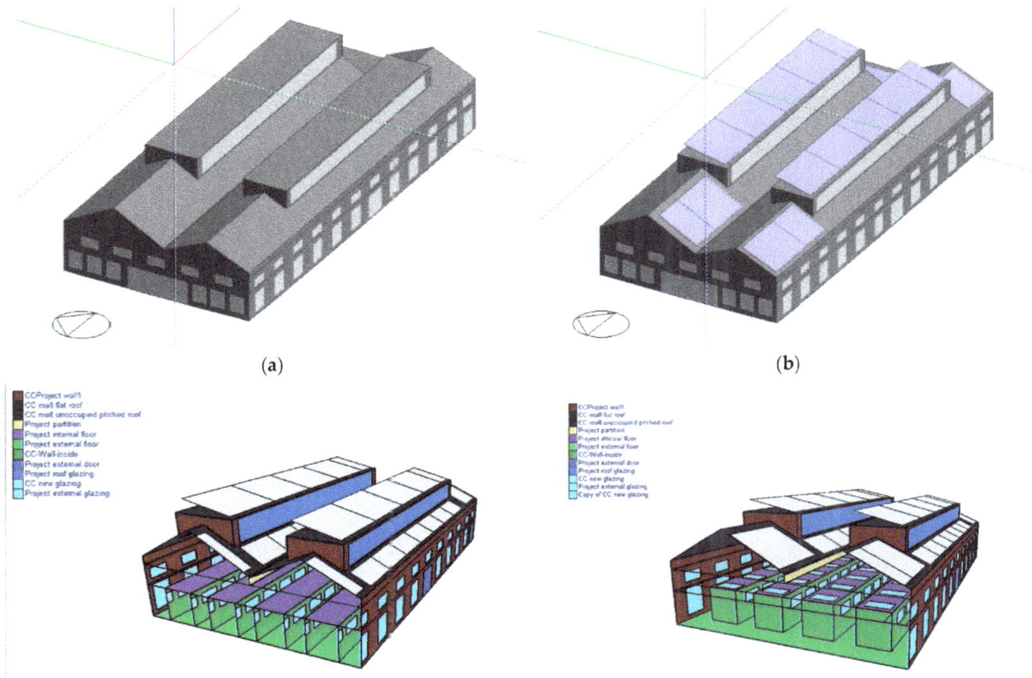

Figure 13. Case 1–4 models built in DesignBuilder software: (**a**) Case 1 model diagram; (**b**) Case 2 model diagram; (**c**) Case 3 model diagram; (**d**) Case 4 model diagram.

4. Results

4.1. Analysis of the Original Model—Current Situation with No Modifications

We input the normal commercial usage requirements, including parameters for lighting, heating, and other factors, into the unrenovated factory to obtain the energy consumption data for the original building (referred to as the Original Model). The calculation results from the Original Model show that during winter, due to the large factory space, the heating system can only maintain a temperature range of indoor temperatures between 5 °C and 18 °C. In summer, due to the lack of proper passive ventilation, mechanical ventilation, and cooling systems, indoor temperatures significantly exceed outdoor levels, peaking at 30 °C. These conditions fail to meet human comfort requirements, making reuse infeasible. The monthly temperature results output by DesignBuilder are shown in Table 13. Additionally, the monthly energy consumption throughout the year, illustrated in Table 14, clearly demonstrates that heating energy consumption far exceeds that of lighting and equipment, with the energy consumption per unit area reaching as high as 19,562.84 kWh/m^2. In this context, the energy consumption attributable to equipment and lighting is almost negligible. Moreover, according to current Chinese building energy consumption standards—which stipulate that the energy consumption per unit area should not exceed 300 kWh/m^2—it is evident that the factory building's energy consumption far exceeds acceptable usage standards.

Table 13. The monthly temperature results output by DesignBuilder.

Date	Relative Humidity (%)	Fanger PMV	Fanger PPD (%)	Air Temperature	Radiant Temperature	Operative Temperature	Outside Dry-Bulb Temperature
Jan	12.11028	−1.83746	53.03131	14.68633	12.92066	13.80349	−15.3989
Feb	12.35588	−1.64383	49.12699	15.3051	14.26906	14.78708	−9.48382
Mar	20.13476	−1.33774	42.95947	16.39322	15.94101	16.16712	−3.31946
Apr	27.02408	−2.21599	68.80729	17.31706	18.46242	17.88974	7.259167
May	35.56598	−1.04275	34.4886	21.04499	22.70672	21.87586	15.05538
Jun	50.80415	2.65×10^{-2}	21.27699	24.16246	25.99894	25.0807	20.20503
Jul	64.01617	0.660134	20.49438	25.74809	27.90072	26.8244	22.94425
Aug	61.84307	0.349352	19.19988	24.87872	26.87755	25.87814	21.30336
Sept	45.78144	−0.798	27.5013	21.51686	23.48149	22.49918	15.21132
Oct	36.45867	−0.67075	22.966	18.53919	19.57963	19.05941	7.616667
Nov	21.47486	−1.39804	43.77863	16.21113	15.43637	15.82375	−5.01031
Dec	16.05555	−1.75822	51.64983	14.96786	13.30296	14.13541	−11.6448

Table 14. The monthly energy consumption throughout the year.

Date/Time	Room Electricity (kWh)	Lighting (kWh)	Heating (kWh)
Jan	134,958.4	114,195.5	1.00×10^8
Feb	121,897.9	103,144.4	6.51×10^7
Mar	134,958.4	114,195.5	4.63×10^7
Apr	130,604.9	110,511.8	4,406,595
May	134,958.4	114,195.5	0
Jun	130,604.9	110,511.8	0
Jul	134,958.4	114,195.5	0
Aug	134,958.4	114,195.5	0
Sept	130,604.9	110,511.8	0
Oct	134,958.4	114,195.5	7,094,450
Nov	130,604.9	110,511.8	5.27×10^7
Dec	134,958.4	114,195.5	8.21×10^7
Annual all	1,589,026	1,344,561	3.58×10^8
Energy consumption per unit area (kWh/m^2)	86.83	73.47	19,562.84

4.2. Analysis of Modeling Results for the "Two Enhancements and Two Reductions" Strategy

As shown on the left side of Figure 14, the indoor temperature for Case 1 is not ideal, particularly in winter, with an average temperature below 20 °C. In comparison, the indoor temperatures for the mall space in Cases 2–4 are relatively similar. However, the maximum temperatures in Cases 3 and 4 are significantly lower than in Case 2, and the annual temperature fluctuation in Case 4 is slightly smaller than in the other two cases. On the right side of Figure 14, which represents the air temperature variation of the month period in the walking space, the overall temperature in Case 2 is relatively high without zoned temperature control, exceeding the normal temperature range for its mall spaces. After implementing zoned temperature control, the minimum temperature requirement was reduced. However, Case 4 still demonstrates better annual temperature fluctuation compared to Case 3.

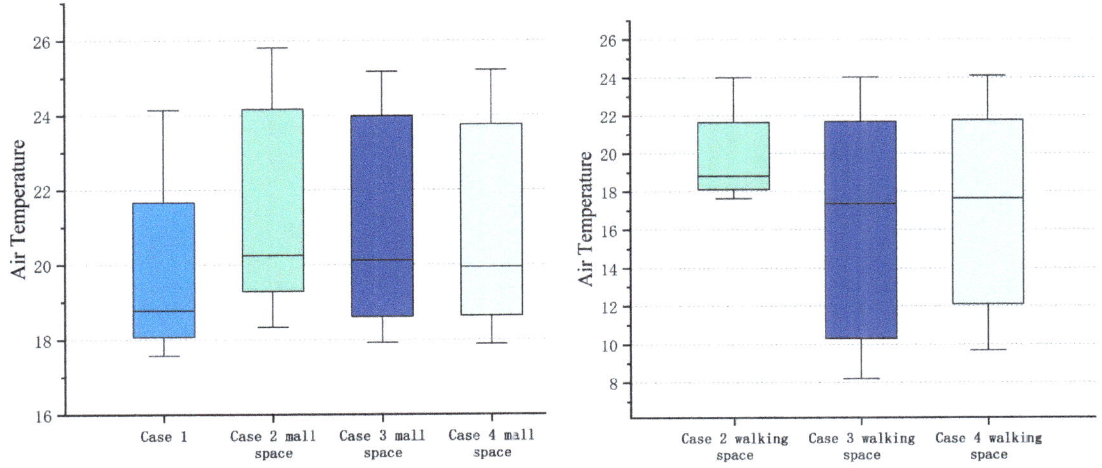

Figure 14. The air temperature variation of the month period in the different spaces of Cases 1–4.

Indoor temperature variation is only a minor aspect of reflecting indoor comfort. According to international standards, the PMV (Predicted Mean Vote) model is adopted as the basis for assessing thermal comfort requirements in building interiors [71]. PMV was developed by Fanger in the 1960s and specifies that the standard indoor comfort range should fall between −0.5 and 0.5 [72]. However, some researchers argue that variables (such as age, health status, body mass index, and other characteristics) also influence human perceptions of comfort [73]. Nevertheless, this metric primarily predicts the average comfort level for the majority of people. In China, the government has similarly used this as a standard, implementing codes as shown in Table 15 [74]. Given the context of this study, the standard is sufficient to hold on the analysis of whether the proposed reuse strategies are effective.

Table 15. Human comfort evaluation standard according to Chinese standard [74].

Level	Evaluation Indicator 1	Evaluation Indicator 2	Explanation
Level I	$-0.5 \leq$ PMV $\leq +0.5$	PPD $\leq$ 10%	Thermal environment in which 90% of the population is satisfied
Level II	$-1 \leq$ PMV < -0.5 or $+0.5 <$ PMV $\leq +1$	10% < PPD $\leq$ 25%	Thermal environment in which 75% of the population is satisfied
Level III	PMV < −1 or PMV < +1	PPD > 25%	Thermal environment in which less than 75% of the population is satisfied

Figure 15 presents the results for the month period, showing consistent abnormal variations in April, primarily related to the heating schedule. These discrepancies can be addressed by adjusting the heating cutoff time based on actual conditions. In the mall space, the PMV values under zoned control are closer to the Class I standard, with Cases 3 and 4 being nearly identical. For the walking space, Case 4 demonstrates better PMV performance. Figure 16 compares the results over the run period. For the mall space, the PMV values for Cases 2–4 all meet the Class I standard, while the PPD values meet the Class II standard, with Case 1 being slightly higher. In contrast, for the walking space, the PMV and PPD results for Case 2, without zoned control, are more favorable compared to the other cases.

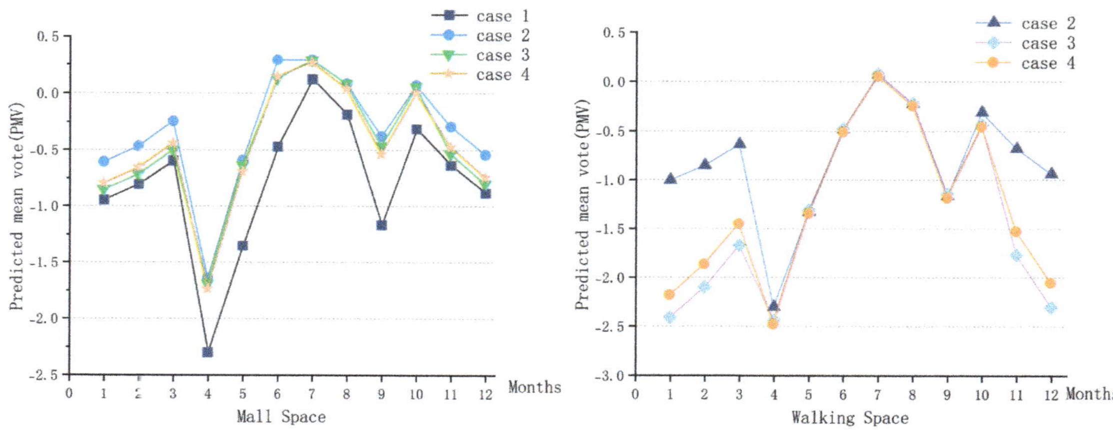

Figure 15. Predicted mean vote (PMV) of the month period in the different spaces of Cases 1–4 from DesignBuilder.

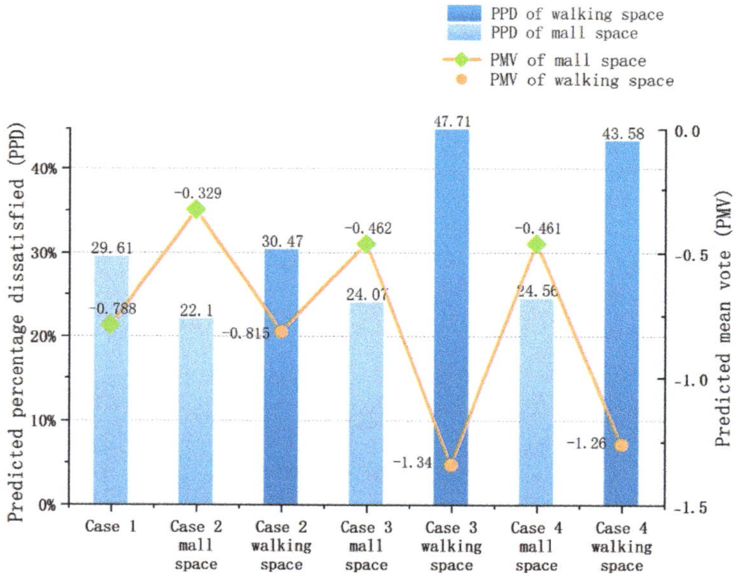

Figure 16. PMV and PPD of the run period in the different spaces of Cases 1–4 from DesignBuilder.

Figures 17 and 18 display the average monthly energy consumption for Cases 1–4 as simulated by DesignBuilder. Additionally, by comparing the heating and cooling energy consumption data for each month, the peak months (January and July) are selected as representative of winter and summer, respectively, for a horizontal comparative analysis of the four cases, as shown in Figures 19 and 20. It is evident that Room Electricity shows minimal variation across cases. For Lighting, after adding interior skylights in Case 4, the monthly energy consumption decreased by approximately 0.169 (kwh/m^2), resulting in an annual reduction of about 2.028 (kwh/m^2). For Heating, after implementing zoned control (Case 3 and Case 4), monthly consumption decreased by approximately 1665 (kwh/m^2), leading to a reduction of about 8325 (kwh/m^2) over the heating season. For Cooling (Electricity), zoned control (Case 3 and Case 4) reduced average summer monthly consumption by approximately 0.35 (kwh/m^2). In Case 4, photovoltaic energy production, combining solar panels and photovoltaic tiles, achieved a maximum monthly energy generation of 25,166 (kWh) and an annual total of approximately 203,946 (kWh). After deducting the energy used for Lighting, the remaining photovoltaic energy per year can supplement Room Electricity consumption or be stored in energy storage facilities for backup use.

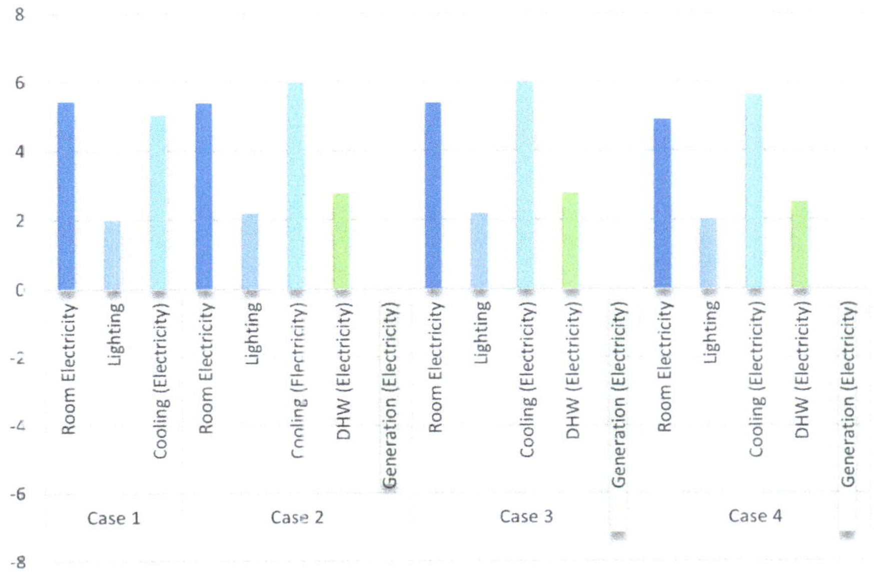

Figure 17. Average energy consumption (excluding heating) per unit area (kWh/m^2) for valid months of DesignBuilder simulation Cases 1–4.

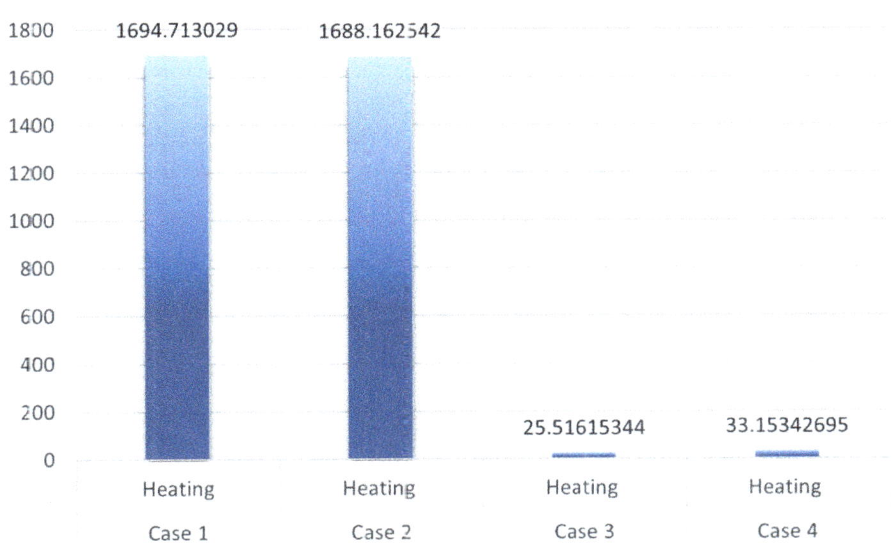

Figure 18. Average heating energy consumption per unit area (kWh/m^2) for valid months of DesignBuilder simulation Cases 1–4.

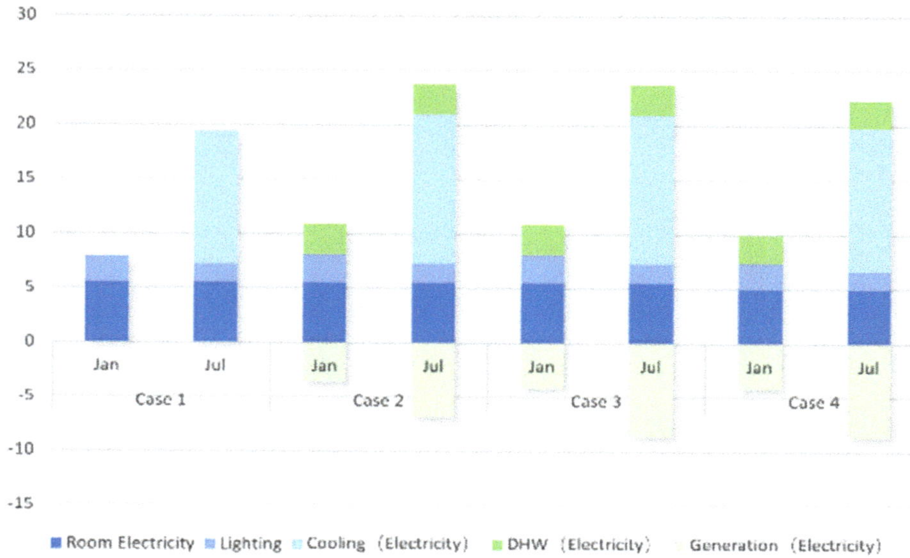

Figure 19. Comparison of energy consumption (excluding heating) per unit area (kWh/m^2) for Cases 1–4 in a typical month by DesignBuilder.

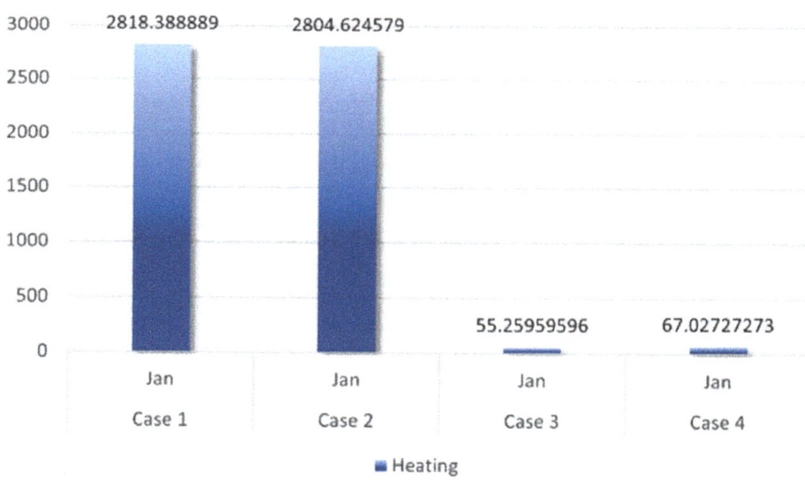

Figure 20. Comparison of heating energy consumption per unit area (kWh/m^2) for Case 1–4 in a typical month by DesignBuilder.

5. Discussion

To comprehensively evaluate the energy-saving of the design strategy proposed in this study, the energy consumption output from DesignBuilder simulations is converted into primary energy, thereby providing a more scientifically accurate reflection of actual energy usage. According to the calculation method for coal-fired heating and the standard coal conversion coefficient specified in the General Rules for Calculation of the Comprehensive Energy Consumption (GB/T 2589-2020) [75] published by China, 1 kg of raw coal, with a lower heating value of 20,934 KJ/kg (5000 kcal/kg), is equivalent to 0.7143 kgce/kg. The formula for calculating primary energy is as follows:

$$PEC = \frac{E_{heat} \times 3600}{Q_{coal} \times \eta} \times f_{coal} \quad (3)$$

Here, *PEC* is Primary Energy Consumption (kgce/tce); E_{heat} is the energy consumption demand (kWh); Q_{coal} is the lower heating value of raw coal (KJ/kg); f_{coal} is the standard coal conversion coefficient for raw coal (kgce/kg); and η is the nominal efficiency of the boiler.

According to the "Limits on Energy Consumption per Unit Product for Coal-Fired Power Generation Units" issued in China, the coal used for power supply must not exceed 300 g of standard coal per kWh. Furthermore, public data from the Chinese Resources Bureau indicate that the maximum energy conversion efficiency of power plants is 38%, with a transmission loss rate of approximately 6%. The primary energy calculation formula for coal-fired power generation is as follows:

$$PEC_{device} = \frac{E_{device}}{\eta_{device}} \times \frac{CEB}{\eta_{plant} \times f_{coal} \times (1 - \eta_{loss})} \quad (4)$$

Here, E_{device} is the energy consumption demand of the equipment (kWh); *CEB* is the nationally prescribed upper limit for coal supply for power; η_{device} is the efficiency of the equipment; η_{plant} is the energy conversion efficiency of the power plant; f_{coal} represents the standard coal conversion coefficient for raw coal; and η_{loss} is the transmission loss rate. The unit for *PEC* is standard coal (kgce/tce).

By consolidating and calculating the data, Table 16 presents the energy consumption and primary energy usage for different project models over one operational cycle. It is evident that the total primary energy consumption for Case 1 and Case 2 reaches as high as 4,353,303 (kgce/tce), with heating energy consumption being the predominant contributor at 4,073,358 (kgce/tce), accounting for 98–99% of the project's total energy consumption and an energy intensity of up to 11,908 (kWh/m^2). Therefore, the primary objective in reducing energy consumption is to decrease the heating load. By employing the "Two Enhancements and Two Reductions" design strategy along with zonal temperature control technology, Case 3 and Case 4 successfully reduced the heating energy consumption to approximately 50–60% of the total energy consumption, lowering the primary energy by about 4 million (kgce/tce) and reducing the energy intensity to around 180–230 (kWh/m^2)—a reduction of roughly 90–95% compared to the first two cases. Additionally, the installation of 30 sets of photovoltaic solar panels is capable of providing approximately 203,946 kWh of actual power generation annually, thereby substituting a portion of the primary energy. Considering the 95% efficiency of the LED lighting system, this substitution can effectively replace all coal-fired electricity. In the Lighting sector, this results in a reduction of about 71,429 (kgce/tce) of primary energy, which accounts for 20% of the total primary energy consumption for this case.

Moreover, to verify whether the updated simulation results using the "Two Enhancements and Two Reductions" design strategy meet the ultra-low energy consumption design standards, we compare the project's simulation outcomes with the regulatory requirements. Unfortunately, there are currently no specific standards in China for ultra-low energy consumption in industrial and architectural heritage. Therefore, we refer to the requirements for energy-saving renovations of existing buildings—which mandate that such renovations should begin with an energy diagnosis, followed by the development of renovation plans, energy-saving indicators, and testing and acceptance procedures—and to the "Ultra-Low Energy Consumption Requirements for Public Buildings in Jilin Province (DB22/T 5128-

2022)" [76] issued by Jilin Province in China. Table 17 presents a comparison between the ultra-low energy requirements for public buildings and the conditions of this project. In order to preserve the historical value of the peripheral structure of the industrial heritage, the "Two Enhancements and Two Reductions" reuse strategy was implemented. While the building envelope's airtightness, comprehensive energy-saving rate, and renewable energy utilization rate meet the requirements for ultra-low-energy public buildings, the energy-saving rate of the building itself is adversely affected by the original enclosure of the industrial heritage and thus does not meet the standards. Given the unique characteristics of industrial heritage buildings, the ultra-low-energy renovation requirements cannot be directly replicated from those of public buildings. As a result, the current simulation results of the renovation strategy essentially satisfy most of the ultra-low-energy public building requirements.

Table 16. The energy consumption and its proportions for Cases 1–4 in the run period.

Category	Indicator	Original Model	Case 1	Case 2	Case 3	Case 4
Room	Energy consumption (kWh)	1,589,026	155,024	153,890	153,890	139,916
	Energy consumption per unit area (kWh/m^2)	86.832	65.49	65.01	65.01	59.11
	Percentage	0.44%	0.5445%	0.5408%	19.0981%	15.5237%
	PEC	1,966,692.91	191,868.86	190,465.34	190,465.34	173,170.11
Lighting	Energy consumption (kWh)	1,344,561	56,926	62,509	62,532	57,713
	Energy consumption per unit area (kWh/m^2)	73.47	24.05	26.41	26.42	24.38
	Percentage	0.37%	0.1999%	0.2196%	7.7604%	6.4032%
	PEC	1,664,125.44	70,455.71	77,365.64	77,394.10	71,429.76
Heating	Energy consumption (kWh)	358,000,000	28,186,467	28,077,519	424,385	551,408
	Energy consumption per unit area (kWh/m^2)	19,562.84	11,908.10	11,862.07	179.29	232.96
	Percentage	99.19%	99.0029%	98.6612%	52.6673%	61.1787%
	PEC	51,736,260.18	4,073,358.63	4,057,614.05	61,329.87	79,686.56
Cooling	Energy consumption (kWh)		71,934	85,452	85,824	80,303
	Energy consumption per unit area (kWh/m^2)		30.39	36.10	36.26	33.93
	Percentage		0.2527%	0.3003%	10.6510%	8.9096%
	PEC		17,620.66	20,931.97	21,023.10	19,670.69
DHW	Energy consumption (kWh)	-	-	79,154	79,154	71,967
	Energy consumption per unit area (kWh/m^2)			33.44	33.44	30.40
	Percentage			0.2781%	9.8232%	7.9847%
	PEC			116,335.44	116,335.4	105,772.5
All	Energy consumption (kWh)	360,933,587	28,470,350	28,458,524	805,784	901,307
	Energy consumption per unit area (kWh/m^2)	19,723.15	12,028	12,023	340.42	380.78
	PEC	55,367,078.53	4,353,303.86	4,346,376.99	350,212.40	343,957.12
Generation (Electricity)		-	-	−163,978	−203,946	−203,946

Table 17. Comparison of energy-saving requirements for ultra-low-energy public buildings and the situation of this project.

Energy Efficiency Requirements for Public Buildings with Low Energy Consumption	Limit Values	This Project
Building envelope's airtightness (N_{50}, h^{-1})	≤1.0	New structures under the "Two Enhancements and Two Reductions" reuse strategy (Case 3,4) Original factory > 1.0
Comprehensive energy-saving rate η	≥50%	92% (Case 3,4)
Energy-saving of the building itself	≥25%	10%
Renewable energy utilization rate	≥10%	14.82% (Case 3,4)

To assess whether the updated simulation using the "Two Enhancements and Two Reductions" design strategy improves occupant comfort, we utilized the indoor comfort data derived from DesignBuilder simulations. These data not only reflect the occupants' experiences but also directly influence their satisfaction, constituting one of the indispensable criteria for sustainable development. According to the results shown in Figures 15 and 16, the comfort levels of the mall space in Cases 2–4 meet Class I of the Chinese standard, achieving a 90% satisfaction rate. However, for the walking space, Case 2 meets the Class II standard, while Case 3 and Case 4, after implementing zoned temperature control, only meet the Class III standard. The walking space refers to usage areas similar to semi-enclosed corridors, semi-enclosed pedestrian streets, and loggias, where the primary goal is to maintain a comfortable level of warmth in winter and a pleasant coolness in summer, providing pedestrians with a more comfortable passage compared to outdoor conditions. Additionally, the walking space serves to seamlessly connect mall areas, creating an integrated environment for both strolling and shopping. Based on the simulation results of the "Two Enhancements and Two Reductions" reuse strategy, the comfort levels in Case 3 and Case 4 are sufficient to meet the usage needs of the walking space.

In comparison to Case 3, Case 4 introduces internal roof skylights, which, while causing some negative impacts on heat loss and other factors, are essential for regions in Northeast China, where natural lighting is in high demand to enhance user comfort by increasing natural daylight, particularly during the winter months. Therefore, even though Case 3 already demonstrates excellent energy reduction results compared to the original factory condition, Case 4 was designed with further improvements in mind, incorporating natural light for user comfort. Ultimately, whether the space is utilized as a cultural and creative commercial street, a shared educational space, or a temporary exhibition area, the design enables flexibility and adaptability to meet the varying needs of users, allowing for seamless transitions between different functions.

In summary, the analysis of the aforementioned results demonstrates that the "Two Enhancements and Two Reductions" industrial heritage reuse strategy significantly facilitates the reuse of industrial heritage within an ultra-low energy consumption framework. The development of an indoor walking space model with zonal control offers the following advantages:

1. Preservation of the external envelope, conserving the historical value of industrial heritage.
2. Utilization of large spaces and high ceilings in factory buildings, transforming the indoor walking space into more than just a transitional commercial area. It also provides an improved walking environment during the five-month winter period in cold regions, with an annual temperature maintained between 8 °C and 26 °C (as shown in Figure 21).

3. Significant reduction in total energy consumption—compared to Case 1, total energy consumption was reduced by approximately 95%, with an annual decrease in primary energy of 4,009,346 (kgce/tce).
4. Ensuring the usability and thermal comfort of the mall space, with an annual temperature maintained between 18 °C and 26 °C (as shown in Figure 21), meeting human comfort standards.

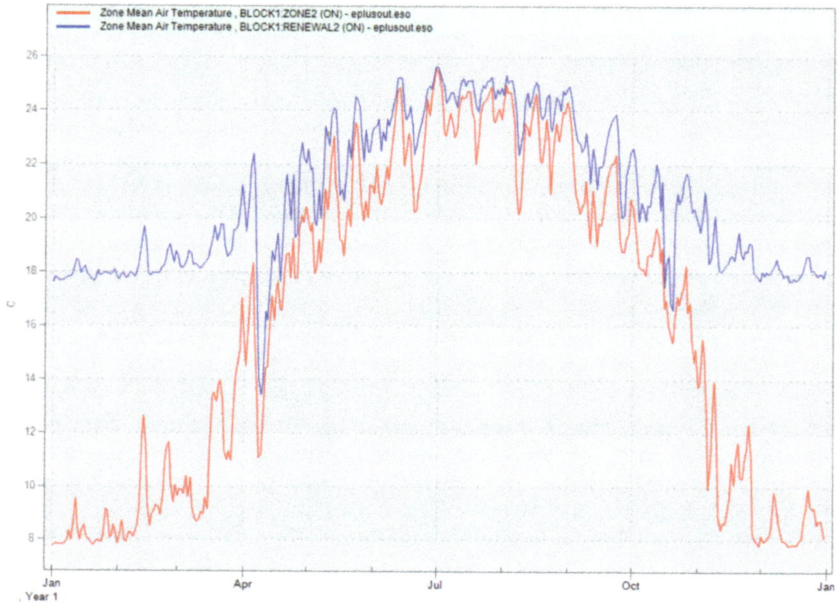

Figure 21. Comparison of internal temperature simulations for Case 3 and Case 4 zones.

Therefore, zoned temperature control not only satisfies human thermal comfort but also optimally regulates temperature requirements for different functional spaces. Additionally, it serves as a crucial step in reducing energy consumption during the adaptive reuse of large industrial heritage factory buildings.

Although it does not possess an absolute advantage in reducing energy consumption compared to newly constructed buildings, industrial heritage—whether viewed from the perspective of historical continuity or urban transformation—demands sustainable development even more critically than new constructions. Based on the simulation results of the representative case in the DesignBuilder software, the "Two Enhancements and Two Reductions" strategy is applicable to most factory-type, high-ceiling, and large-space industrial heritage structures. During the investigation of industrial heritage in Northeast China, it was found that these types of structures were originally used for heavy industry and mechanical production. Since heavy industry accounts for approximately 70% of the "156 Project" in China, these data increase the universal applicability of the "Two Enhancements and Two Reductions" industrial heritage reuse design strategy in China.

However, the application of the "Two Enhancements and Two Reductions" industrial heritage reuse design strategy has certain limitations. This strategy may not be universally applicable to multi-story industrial buildings left by light industry. In addition, DesignBuilder's energy consumption simulations still have certain limitations in terms of accuracy, particularly when it comes to the continuous changes over a building's lifecycle (e.g., building deterioration). It cannot provide precise simulations of energy consumption under these evolving conditions, which restricts the ability to model and calculate sustainable low-energy performance for reused industrial heritages. Additionally, this study lacks comparisons with simulation data from other cities in Northeast China. Different regions may have varying conditions, but overall, the approach can still achieve relatively better energy consumption reduction results.

6. Conclusions

The perfect conservation and development of industrial heritage is not merely about preserving historical buildings but also about reusing them while maintaining protection. Balancing the tension between conservation and reuse is essential for the sustainable development of industrial heritage. At the same time, in the face of the global ecological crisis, industrial heritage, as part of the built environment, also faces the issue of achieving low energy consumption. Therefore, to realize a successful industrial heritage reuse project, it is necessary to balance the relationship between energy consumption, conservation, and reuse. This study integrates quantitative energy modeling with heritage conservation goals, addressing both energy consumption and cultural sustainability in the context of China's industrial heritage. Additionally, this study identifies two main advantages of industrial heritage in Northeast China:

1. The high ceiling and large space provide a high level of flexibility for reuse, accommodating diverse functional needs.
2. Most are located in the city center, offering excellent foot traffic and geographical location advantages.

However, the disadvantages of these industrial heritage buildings are more pronounced:

1. The large spaces result in excessively high demands for heating and cooling.
2. To preserve the historical value of the industrial heritage, the factory's envelope structure is retained, but these structures have poor thermal insulation properties.
3. The large spans and multiple bays of these factories also impose significant limitations on natural daylight.

This study leverages the advantages of this industrial heritage using Factory No. 2 of Changchun Tractor Factory as a case example. It analyzes the architectural form, structural form, spatial configuration, and available space characteristics. By employing DesignBuilder software for simulation and calculation, this study integrates ultra-low energy consumption building theories, industrial heritage conservation requirements, and the "Two Enhancements and Two Reductions" strategy for the reuse of industrial heritage. This study approaches the problem from multiple angles, dimensions, and perspectives, constructing four comparative models to comprehensively address the building's deficiencies. The final Case 4 results in a reduction of about 11,648 (kWh/m^2) over the heating season. With the addition of photovoltaic solar panels, the total lighting energy consumption for the entire year is fully compensated. Although there is an increase in domestic hot water (DHW) and equipment energy consumption, the total primary energy was reduced by about 4 million (kgce/tce), and the total energy consumption still decreased by approximately 95%. This verifies the ultra-low energy consumption advantages of the "Two Enhancements and Two Reductions" reuse design strategy for factory-type industrial heritage.

In conclusion, the "Two Enhancements and Two Reductions" reuse design strategy for industrial heritage proposed in this study breaks through the limitations of traditional reuse perspectives. It provides a new approach to reducing energy consumption in factory-type industrial heritage reuse in cold regions while ensuring the preservation of historical value. This strategy proactively aligns with ultra-low energy consumption building design requirements, filling a gap in the field of industrial heritage reuse. The innovative "Two Enhancements and Two Reductions" industrial heritage reuse design strategy for industrial heritage reuse not only plays a regulatory role in the future development of industrial heritage reuse but also provides valuable reference for architects and scholars. In the future, we will consider applying this strategy to different cases, seeking a balance between commonalities and uniqueness, and continuously exploring the ultra-low energy consumption potential of industrial heritage reuse. This ongoing effort will further refine and perfect the reuse strategy. Of course, this study has certain limitations, as it does not conduct a universal analysis across more projects. However, the findings still hold significant guidance for industrial heritage reuse, particularly in cold regions, and are applicable to factory-type industrial heritage worldwide.

Author Contributions: Conceptualization: S.Y.; investigation: N.L.; methodology: S.Y.; funding acquisition: H.M.; software: S.Y.; validation: H.M., S.X.; writing—original draft: S.Y.; review and editing: H.M., S.X. and F.G. All authors have read and agreed to the published version of the manuscript.

Funding: Research on Paradigm Innovation and Practical Exploration of VR Virtual Reality Technology Intervention in Art and Design Teaching under the Background of Digital Education (Subject No. 2023RY051).

Data Availability Statement: The original contributions presented in the study are included in the article, further inquiries can be directed to the corresponding author.

Acknowledgments: We would like to thank the anonymous reviewers for their constructive and supportive feedback.

Conflicts of Interest: Author Na Li was employed by the Design Institute of Civil Engineering & Architecture of Dalian University of Technology Co., Ltd. The remaining authors declare that the research was conducted in the absence of any commercial or financial relationships that could be construed as a potential conflict of interest.

References

1. IEA. Global ABC Roadmap for Buildings and Construction. 2020–2050. Available online: https://www.iea.org/reports/globalabc-roadmap-for-buildings-and-construction-2020-2050 (accessed on 15 October 2024).
2. CPC Central Committee State Council. Oponions on Accelerating Comprehensive Green Transformation of Economic and Social Development. Available online: https://www.gov.cn/gongbao/2024/issue_11546/202408/content_6970974.html (accessed on 15 October 2024).
3. Alsabry, A.; Szymański, K.; Backiel-Brzozowska, B. Analysis of the Energy, Environmental and Economic Efficiency of Multi-Family Residential Buildings in Poland. *Energies* **2024**, *17*, 2057. [CrossRef]
4. Spandagos, C.; Ng, T.L. Equivalent full-load hours for assessing climate change impact on building cooling and heating energy consumption in large Asian cities. *Appl. Energy* **2017**, *189*, 352–368. [CrossRef]
5. Blagojević, M.R.; Tufegdžić, A. The new technology era requirements and sustainable approach to industrial heritage renewal. *Energ Build.* **2016**, *115*, 148–153. [CrossRef]
6. Douet, J. *Industrial Heritage Re-Tooled: The TICCIH Guide to Industrial Heritage Conservation*; Routledge: London, UK, 2016.
7. Misirlisoy, D.; Günçe, K. Adaptive reuse strategies for heritage buildings: A holistic approach. *Sustain. Cities Soc.* **2016**, *26*, 91–98. [CrossRef]
8. Vardopoulos, I. Critical sustainable development factors in the adaptive reuse of urban industrial buildings. A fuzzy DEMATEL approach. *Sustain. Cities Soc.* **2019**, *50*, 101684. [CrossRef]
9. Scaffidi, F. Average social and territorial innovation impacts of industrial heritage regeneration. *Cities* **2024**, *148*, 104907. [CrossRef]
10. Guidetti, E.; Ferrara, M. Embodied energy in existing buildings as a tool for sustainable intervention on urban heritage. *Sustain. Cities Soc.* **2023**, *88*, 104284. [CrossRef]
11. Wang, S.; Zhang, J.; Wang, F.; Dong, Y. How to achieve a balance between functional improvement and heritage conservation? A case study on the renewal of old Beijing city. *Sustain. Cities Soc.* **2023**, *98*, 104790. [CrossRef]
12. Banti, N. Existing industrial buildings—A review on multidisciplinary research trends and retrofit solutions. *J. Build. Eng.* **2024**, *84*, 108615. [CrossRef]

13. Yue, J.; Lu, J.; Yan, C.; Liu, J.; Pan, Y. Dynamic Simulation Analysis of Influencing Factors of Energy Consumption of Public Building Envelope in Central China. *Sustainability* **2023**, *15*, 14982. [CrossRef]
14. Harish, V.S.K.V.; Kumar, A. A review on modeling and simulation of building energy systems. *Renew. Sustain. Energy Rev.* **2016**, *56*, 1272–1292. [CrossRef]
15. Gan, V.J.L.; Lo, I.M.C.; Ma, J.; Tse, K.T.; Cheng, J.C.P.; Chan, C.M. Simulation optimisation towards energy efficient green buildings: Current status and future trends. *J. Clean. Prod.* **2020**, *254*, 120012. [CrossRef]
16. Tian, Z.C.; Zhang, X.K.; Jin, X.; Zhou, X.; Si, B.H.; Shi, X. Towards adoption of building energy simulation and optimization for passive building design: A survey and a review. *Energy Build.* **2018**, *158*, 1306–1316. [CrossRef]
17. Zhang, Y.; Wang, H.; Gao, W.; Wang, F.; Zhou, N.; Kammen, D.M.; Ying, X. A survey of the status and challenges of green building development in various countries. *Sustainability* **2019**, *11*, 5385. [CrossRef]
18. Cucuzzella, C. *Analyzing Eco-Architecture Beyond Performance*; Editions JFD: Montreal, QC, Canada, 2020.
19. Esbensen, T.V.; Korsgaard, V. Dimensioning of the solar heating system in the zero energy house in Denmark. *Sol. Energy* **1977**, *19*, 195–199. [CrossRef]
20. Ding, Z.; Fan, Z.; Tam, V.W.Y.; Bian, Y.; Li, S.; Illankoon, I.M.C.S.; Moon, S. Green building evaluation system implementation. *Build. Environ.* **2018**, *133*, 32–40. [CrossRef]
21. Bere, J. *An introduction to Passive House*; Riba Publishing: London, UK, 2019.
22. Martiskainen, M.; Kivimaa, P. Creating innovative zero carbon homes in the United Kingdom—Intermediaries and champions in building projects. *Environ. Innov. Soc. Transit* **2018**, *26*, 15–31. [CrossRef]
23. Dang, M.; van den Dobbelsteen, A.; Voskuilen, P. A Parametric Modelling Approach for Energy Retrofitting Heritage Buildings: The Case of Amsterdam City Centre. *Energies* **2024**, *17*, 994. [CrossRef]
24. Xiaoxiang, Q.; Junjia, Y.; Haron, N.A.; Alias, A.H.; Law, T.H.; Abu Bakar, N. Status, Challenges and Future Directions in the Evaluation of Net-Zero Energy Building Retrofits: A Bibliometrics-Based Systematic Review. *Energies* **2024**, *17*, 3826. [CrossRef]
25. Zhang, Y.; Wang, J.; Hu, F.; Wang, Y. Comparison of evaluation standards for green building in China, Britain, United States. *Renew. Sustain. Energy Rev.* **2017**, *68*, 262–271. [CrossRef]
26. *Collection of Building Energy Effciency Standards*; China Architecture & Building Press: Beijing, China, 2017.
27. CECC. National Standard E-Book Library. Available online: https://ebook.chinabuilding.com.cn/zbooklib/index?SiteID=1&IsSiteUrl=Y (accessed on 31 October 2024).
28. Francioni, F.; Lenzerini, F. *The 1972 World Heritage Convention: A Commentary*; Oxford University Press: Oxford, UK, 2023.
29. Unesco. World Heritage Convention. Available online: https://whc.unesco.org/ (accessed on 10 October 2024).
30. Chen, J.; Judd, B.; Hawken, S. Adaptive reuse of industrial heritage for cultural purposes in Beijing, Shanghai and Chongqing. *Struct. Surv.* **2016**, *34*, 331–350. [CrossRef]
31. Bandarin, F.; Van Oers, R. *The Historic Urban Landscape: Managing Heritage in an Urban Century*; John Wiley & Sons: Hoboken, NJ, USA, 2012
32. Mo, C.; Wang, L.; Rao, F. Typology, preservation, and regeneration of the post-1949 industrial heritage in China: A case study of Shanghai. *Land* **2022**, *11*, 1527. [CrossRef]
33. Alfrey, J.; Putnam, T. *The Industrial Heritage: Managing Resources and Uses*; Routledge: London, UK, 2003.
34. Jankovich. B.; Puccetti, P. Building rehabilitation with retrofit of passive and low energy systems. In *Passive and Low Energy Architecture*; Yannas, S., Ed.; Pergamon: Oxford, UK, 1983; pp. 177–181.
35. Attia, S. *State of the Art of Existing Early Design Simulation Tools for Net Zero Energy Buildings: A Comparison of Ten Tools*; Architecture et climat: Ottignies-Louvain-la-Neuve, Belgium, 2011.
36. Sagdiçoglu, M.S.; Yenice, M.S.; Tel, M.Z. The Use of Energy Simulations in Residential Design: A Systematic Literature Review. *Sustainability* **2024**, *16*, 8138. [CrossRef]
37. Wang, H.D.; Zhai, Z. Advances in building simulation and computational techniques: A review between 1987 and 2014. *Energy Build.* **2016**, *128*, 319–335. [CrossRef]
38. Bevilacqua, P.; Bruno, R.; Szyszka, J.; Cirone, D.; Rollo A. Summer and winter performance of an innovative concept of Trombe wall for residential buildings. *Energy* **2022**, *258*, 124798. [CrossRef]
39. Rahman, M.M.; Rasul, M.; Khan, M.M.K. Energy conservation measures in an institutional building in sub-tropical climate in Australia. *Appl. Energy* **2010**, *87*, 2994–3004. [CrossRef]
40. Liu, Y.; Chen, H.; Zhang, L.; Feng, Z. Enhancing building energy efficiency using a random forest model: A hybrid prediction approach. *Energy Rep.* **2021**, *7*, 5003–5012. [CrossRef]
41. Goenaga-Pérez, A.; Álvarez-Sanz, M.; Terés-Zubiaga, J.; Campos-Celador, A. Cost-effectiveness and minimum requirements of nZEB for residential buildings under the new Spanish Technical Building Code. *Energ Build.* **2023**, *287*, 112986. [CrossRef]
42. Hosseini, S.M.A.; Farahzadi, L.; Pons, O. Assessing the sustainability index of different post-disaster temporary housing unit configuration types. *J. Build. Eng.* **2021**, *42*, 102806. [CrossRef]

43. Martinaitis, V.; Zavadskas, E.K.; Motuzienė, V.; Vilutienė, T. Importance of occupancy information when simulating energy demand of energy efficient house: A case study. *Energy Build.* **2015**, *101*, 64–75. [CrossRef]
44. Zhang, A.; Bokel, R.; van den Dobbelsteen, A.; Sun, Y.; Huang, Q.; Zhang, Q. An integrated school and schoolyard design method for summer thermal comfort and energy efficiency in Northern China. *Build. Environ.* **2017**, *124*, 369–387. [CrossRef]
45. Development and Reform Commission. Circular on the Issuance of the National Plan for the Adjustment and Transformation of Old Industrial Bases. Available online: https://www.gov.cn/gongbao/content/2013/content_2441018.htm (accessed on 15 October 2024).
46. *GB 50176-2016*; Code for Thermal Design of Civil Building. China Architecture & Building Press: Beijing, China, 2016.
47. Zhu, G. Resource Conservation Study of Changchun Tractor Works Industrial Heritage Under the Concept of Urban Micro Renewal. Master's Thesis, Jilin Jianzhu University, Changchun, China, 2022.
48. Anderson, K. *Design Energy Simulation for Architects: Guide to 3D Graphics*, 1st ed.; Routledge: London, UK, 2014.
49. Wilcox, S. *Users Manual for TMY3 Data Sets*; NREL/TP-581-43156; National Renewable Energy Laboratory: Golden, CO, USA, 2008.
50. *DB22/T 5160-2024*; Design Standard for Energy Efficiency of Public Buildings (Energy Efficiency 72%). Housing and Urban-Rural Development Department of Jilin Province: Changchun, China, 2024.
51. *DB22/T 5055-2021*; Design standard for Green Building. Housing and Urban-Rural Development Department of Jilin Province: Changchun, China, 2021.
52. Han, R.; Yang, S. A study on industrial heritage renewal strategy based on hybrid Bayesian network. *Sustainability* **2023**, *15*, 10707. [CrossRef]
53. Felipe, C.A.S.; Rocha, S.C.S. Prediction of minimum fluidization velocity of gas–solid fluidized beds by pressure fluctuation measurements—Analysis of the standard deviation methodology. *Powder Technol.* **2007**, *174*, 104–113. [CrossRef]
54. Welch, S.; Obonyo, E.; Memari, A.M. A review of the previous and current challenges of passive house retrofits. *Build. Environ.* **2023**, *245*, 110938. [CrossRef]
55. Hensley, J.E.; Aguilar, A. Preservation Brief 3: Improving Energy Efficiency in Historic Buildings, National Park Service. 2011. Available online: https://www.nps.gov/orgs/1739/index.htm (accessed on 15 October 2024).
56. Kertsmik, K.-A.; Arumägi, E.; Hallik, J.; Kalamees, T. Low carbon emission renovation of historical residential buildings. *Energy Rep.* **2024**, *11*, 3836–3847. [CrossRef]
57. Menconi, M.; Painting, N.; Piroozfar, P. Modelling and simulation of low-risk energy retrofit measures for Traditional Listed Dwellings in the UK. *J. Build. Eng.* **2024**, *82*, 108346. [CrossRef]
58. Oki, R.; Tsuneoka, Y.; Yamaguchi, S.; Sugano, S.; Watanabe, N.; Akimoto, T.; Hayashi, Y.; Wakao, S.; Tanabe, S.-i. Renovating a house to aim for net-zero energy, thermal comfort, energy self-consumption and behavioural adaptation: A method proposed for ENEMANE HOUSE 2017. *Energy Build.* **2019**, *201*, 183–193. [CrossRef]
59. Hao, L.; Chengyuan, Z. Regional Linkage, Strategic Reserve, and the Dual Track of Peacetime and Wartime: Reflections on the Architectural Design of Contemporary Infectious Disease Hospitals Based on Historical Study and the Current Outbreak of Pandemics. *Archit. J.* **2020**, *3*, 41–48. [CrossRef]
60. Chwieduk, D. Towards sustainable-energy buildings. *Appl. Energy* **2003**, *76*, 211–217. [CrossRef]
61. Liu, Z.; Yu, Z.; Yang, T.; Qin, D.; Li, S.; Zhang, G.; Haghighat, F.; Joybari, M.M. A review on macro-encapsulated phase change material for building envelope applications. *Build. Environ.* **2018**, *144*, 281–294. [CrossRef]
62. Brigante, D. *New Composite Materials: Selection, Design, and Application*; Springer: Berlin/Heidelberg, Germany, 2014. [CrossRef]
63. Sah, T.P.; Lacey, A.W.; Hao, H.; Chen, W. Prefabricated concrete sandwich and other lightweight wall panels for sustainable building construction: State-of-the-art review. *J. Build. Eng.* **2024**, *89*, 109391. [CrossRef]
64. Tripathi, B.M.; Shukla, S.K. A comprehensive review of the thermal performance in energy efficient building envelope incorporated with phase change materials. *J. Energy Storage* **2024**, *79*, 110128. [CrossRef]
65. Sharafi, P.; Samali, B.; Ronagh, H.; Ghodrat, M. Automated spatial design of multi-story modular buildings using a unified matrix method. *Autom. Constr.* **2017**, *82*, 31–42. [CrossRef]
66. Building Energy Efficiency Research Center, Tsinghua University. *China Annual Development Report on Building Energy Efficiency 2022 (Public Building Edition)*; China Building Industry Press: Beijing, China, 2022.
67. Zhao, Y.; Sun, Q.; Wei, W.; Xiao, C.; Ren, J.; Wang, Z.; Gao, S.; Dai, Y. Research on the impact of building structure type and scheme on carbon emission. *Build. Struct.* **2023**, *53*, 14–18.
68. *JGJ/T 449-2018*; Standard for Green Performance Calculation of Civil Buildings. China Architecture & Building Press: Beijing, China, 2018.
69. *GB 55015-2021*; General Code for Energy Efficiency and Renewable Energy Application in Buildings. China Architecture & Building Press: Beijing, China, 2021.
70. *GB 21455-2019*; Minimum Allowable Values of the Energy Efficiency and Energy Efficiency Grades for Room Air Conditioners. State Administration for Market Regulation of the People's Republic of China: Beijing, China, 2019.

71. Arakawa Martins, L.; Soebarto, V.; Williamson, T. A systematic review of personal thermal comfort models. *Build. Environ.* **2022**, *207*, 108502. [CrossRef]
72. Fanger, P. *Thermal Comfort: Analysis and Applications in Environmental Engineering*; Danish Technical Press: Copenhagen, Denmark, 1970.
73. Kim, J.; Schiavon, S.; Brager, G. Personal comfort models—A new paradigm in thermal comfort for occupant-centric environmental control. *Build. Environ.* **2018**, *132*, 114–124. [CrossRef]
74. *GB/T 50785-2012*; Evaluation Standard for Indoor Thermal Environment in Civil Buildings. China Architecture & Building Press: Beijing, China, 2012.
75. *GB/T 2589-2020*; General Rules for Calculation of the Comprehensive Energy Consumption. State Administration for Market Regulation of the People's Republic of China: Beijing, China, 2020.
76. *DB22/T 5128-2022*; Ultra-Low Energy Consumption Requirements for Public Buildings in Jilin Province. Housing and Urban-Rural Development Department of Jilin Province: Changchun, China, 2022.

energies

MDPI

Article

Environmental and Economic Benefits of Waste Heat Recovery as a Symbiotic Scenario in Sweden

María Dolores Mainar-Toledo [1,*], Irene González García [1], Hector Leiva [1], Jack Fraser [2], Danna Persson [2,3] and Thomas Parker [2,3]

1 CIRCE—Technology Center for Energy Resources and Consumption, Parque Empresarial Dinamiza, Avda. Ranillas 3D, 1st Floor, 50018 Zaragoza, Spain
2 WA3RM AB, 211 41 Malmö, Sweden
3 Department of Management and Engineering, Division of Environmental Technology and Management, Linköping University (LiU), 581 83 Linköping, Sweden
* Correspondence: mdmainar@fcirce.es

Abstract: This study pioneers a combined Life Cycle Assessment (LCA) and Life Cycle Costing (LCC) evaluation of an industrial symbiosis (IS) case involving waste heat recovery from a pulp and paper mill to a tomato greenhouse in Sweden. Unlike previous studies that assess environmental or economic aspects separately, this research provides a holistic assessment quantifying both environmental burdens and economic feasibility. A comparative analysis framework is applied, evaluating a symbiotic real case of waste heat recovery versus conventional greenhouse tomato production in the Netherlands and subsequent import to Sweden. LCA examines greenhouse gas emissions, eutrophication, toxicity, land use, and resource depletion, while LCC assesses total ownership costs, including Capital and Operational Expenditures. The findings demonstrate that the IS scenario significantly reduces greenhouse gas emissions and operational costs while enhancing energy efficiency. This work fills a gap in IS literature, offering a replicable framework for sustainable greenhouse operations. The results highlight the potential of IS to improve resource efficiency, promote circular economy strategies, and foster sustainable practices in the agri-food sector.

Keywords: greenhouse; industrial symbiosis; life cycle assessment; life cycle costing; waste heat recovery

1. Introduction

The agroindustry is a major consumer of energy, with greenhouse production being particularly energy-intensive due to the need for controlled environments to optimize crop yields, primarily for heating [1]. Greenhouse farming enhances environmental conditions for crop growth by utilizing controlled microclimates to boost production [2,3]. The integration of intelligent decision systems and real-time monitoring can complement energy efficiency strategies in agricultural practices [4] and reduce operational costs in controlled environments [5].

In the context of tomato production, greenhouses are predominantly located in regions with favorable climatic conditions and access to energy resources. However, greenhouse tomato production in Europe exhibits significant regional disparities, particularly between northern and southern regions. Northern European greenhouses rely heavily on high-energy systems for heating and cooling, resulting in greater energy consumption compared to their southern counterparts, which benefit from greater natural sunlight. The primary energy sources used in greenhouse tomato production include natural gas, electricity, and, increasingly, renewable energy sources such as biomass and solar energy [5]. The dependence on fossil fuels for heating during winter months leads to elevated greenhouse gas emissions, as evidenced by studies illustrating that countries like Austria experience higher emissions due to intensive heating systems [6,7], which are crucial for greenhouse operations. This discrepancy in energy use highlights the need for efficient energy management in greenhouses to enhance productivity and reduce operational costs [1]. There is a growing trend toward integrating renewable energy sources to reduce reliance on fossil fuels and lower emissions. In this sense, the Netherlands exemplifies the adoption of advanced greenhouse technologies, emphasizing energy efficiency and renewable sources to reduce its dependence on fossil fuel resources [8,9]. Additionally, research indicates a substantial reduction in climate impact due to the transition to renewable energy, with a 61% decrease in the climate footprint of Finnish greenhouse vegetables from 2004 to 2017 [7]. An extensive review developed by Cuce et al. (2016) on the renewable and sustainable energy-saving strategies of greenhouse systems analyzes different technologies applied in greenhouses, going from photovoltaic modules to phase change materials, passing through heat pumps, and others; it concludes that renewable and sustainable energy technologies are required to reach energy and greenhouse emission savings in the greenhouse sector; the investment can obtain 4 to 8 years of payback, depending on the technology chosen, climate zone, and crop type [10]. Despite these promising developments, achieving consistent energy efficiency across diverse greenhouse systems remains a challenge, particularly for maintaining optimal growing conditions without excessive energy consumption [8].

In recent years, industrial symbiosis (IS) has gained momentum in Europe as a strategy to improve resource efficiency and sustainability. IS involves the exchange of materials, energy, water, or other by-products between industries, fostering mutually beneficial relationships. This approach plays a crucial role in enhancing emission reductions in industrial settings, by promoting resource efficiency and sustainability in industrial processes [11,12].

Life Cycle Assessment (LCA) and Life Cycle Costing (LCC) are well-established methods advocated by the EU Commission to assess the environmental footprint and economic performance of products and organizations throughout their life span to promote sustainable and cost-effective choices [13,14]. Interestingly, while LCA is widely used in the IS literature to assess the environmental performance of existent or future symbiotic developments [11,15,16], economic performance is seldom studied in-depth in combination with LCA.

Through IS, industries can exchange materials, energy, and by-products, thereby contributing to a circular economy. The integration of IS within LCA and LCC frameworks enables a comprehensive evaluation of environmental impacts and economic benefits throughout the product life cycle. By utilizing waste streams as feedstock, industries can significantly reduce their environmental footprint [17]. LCA and LCC incorporate both economic and environmental metrics, allowing for a holistic assessment of IS impacts [18,19]. Examples from biopolymer production illustrate how IS can enhance sustainability through effective waste valorization [19], while LCC analyses show that initial investments in symbiotic relationships can yield significant long-term savings and environmental benefits [20]. However, challenges such as limited collaboration and resource availability can hinder the effectiveness of IS [21]. Overcoming these barriers may require significant changes in organizational practices and stakeholder engagement [22]. IS not only reduces waste but also lowers operational costs and mitigates environmental impacts [18].

Although industrial symbiosis (IS) has been widely studied in various industrial contexts [21], few studies integrate both environmental and economic assessments within a real case of waste heat recovery applied to greenhouse farming [18,23]. Previous works have primarily focused on energy recovery in heavy industries [12,24–26]. Waste heat recycling is highlighted as an important measure to achieve GHG emission reduction targets in industrial processes, for instance, by reutilizing waste heat in the iron and steel industry for municipal heating [24], recovering low-grade waste heat from the flue gases in a coking plant for preheating coal in adjacent processes [25], reusing waste heat generated from an iron and steel plant and ironmaking to supply a thermal power plant with heat [12], or replacing district heating with waste heat from a hypothetical urban vertical farm integrated with a host building and other local companies [27], neglecting the agricultural sector as a potential beneficiary of symbiotic heat recovery solutions [27]. Nevertheless, these studies do not elaborate on estimations of several environmental impact reduction potentials from waste heat recovery measures, nor do these studies provide any economic assessments of the proposed symbiotic developments.

In the case of the greenhouse under study, waste heat from a nearby pulp and paper mill is utilized to heat the greenhouse, exemplifying a successful IS application. The pulp and paper industry is known for its high energy consumption and waste generation, producing waste streams such as waste heat, solid waste, wastewater containing organic matter and chemicals, and air emissions. By redirecting waste heat to the greenhouse, the environmental footprint of both the pulp and paper mill and the greenhouse is significantly reduced [1].

While improving energy efficiency and sustainability in greenhouse production is essential, it is also crucial to consider potential challenges, such as the initial investment costs required for renewable technologies and the necessary infrastructure to support IS.

This study aims to assess the environmental and economic benefits of using waste heat recovery for greenhouse heating in Sweden, optimizing resource use and reducing costs [23]. The study aims to answer the following research questions:

1. Is the use of low-grade waste heat for greenhouse heating in Sweden economically comparable with a BAU greenhouse using fossil fuels or biomass as a heat energy source?
2. What financial and environmental benefits arise from utilizing low-grade waste heat for greenhouse heating in Sweden?
3. How does the real case of waste heat reuse in Frövi for a tomato greenhouse compare to various business-as-usual (BAU) scenarios where greenhouses and systems operate under standard conditions, in terms of environmental and economic performance?

The innovative aspect of this research lies in comparing the environmental and economic analysis of the symbiotic scenario of local tomato production in Sweden using waste heat, generated in a pulp and paper mill from the CHP process, with the BAU scenario of importing tomatoes produced in a regular greenhouse located in the Netherlands. The novelty of this study lies in applying LCA and LCC methodologies in tandem, to assess the environmental and economic effects of waste heat recycling, exploring a real IS case, and providing insights into estimations on the environmental burden reduction and environmental benefits, which have been identified as a gap in the current IS literature addressing waste heat recovery. The findings are expected to contribute to the scientific community by providing a model for enabling competitive greenhouse operations in Sweden, potentially reducing dependence on imported tomatoes from the Netherlands. This, in turn, would enhance local food production and sustainability by leveraging low-grade waste heat recovery while lowering greenhouse gas emissions associated with food transportation, among other environmental benefits.

2. Methodologies

This section outlines the process followed to obtain the LCA and LCC results for the IS demonstration case in Sweden, specifically located in Frövi. It also describes the scenarios against which it is compared.

2.1. Demo Case Demonstrator Description

The Frövi demonstration case, part of the CORALIS project initiative, funded by the European Commission, features a 10-hectare tomato greenhouse in Frövi, Sweden. This greenhouse utilizes industrial waste heat from a nearby pulp and paper facility. Construction began in 2022 and was completed by mid-2024, with planning having started much earlier. The greenhouse is notable for its innovative energy and resource optimization strategies, including waste heat reuse, rainwater and snowmelt collection, thermal and light screens, water filtration and reuse, and efficient LED lighting. It is the first large-scale vegetable greenhouse in Sweden to repurpose low-grade industrial waste heat [28].

The data used in this study were collected from key project partners; they included business and production information from the tomato producer, data from materials and quantities, and costs from the project developer, the grower, the greenhouse provider, commercial databases, scientific and technical literature, and detailed engineering drawings and specifications. The data were reviewed by WA3RM and CIRCE to ensure high-quality, site-specific information following a data quality approach connecting the SimaPro, the LCA, and the LCC [29,30].

Key partners included the greenhouse and materials supplier, the tomato producer, the special purpose vehicle entity, the project developer (WA3RM), the on-site management team, and the construction contractor.

LCC assessment followed the same system as for the LCA assessment processes, considering Capital Expenditure (CapEx) and Operational Expenditure (OpEx).

- CapEx data covered building materials, heat and water systems, lighting, land preparation, construction, transport, and fuel.
- OpEx data included heat, electricity, water, fertilizers, substrate, packaging, and waste consumption.

2.2. Life Cycle Inventory (LCI)

The LCI is essential for evaluating the environmental impact and sustainability of the system, providing insights to enhance resource efficiency. This analysis encompasses all inputs, outputs, energy, materials, and expenses within the system boundaries, as defined in the processes.

The LCI involves reviewing previous inventories, collecting primary and secondary data, normalizing inputs and outputs, and modeling the inventory using Ecoinvent v3.8. The aggregated LCI data used for LCA assessment are provided in Appendix A.

2.3. Life Cycle Assessment

LCA is a standardized, comprehensive tool for systematically evaluating the environmental impacts of different sectors, processes, and products throughout their entire life cycle, from resource extraction to disposal. It assesses value chain flows to identify strategic routes for environmental improvement. The LCA framework follows the standardized methodology outlined in ISO 14040:2006 and 14044:2006 [31,32], ensuring systematic assessment of environmental impacts, and consistency and quality [33]. LCA studies consist of four stages: goal and scope definition, inventory analysis, impact assessment, and interpretation (Figure 1) [34]. The methodology allows for the identification of hotspots in product life cycles, facilitating targeted improvements [35].

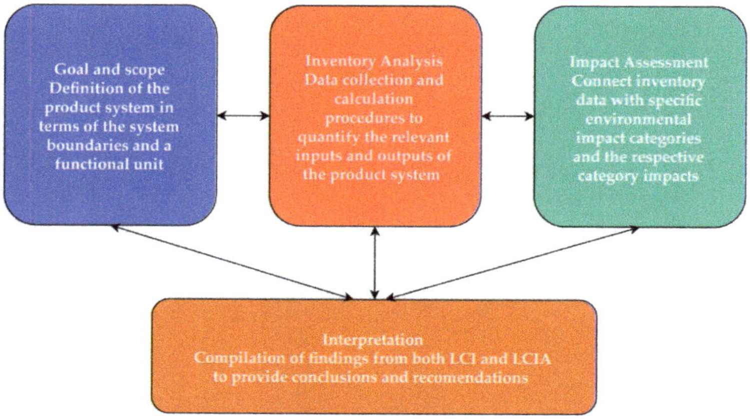

Figure 1. Four phases of a life cycle assessment (own preparation).

2.3.1. Goal and Scope Definition

This stage establishes the scope, system boundary, and functional unit (FU) of the study. The main objective is to assess the environmental performance of CORALIS scenarios by comparing the baseline scenario with the IS scenarios.

A cradle-to-gate approach is applied, evaluating environmental impacts from raw material extraction to the production facility's gate, excluding use and disposal phases. The FU is calculated based on the expected lifetime of the greenhouses (30 years), in line with previous studies [36,37] and the total tomato production during that period. Consequently, the assessment considers 1 kg of tomatoes produced as the FU.

2.3.2. System Description and Boundaries

The study evaluates multiple scenarios, summarized in Table 1.

Table 1. Scenarios considered for the study of Frövi's demo case demonstrator.

Scenario	Assumptions	Data Included and Data Sources
Scenario 1.1—Symbiotic scenario in Frövi's greenhouse	• All electricity comes from local hydropower. • All heat is supplied by waste heat from the pulp and paper mill. • CO_2 (liquified) is supplied by an external provider. • Fertilizers and chemicals used follow expected values provided by the tomato producer.	• All production data (water, heat, electricity, waste, substrate, plants, tomatoes, plant protection products, fertilizers, packaging) according to their expertise, based on assumptions. • All building data from different suppliers (steel, glass, aluminum, machinery, lighting, land preparation, heat recovery system) were collected in 2023 and 2024. • Land use change (from forest to industrial land).
Scenario 1.2—Frövi's greenhouse, non-symbiotic solution, heat from biomass	• All electricity is average Swedish grid. • All heat is from biomass heat. • All water from municipal supply, lake, and rain. • CO_2 from external supplier. • Fertilizers and chemicals use is the average use, information taken from the tomato producer.	• All production data from tomato production (water, heat, electricity, waste, substrate, plants, tomatoes, plant protection products, fertilizers, packaging), according to their expertise, based on assumptions. • All building data from different suppliers (steel, glass, aluminum, machinery, lighting, land preparation), collected in 2023 and 2024. • Land use change (from forest to industrial land).

Table 1. *Cont.*

Scenario	Assumptions	Data Included and Data Sources
Scenario 1.3—Frövi's greenhouse, non-symbiotic solution, heat from natural gas	• All electricity is average Swedish grid. • All heat is from natural gas heat. • All water from municipal supply, lake, and rain. • CO_2 from external supplier. • Fertilizer and chemical use is the average use; information taken from the tomato producer.	• All production data from tomato producers (water, heat, electricity, waste, substrate, plants, tomatoes, plant protection products, fertilizers, packaging). • All building data from different suppliers (steel, glass, aluminum, machinery, lighting, land preparation), according to their expertise, based on assumptions. • Land use change (from forest to industrial land).
Scenario 2—Generic greenhouse in Sweden	• All electricity is average Swedish grid. • All heat is from biomass combustion. • All water consumption is from generic irrigation as in Scenario 3. • CO_2 from external supplier. • Fertilizer and chemical use is the average use; information taken from the tomato producer, according to their expertise, based on assumptions.	• Generic greenhouse in the Netherlands, included in the Ecoinvent dataset (greenhouse, glass walls, and roof {GLO market for greenhouse, glass walls, and roof, Cut-off, U}).
Scenario 3—Generic greenhouse in the Netherlands	• Electricity is from the average Netherlands grid system. • All heat is from natural gas. • All water consumption is from generic irrigation (drip (6%), sprinkler (24%), surface (70%)). • It includes tomato transport from the Netherlands to Sweden (average of 1580 km). • CO_2 is from emissions of the CHP system onsite. • Fertilizer and chemical use is the average use; information taken from the tomato producer, according to their expertise, based on assumptions.	• Generic greenhouse in the Netherlands, included in the Ecoinvent dataset (greenhouse, glass walls, and roof {GLO market for greenhouse, glass walls, and roof, Cut-off, U}).

Scenario 1—The real Frövi greenhouse in different conditions:

- **Scenario 1.1**: The symbiotic scenario demonstrated in the Frövi demo case, using 100% hydropower electricity from Linde Energi and heat recovered from a pulp and paper mill.
- **Scenario 1.2**: The Frövi greenhouse operates with biomass heating and the Swedish electricity mix but without the symbiotic heat connection.
- **Scenario 1.3**: Similar to Scenario 1.2, but with natural gas as the primary heat source.

Scenario 2—A generic Swedish greenhouse:

- A BAU greenhouse in Sweden, using the Swedish electricity mix, biomass for heating, and generic irrigation methods.

Scenario 3—A generic Dutch greenhouse:

- A BAU greenhouse in the Netherlands, producing tomatoes for export to Sweden.
- Uses the Netherlands electricity mix and natural gas for heating, along with generic irrigation methods.
- Tomatoes are transported to Sweden by road and sea, covering an average distance of 1580 km.

The study follows Product Category Rule (PCR) guidelines for arable vegetables [38] to define the scope, boundaries, and data requirements for the LCI and LCA of the different scenarios. Both LCA and LCC assessments consider the same system boundaries, including material, water, energy, and waste flows, as detailed in Table 2.

Table 2. Boundaries considered for the scenarios.

Type of Process	Processes Included	Temporality
Tomato production	Upstream processes ○ Chemical and fertilizer production and transport ○ Tomato starter plant production and transport ○ Substrate production and transport ○ Plant support equipment production and transport ○ Electricity production and transport ○ Heat source production and transport (waste heat, biomass heat, or natural gas heat, depending on the scenario analyzed) ○ Water pumping ○ CO_2 liquefying and transport ○ Packaging production and transport ○ Waste generation and treatment	Yearly
	Downstream processes ○ Emissions due to fertilizing (fertilizer application) ○ Emissions due to the combustion of fuels (mainly diesel)	
Greenhouse production and operation	Upstream processes ○ Structure materials production and transport for the greenhouse ○ Auxiliary building materials production and transport ○ Grow/cultivation system materials production and transport ○ Shading system materials production and transport ○ Heating system materials production and transport ○ Heat recovery system materials (for the IS scenario) (including heat exchangers, heat pumps, heat pipes production and transport) ○ Lighting system materials production and transport ○ Irrigation system materials production and transport ○ Land transformation and preparation ○ Land use change ○ Waste generation, transport, and treatment	30 years
	Core processes ○ Maintenance: pump replacement ○ Maintenance: heat exchanger ○ Maintenance: boilers ○ Maintenance: LEDs	15 years (reference from manufacturers)
	Downstream processes ○ Emissions due to the combustion of fuels (diesel mainly)	Yearly

2.3.3. Cut-Off Criteria

The cut-off criteria applied in this study focus on the relative contribution of mass and energy to the FU, as well as generated waste (air emissions, water waste, and solid waste). The following criteria are applied: (i) Materials: flows contributing less than 1% of the cumulative mass were excluded, provided their environmental relevance was negligible. However, the sum of neglected material flows did not exceed 5% of the mass or environmental relevance. (ii) Energy: flows contributing less than 1% of cumulative energy were excluded from the analysis.

2.3.4. Impact Assessment

Impact assessment evaluates the environmental impact of value chains using SimaPro Analyst 9.3.0.3., complemented by in-house databases and Ecoinvent 3.8. The ReCiPe 2016 v1.1 midpoint method [39] was used for data analysis.

Midpoint analysis identifies the most impactful categories, while characterization determines the contribution of different processes. Table 3 lists the selected environmental impact indicators.

Table 3. Environmental impact indicators were selected for this study and respective units.

ReCiPe 2016 Midpoint	Abbr.	Unit
Global warming	GWP	kg $CO_{2\,eq}$
Stratospheric ozone depletion	ODP	kg $CFC11_{eq}$
Ionizing radiation	IRP	kBq Co-60_{eq}
Ozone formation, human health	HOFP	kg $NO_{x\,eq}$
Fine particulate matter formation	PMFP	kg $PM2.5_{eq}$
Ozone formation, terrestrial ecosystems	EOFP	kg $NO_{x\,eq}$
Terrestrial acidification	TAP	kg $SO_{2\,eq}$
Freshwater eutrophication	FEP	kg P_{eq}
Marine eutrophication	MEP	kg N_{eq}
Terrestrial ecotoxicity	TETP	kg 1,4-DCB
Freshwater ecotoxicity	FETP	kg 1,4-DCB
Marine ecotoxicity	METP	kg 1,4-DCB
Human carcinogenic toxicity	HTP_c	kg 1,4-DCB
Human non-carcinogenic toxicity	HTP_{nc}	kg 1,4-DCB
Land use	LOP	m2a $crop_{eq}$
Mineral resource scarcity	SOP	kg $Cu_{\,eq}$
Fossil resource scarcity	FFP	kg $oil_{\,eq}$
Water consumption	WCP	m^3

2.3.5. Interpretation

This stage analyzes impact assessment results and suggests improvements. It compares life cycle stages, and between baseline and the IS scenario.

2.4. Life Cycle Costing

LCC [14] evaluates the costs incurred throughout a product's life cycle, following the same principles as LCA. It includes both direct costs (CapEx) and indirect costs (OpEx).

Due to limitations in gathering baseline cost information, the LCC analysis focuses on direct and indirect costs rather than predicting long-term economic trends. When specific cost data are unavailable, estimated cost factors based on Peters (2011) [40] are used.

The lifespan of the Frövi greenhouse is assumed to be 30 years, with CapEx calculations incorporating data from 2022, 2023, and 2024, while OpEx inputs are exclusively from 2024. Some inputs are estimates rather than actual data, particularly for future years.

The total LCC is calculated as follows:

$$LCC = C_{CapEx} + \sum C_{OpEx,\,t}, \quad (1)$$

where

- C_{CapEx} represents initial capital costs, including land acquisition, civil works, engineering, greenhouse construction materials, investment in irrigation systems, electricity and heat generation/distribution, and waste operations.
- $C_{OpEx,\,t}$ represents annual operational costs, including insurance, harvesting, personnel costs, water and energy use, fertilizers, CO_2 supply, and greenhouse maintenance.
- t is the number of years considered in the analysis (30 years).

  Certain costs are excluded from the calculation, such as the cost of the cultivation system, as cultivation is performed manually. The estimated annual crop sale value is based on average figures provided by the tomato producer.

  Additionally, the Levelized production cost (€/kg of tomatoes) is estimated as follows:

$$C_{prod} = \frac{C_{total}}{Q_{tomato}}, \quad (2)$$

where

- C_{total} = LCC represents the total life cycle cost.
- Q_{tomato} is the total number of tomatoes produced over 30 years.

This analysis focuses on comparing different heating scenarios rather than making long-term financial predictions.

3. Results

3.1. LCA Evaluation

The LCA evaluation assesses the environmental impacts associated with all stages of the greenhouse system, from resource extraction to operation and waste management.

Considering the 30-year life expectancy of the greenhouse (in alignment with previous studies) and using a BAU greenhouse from the Ecoinvent database as a reference (minimum lifespan of 25 years), the most and least impactful scenarios per kg of tomatoes analyzed were identified (see Figure 2):

- Scenario 3—The generic Dutch greenhouse exhibits the highest environmental impact in 14 out of 18 impact categories.
- Scenario 1.1—The symbiotic scenario being demonstrated in the Frövi greenhouse has the lowest environmental impact in 17 out of 18 impact categories. However, for water consumption, Scenario 1.1 ranks second highest due to the environmental burden associated with Linde Energi hydropower plant construction and materials production (notably gravel and steel).
- The remaining scenarios fall between these two extremes.

A comparative analysis was conducted between Scenario 1.1 and the remaining scenarios (Table 4). The key findings are as follows:

- Significant environmental burden reductions were observed across multiple impact categories.
- Total avoided GHG emissions in the symbiotic scenario (Scenario 1.1) compared to the BAU scenario amount to 3.382 kg CO_2 equivalent per kg of tomatoes produced, translating to a yearly reduction of 29,270 tons of CO_2 equivalent. While GHG reductions are lower when compared with other scenarios, Scenario 1.1 remains the least impactful in terms of Global Warming Potential (GWP).
- Water consumption per kg of tomatoes is 9% higher (0.003 m^3/kg) in Scenario 1.1 compared to the reference scenario.
- Fossil resource consumption per kg of tomatoes is reduced by 1.07 kg of oil equivalent (87%) compared to the reference scenario.

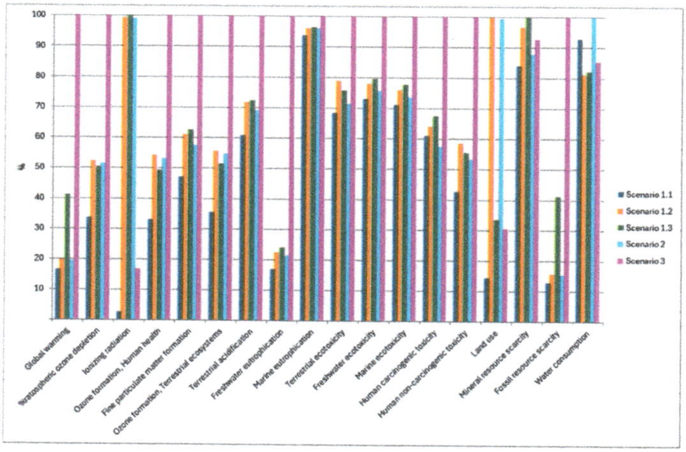

Figure 2. Life cycle assessment for the different scenarios.

Table 4. Life cycle assessment results for comparison of the different scenarios, total impact per FU.

Impact Category	Unit	Scenario 1.1	Scenario 1.2	Scenario 1.3	Scenario 2	Scenario 3
GWP	kg $CO_{2\ eq}$	0.6749	0.8085	1.6699	0.7948	4.0565
ODP	kg CFC11 $_{eq}$	0.0000	0.0000	0.0000	0.0000	0.0000
IRP	kBq Co-60 $_{eq}$	0.0396	1.4506	1.4600	1.4482	0.2447
HOFP	kg $NO_{x\ eq}$	0.0018	0.0030	0.0028	0.0030	0.0056
PMFP	kg PM2.5 $_{eq}$	0.0012	0.0016	0.0016	0.0015	0.0026
EOFP	kg $NO_{x\ eq}$	0.0021	0.0033	0.0031	0.0033	0.0050
TAP	kg $SO_{2\ eq}$	0.0060	0.0071	0.0072	0.0068	0.0099
FEP	kg P_{eq}	0.0003	0.0004	0.0004	0.0003	0.0016
MEP	kg N_{eq}	0.0012	0.0012	0.0012	0.0012	0.0013
TETP	kg 1,4-DCB	10.5769	12.2007	11.7099	11.0542	15.4524
FETP	kg 1,4-DCB	0.1757	0.1869	0.1911	0.1820	0.2400
METP	kg 1,4-DCB	0.2197	0.2350	0.2400	0.2278	0.3088
HTP_c	kg 1,4-DCB	0.1184	0.1241	0.1309	0.1115	0.1938
HTP_{nc}	kg 1,4-DCB	1.3790	1.8810	1.7844	1.7166	3.2148
LOP	m2a crop $_{eq}$	0.0454	0.3172	0.1066	0.3156	0.0967
SOP	kg Cu_{eq}	0.0079	0.0091	0.0094	0.0082	0.0087
FFP	kg oil $_{eq}$	0.1583	0.1932	0.5066	0.1900	1.2266
WCP	m^3	0.0402	0.0351	0.0355	0.0433	0.0369

An environmental impact analysis was conducted for each scenario. The findings for Scenario 1.1 (symbiotic waste heat scenario in Frövi's greenhouse) are presented in Figure 3 and Table 5:

- Greenhouse construction and building processes contribute significantly to environmental impact and warrant further examination, being the highest contributor in human non-carcinogenic and mineral resource scarcity impact categories. This process will be analyzed in more detail.
- Fertilizer and chemical production are among the most impactful processes across all impact categories, particularly in the stratospheric ozone depletion, terrestrial acidification, marine eutrophication, and land use categories due to the application of macronutrients. Emissions from the application of fertilizers appear as an important contributor in the terrestrial acidification and marine eutrophication. These processes are analyzed further in the next section.
- Carbon dioxide represents the first impacting process in the global warming and ionizing radiation impact categories due to its nature, which mainly impacts in those impact categories.
- Substrate use in the hydroponic base is the first impacting process in the ozone formation (human and terrestrial), terrestrial ecotoxicity, and fossil resource scarcity impacting categories.
- Electricity consumption from Linde Energi purchased for tomato cultivation is also identified as a key contributor.

When analyzing the total IS greenhouse's structure and materials—excluding tomato production processes, inputs, and outputs, and land preparation and construction and land use and occupation—the greatest environmental impact comes from the materials used in the greenhouse's construction. The most significant contributors to the global warming impact category (measured in kg $CO_{2\ eq}$) are electrical components, glass panels and aluminum structures. This is illustrated in Figures 4 and 5.

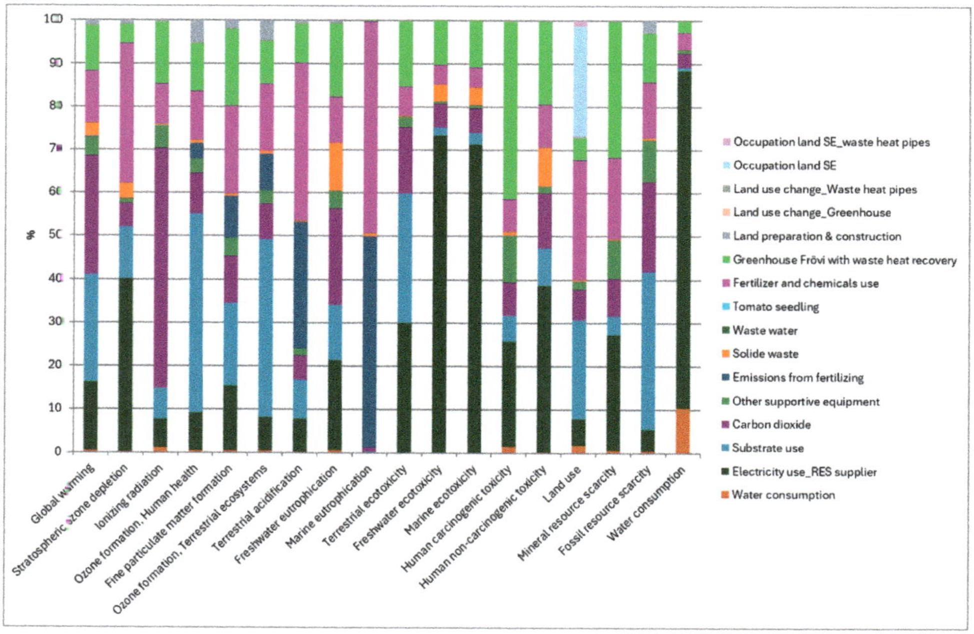

Figure 3. Life cycle assessment for Scenario 1.1, showing the contribution of different processes and activities to the different impact assessment categories.

Table 5. Life cycle assessment results (main impacting processes) for Scenario 1.1 per FU.

Impact Category	Unit	Electricity Use RES Supplier	Substrate Use	Carbon Dioxide	Emissions from Fertilizing	Fertilizer and Chemical Use	Greenhouse Frövi with Waste Heat Recovery
GWP	kg $CO_{2\ eq}$	0.1065	0.1672	0.1856	0.0013	0.0826	0.0699
ODP	kg CFC11 $_{eq}$	0.0000	0.0000	0.0000	0.0000	0.0000	0.0000
IRP	kBq Co-60 $_{eq}$	0.0025	0.0028	0.0219	0.0000	0.0038	0.0056
HOFP	kg $NO_{x\ eq}$	0.0002	0.0008	0.0002	0.0001	0.0002	0.0002
PMFP	kg PM2.5 $_{eq}$	0.0002	0.0002	0.0001	0.0001	0.0003	0.0002
EOFP	kg $NO_{x\ eq}$	0.0002	0.0009	0.0002	0.0002	0.0003	0.0002
TAP	kg $SO_{2\ eq}$	0.0005	0.0005	0.0003	0.0018	0.0022	0.0005
FEP	kg P_{eq}	0.0001	0.0000	0.0001	0.0000	0.0000	0.0000
MEP	kg N_{eq}	0.0000	0.0000	0.0000	0.0006	0.0006	0.0000
TETP	kg 1,4-DCB	3.1535	3.1671	1.6177	0.0000	0.7357	1.5972
FETP	kg 1,4-DCB	0.1286	0.0033	0.0096	0.0000	0.0080	0.0180
METP	kg 1,4-DCB	0.1564	0.0060	0.0124	0.0000	0.0105	0.0235
HTP_c	kg 1,4-DCB	0.0288	0.0070	0.0091	0.0000	0.0088	0.0488
HTP_{nc}	kg 1,4-DCB	0.5287	0.1200	0.1748	0.0000	0.1371	0.2671
LOP	m^2a crop $_{eq}$	0.0028	0.0105	0.0033	0.0000	0.0128	0.0023
SOP	kg Cu_{eq}	0.0021	0.0003	0.0007	0.0000	0.0015	0.0025
FFP	kg oil $_{eq}$	0.0079	0.0576	0.0330	0.0000	0.0206	0.0180
WCP	m^3	0.0314	0.0003	0.0014	0.0000	0.0016	0.0011

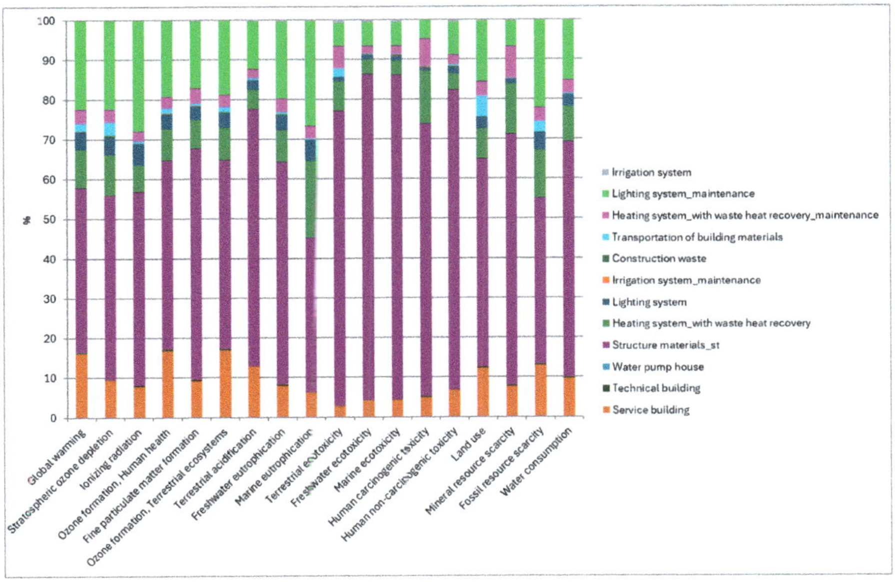

Figure 4. Life cycle assessment for the total amount of materials and inputs for Scenario 1.1: Frövi greenhouse with symbiotic waste heat recovery per FU.

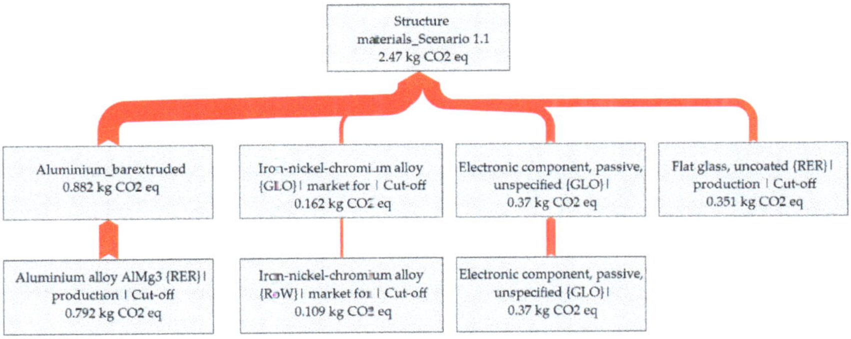

Figure 5. Life cycle assessment for the total quantity of structure materials in kg $CO_{2\,eq}$ produced (global warming impact category for Scenario 1.1) per FU.

The environmental impact per FU was also analyzed across different greenhouse scenarios (Figure 6, Table 6) including the BAU greenhouses (Scenario 2 and Scenario 3), Frövi's greenhouse without symbiotic heat recovery (Scenario 1.2 and Scenario 1.3), and Frövi's greenhouse with symbiotic waste heat recovery system (Scenario 1.1, developed within the CORALIS project).

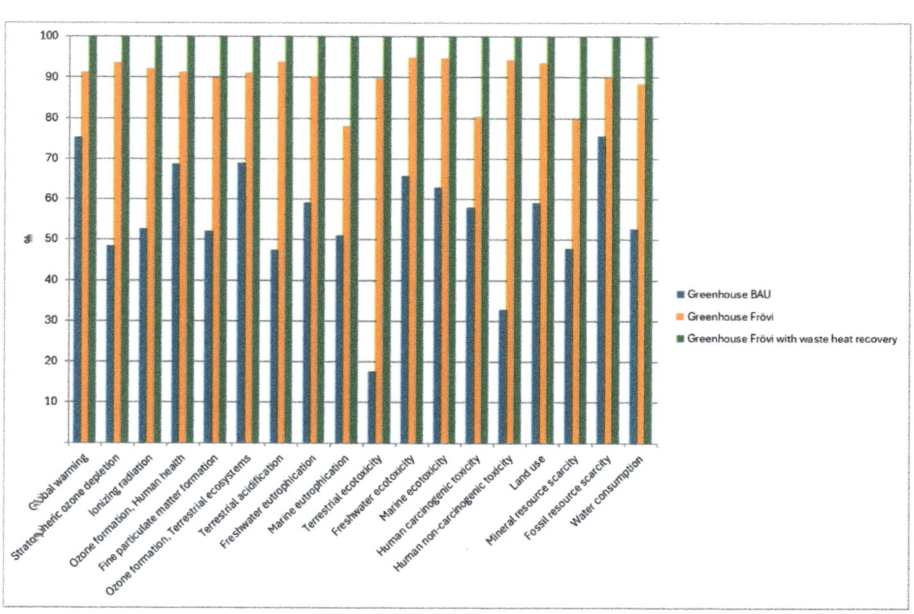

Figure 6. LCA comparison between greenhouses per FU.

Among these results, Scenario 1.1 has the highest environmental impact per FU, mainly because the waste heat recovery system is included within its system boundaries. In contrast, in other scenarios, the heating sources (natural gas or biomass) are outside their boundaries, as they are not part of the greenhouse infrastructure, and are part of the operational processes.

For BAU greenhouse, the most environmentally impactful materials and processes are the production and transport of the aluminum, the flat glass panels, and the steel used in structure components. These materials contribute significantly to Global Warming Potential (kg $CO_{2\,eq}$) (see Figure 7).

Table 6. Life cycle assessment results for the comparison between the different greenhouses contemplated in the 5 scenarios per FU.

Impact Category	Unit	BAU Greenhouse	Frövi Greenhouse	Frövi Greenhouse with Waste Heat Recovery
GWP	kg $CO_{2\,eq}$	0.0528	0.0639	0.0701
ODP	kg CFC11 $_{eq}$	0.0000	0.0000	0.0000
IRP	kBq Co-60 $_{eq}$	0.0030	0.0052	0.0057
HOFP	kg $NO_{x\,eq}$	0.0001	0.0002	0.0002
PMFP	kg PM2.5 $_{eq}$	0.0001	0.0002	0.0002
EOFP	kg $NO_{x\,eq}$	0.0001	0.0002	0.0002
TAP	kg $SO_{2\,eq}$	0.0003	0.0005	0.0005
FEP	kg P_{eq}	0.0000	0.0000	0.0000
MEP	kg N_{eq}	0.0000	0.0000	0.0000
TETP	kg 1,4-DCB	0.2827	1.4358	1.6020
FETP	kg 1,4-DCB	0.0119	0.0171	0.0180
METP	kg 1,4-DCB	0.0149	0.0223	0.0236
HTP_c	kg 1,4-DCB	0.0284	0.0394	0.0490
HTP_{nc}	kg 1,4-DCB	0.0881	0.2524	0.2680
LOP	m^2a crop $_{eq}$	0.0014	0.0022	0.0023
SOP	kg Cu_{eq}	0.0012	0.0020	0.0025
FFP	kg oil $_{eq}$	0.0136	0.0163	0.0181
WCP	m^3	0.0006	0.0010	0.0011

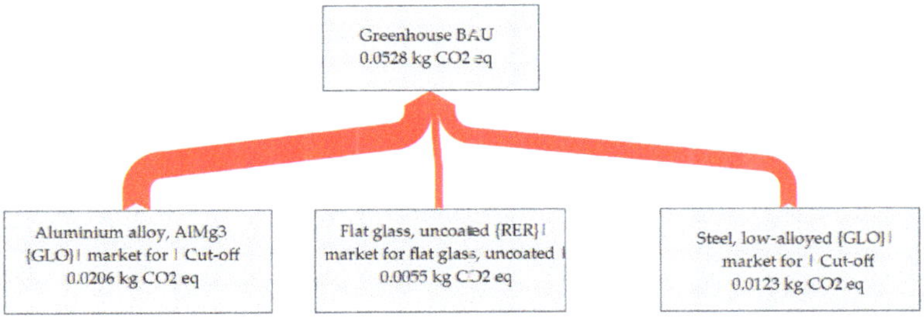

Figure 7. Tree analysis of the greenhouse BAU, for the impact category global warming, per FU, in kg $CO_{2\ eq}$, for Scenario 2 and 3.

The analysis of the fertilizer production and application shows that the main contributor to environmental impact is the production of ammonia, which is utilized to synthesize nitric acid that is later used in the manufacturing of calcium nitrate. This applies to all scenarios (see Figures 8 and 9 for marine eutrophication impact category).

For the case of the emissions due to the process of fertilizing, the main contributor for the marine eutrophication impact category is the emissions coming from the calcium nitrate application due to its content of Nitrogen (0.0153 kg N $_{eq}$ per kg of this emission and 0.000529 kg N $_{eq}$ per kg of tomato production (Figure 10)).

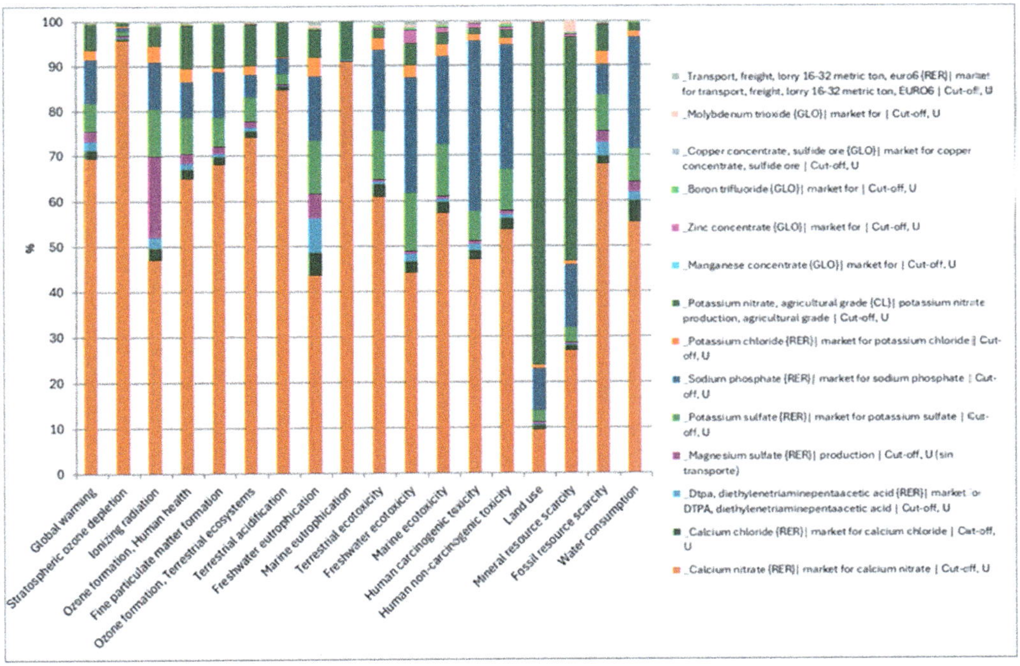

Figure 8. Life cycle assessment for the fertilizer production and application, per FU, for all the scenarios.

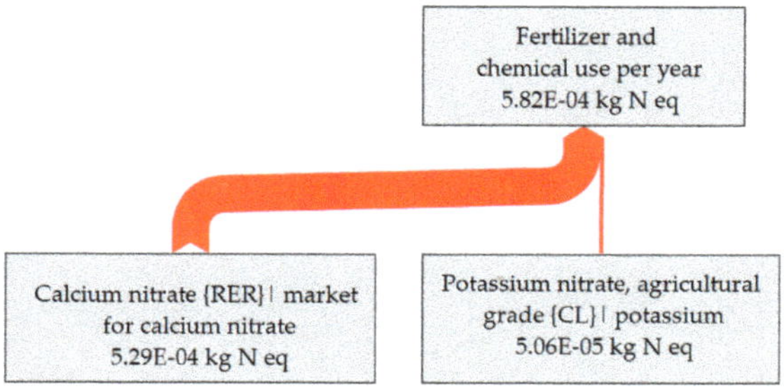

Figure 9. Tree analysis of the fertilizer and chemical use production (for marine eutrophication impact category per FU, in kg N_{eq}, for all the scenarios studied).

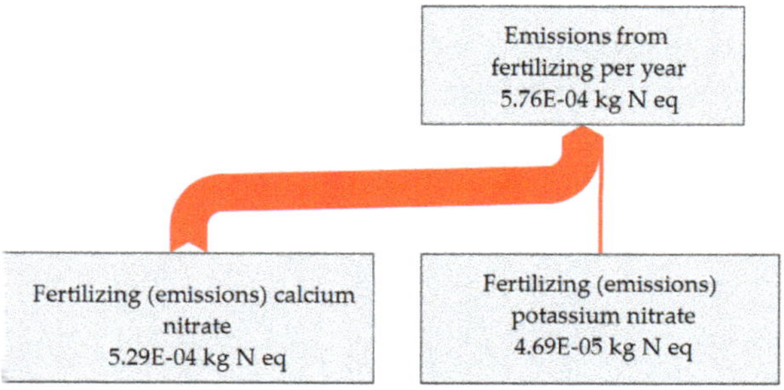

Figure 10. Tree analysis of the emissions from the fertilizing process (for marine eutrophication impact category per FU, in kg N_{eq}, for all the scenarios studied).

The main environmental impact in substrate use (Figure 11) comes from transportation, as the substrate is sourced from India and transported mainly by road to its final destination.

Figure 11. Tree analysis of the substrate use per FU (for the impact category of global warming in kg $CO_{2\ eq}$), for Scenario 1.1, 1.2, and 1.3.

For the electricity input, the energy supplier mix is 100% hydropower, 12.71 kWh is consumed per kg of tomatoes, and 0.0921 m^3 of water is consumed in the electricity high voltage, due to the gravel extraction and steel production for the hydropower plant construction (Figure 12).

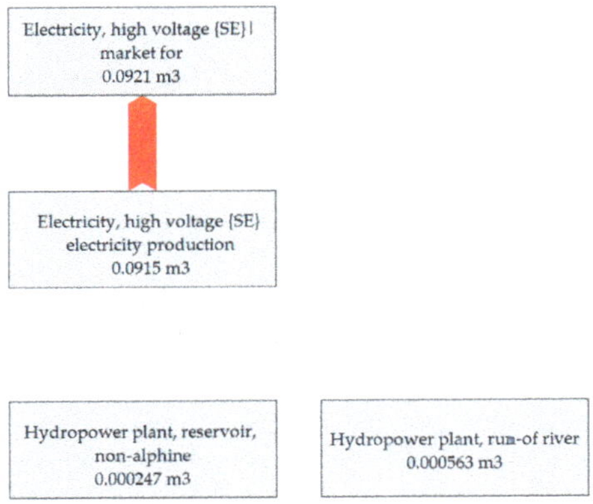

Figure 12. Tree analysis of the electricity high-voltage production used in Scenario 1.1, per FU (for the impact category of water consumption in m^3).

When analyzing Scenario 1.2 (Figure 13, Table 7. Life cycle assessment results (energy processes) for Scenario 1.2 per FU) and Scenario 1.3 (Figure 14, Table 8. Life cycle assessment results for Scenario 1.3 (heat natural gas process) per FU), in addition to the main impact contributors already identified in Scenario 1.1, the heat produced by biomass (Scenario 1.2) and by natural gas (Scenario 1.3) emerges as a major factor in the Global Warming Potential impact category. However, using biomass as a heat source results in lower environmental impact than using natural gas in this category.

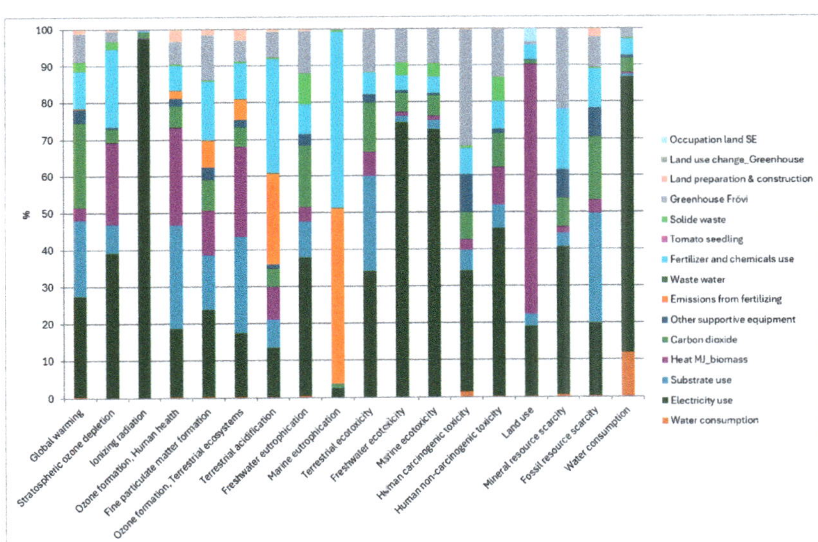

Figure 13. Life cycle assessment for Scenario 1.2 of Frövi's greenhouse with heat from biomass per FU.

Table 7. Life cycle assessment results (energy processes) for Scenario 1.2 per FU.

Impact Category	Unit	SE Electricity Mix (Same as Scenario 2)	Heat Biomass (Same as Scenario 2)
GWP	kg $CO_{2\ eq}$	0.2178	0.0284
ODP	kg $CFC11_{eq}$	0.0000	0.0000
IRP	kBq Co-60 $_{eq}$	1.4120	0.0020
HOFP	kg $NO_{x\ eq}$	0.0006	0.0008
PMFP	kg PM2.5 $_{eq}$	0.0004	0.0002
EOFP	kg $NO_{x\ eq}$	0.0006	0.0008
TAP	kg $SO_{2\ eq}$	0.0009	0.0006
FEP	kg P_{eq}	0.0001	0.0000
MEP	kg N_{eq}	0.0000	0.0000
TETP	kg 1,4-DCB	4.1312	0.8119
FETP	kg 1,4-DCB	0.1391	0.0017
METP	kg 1,4-DCB	0.1702	0.0028
HTP_c	kg 1,4-DCB	0.0405	0.0036
HTP_{nc}	kg 1,4-DCB	0.8535	0.1927
LOP	m^2a crop $_{eq}$	0.0594	0.2154
SOP	kg Cu_{eq}	0.0036	0.0002
FFP	kg oil $_{eq}$	0.0376	0.0070
WCP	m^3	0.0262	0.0002

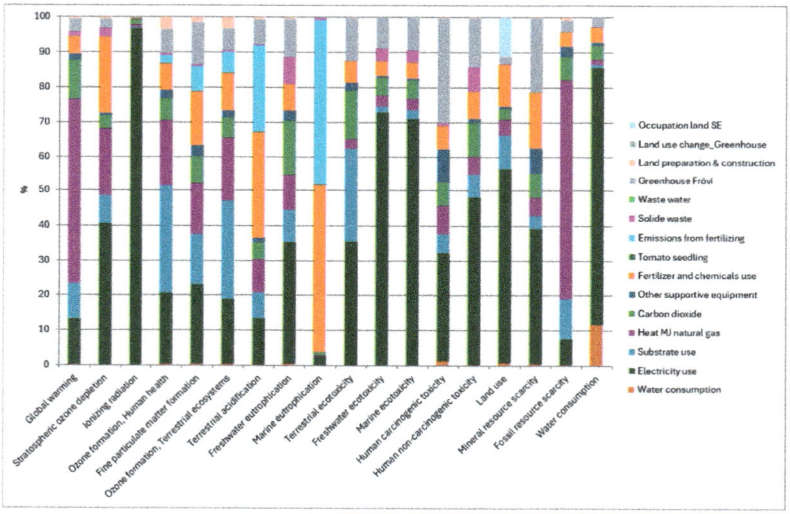

Figure 14. Life cycle assessment for Scenario 1.3 for Frövi's greenhouse with heat from natural gas per FU.

Scenario 2 (Figure 15) indicates that the Swedish electricity mix (Table 7) has a lower environmental impact compared to the Dutch electricity mix. Meanwhile, the transport of tomatoes from the Netherlands to Sweden in Scenario 3 (Figure 16 and Table 9) represents a considerable impact, which is avoided in all other scenarios. However, in Scenario 1.1, the waste heat recovery system increases the impact associated with the greenhouse infrastructure. Despite this, its overall impact is offset by the avoided heat production from biomass (Scenario 1.2) and natural gas (Scenario 1.3).

Table 8. Life cycle assessment results for Scenario 1.3 (heat natural gas process) per FU.

Impact Category	Unit	Heat Natural Gas (Same as Scenario 3)
GWP	kg $CO_{2\,eq}$	0.8898
ODP	kg CFC11 $_{eq}$	0.0000
IRP	kBq Co-60 $_{eq}$	0.0114
HOFP	kg $NO_{x\,eq}$	0.0005
PMFP	kg PM2.5 $_{eq}$	0.0002
EOFP	kg $NO_{x\,eq}$	0.0006
TAP	kg $SO_{2\,eq}$	0.0007
FEP	kg P $_{eq}$	0.0000
MEP	kg N $_{eq}$	0.0000
TETP	kg 1,4-DCB	0.3209
FETP	kg 1,4-DCB	0.0058
METP	kg 1,4-DCB	0.0078
HTP_c	kg 1,4-DCB	0.0104
HTP_{nc}	kg 1,4-DCB	0.0960
LOP	m^2a crop $_{eq}$	0.0047
SOP	kg Cu $_{eq}$	0.0005
FFP	kg oil $_{eq}$	0.3204
WCP	m^3	0.0005

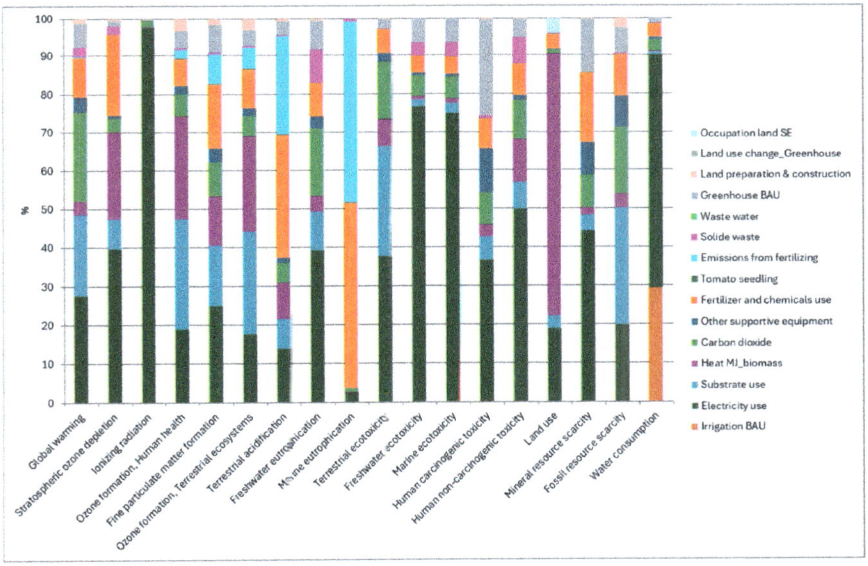

Figure 15. Life cycle assessment for Scenario 2 for BAU greenhouse in Sweden, per FU.

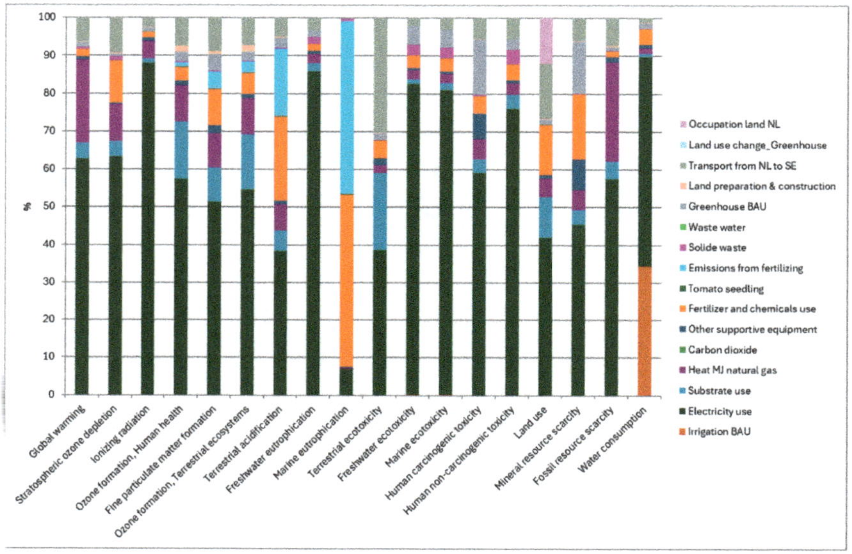

Figure 16. Life cycle assessment for Scenario 3 for BAU greenhouse in the Netherlands, per FU.

Table 9. Life cycle assessment results for Scenario 3 (NL electricity mix and transport from NL to SE) per FU.

Impact Category	Unit	NL Electricity Mix	Transport from NL to SE
GWP	kg $CO_{2\,eq}$	2.5470	0.2568
ODP	kg CFC11 $_{eq}$	0.0000	0.0000
IRP	kBq Co-60 $_{eq}$	0.2149	0.0061
HOFP	kg $NO_{x\,eq}$	0.0032	0.0004
PMFP	kg PM2.5 $_{eq}$	0.0013	0.0002
EOFP	kg $NO_{x\,eq}$	0.0032	0.0004
TAP	kg $SO_{2\,eq}$	0.0038	0.0005
FEP	kg P $_{eq}$	0.0013	0.0001
MEP	kg N $_{eq}$	0.0001	0.0000
TETP	kg 1,4-DCB	5.9513	4.6868
FETP	kg 1,4-DCB	0.1976	0.0051
METP	kg 1,4-DCB	0.2493	0.0093
HTP_c	kg 1,4-DCB	0.1146	0.0105
HTP_{nc}	kg 1,4-DCB	2.4440	0.1792
LOP	m^2a crop $_{eq}$	0.0406	0.0138
SOP	kg Cu $_{eq}$	0.0039	0.0005
FFP	kg oil $_{eq}$	0.7051	0.0886
WCP	m^3	0.0204	0.0004

3.2. LCC Evaluation

The most significant Capital Expenditures (CapEx) are related to infrastructure and functionality, while the main Operational Expenditures (OpEx) are associated with personnel and energy costs.

Figures 17 and 18 present the financial structure of the project, detailing the distribution of Capital Expenditures (CapEx) and Operational Expenditures (OpEx).

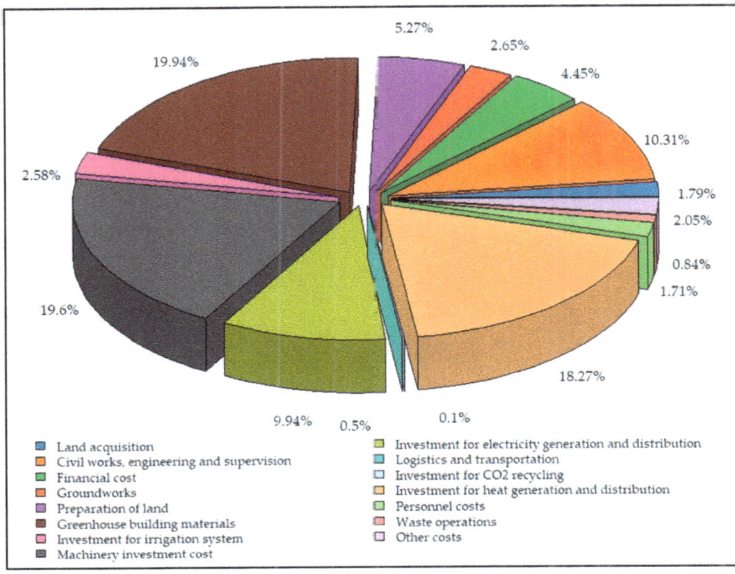

Figure 17. Capital Expenditures—CapEx for Frövi Greenhouse.

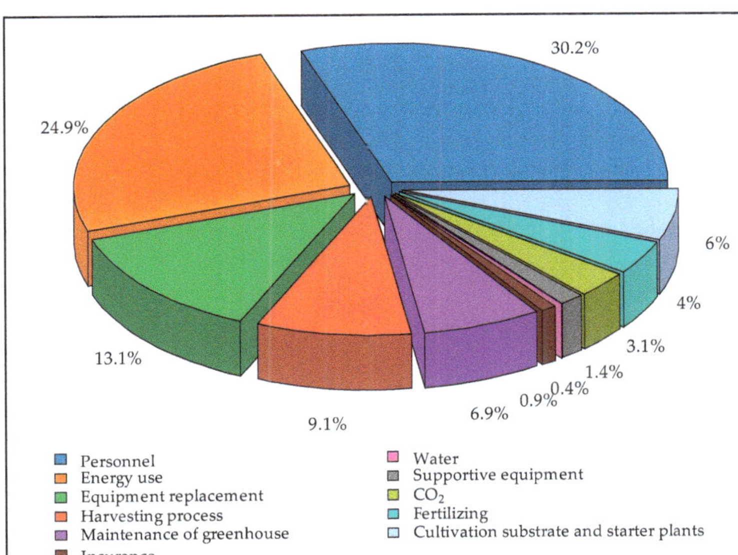

Figure 18. Annual Operating Expenditures—OpEx for Frövi Greenhouse.

Figure 19 illustrates the differences in production costs per kg of tomatoes across the analyzed scenarios. The results indicate that production costs range from €1.149/kg (Scenario 2) to €1.584/kg (Scenario 3), highlighting the economic trade-offs between sustainable and conventional greenhouse practices.

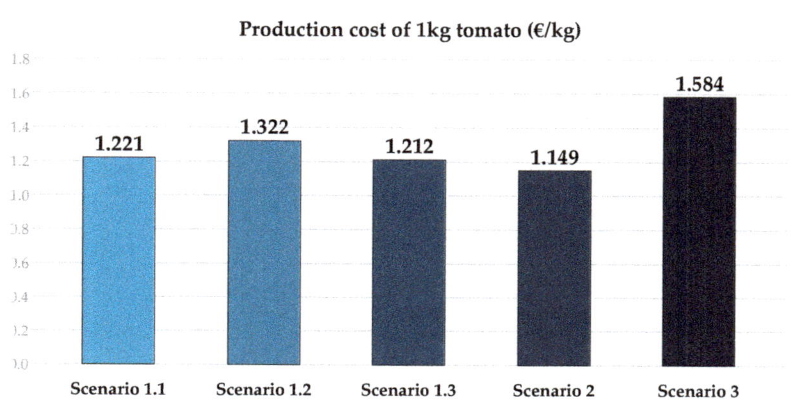

Figure 19. Production cost of 1 kg of tomatoes (€/kg) in different scenarios.

Scenario 1.1 (€1.221/kg) has a relatively low production cost, benefiting from 100% hydropower electricity supplied locally and waste heat from the pulp and paper mill. However, it is slightly more expensive than Scenario 1.3 (€1.212/kg), due to Capital Expenditures, and Scenario 2 (€1.149/kg), which achieve lower costs due to their specific energy configurations. Scenario 1.2 (€1.322/kg) shows a moderate cost increase due to its dependence on the Swedish energy mix (SE mix), which combines renewable and non-renewable sources. Scenario 1.3, which relies on natural gas for heating, is more cost-efficient than Scenario 1.1.

Scenario 2 (€1.149/kg), representing a typical Swedish greenhouse, closely aligns with Scenario 1.3, reflecting similar energy dependencies and efficiencies. Scenario 3 (€1.584/kg), the most expensive scenario, represents the Dutch greenhouse model, which relies on energy-intensive operations, mainly using natural gas, leading to higher economic costs and environmental impacts.

These findings highlight that while Scenario 1.1 benefits from renewable and waste-based energy sources, Scenario 1.3 and Scenario 2 offer lower costs for greenhouse tomato production in Sweden.

4. Discussion

The present work has been dedicated to analyzing the environmental and economic impact of the CORALIS project through the study of a demo case where different industrial symbiotic solutions have been tested, with a focus on the demonstration of recycling of low-grade waste heat for ambient heating in a tomato greenhouse.

The innovative tomato greenhouse demo case constructed in Frövi, Sweden shows that there are both environmental and economic benefits when comparing the industrial symbiotic scenario with the rest of the scenarios included in the study. The main findings from this study are the following:

- The LCA results demonstrate significant reductions in environmental burdens across key impact categories (in fact, 17 out of 18 environmental impact categories are included in the LCA, when the symbiotic scenario (1.1) is compared with the Netherlands BAU scenario (3)).
- The total avoided GHG emissions of the symbiotic scenario (Scenario 1.1), compared with the Netherlands BAU scenario (Scenario 3) are 3.382 kg CO_2 equivalents per kg of tomatoes produced, or 29,270 tons of CO_2 equivalents yearly when considering the 30-year estimated lifespan of the greenhouse. When compared with the rest of the

scenarios the GHG emissions avoided are less, but still the symbiotic scenario is the least impacting with respect to the Global Warming Potential impact.

- The total water consumption per kg of tomatoes is 9% higher (0.003 m^3/kg tomatoes) when the symbiotic scenario (Scenario 1.1) is compared to the Netherlands BAU scenario (Scenario 3).
- The total fossil resources used per kg of tomatoes is 87% lower (1.07 kg of oil equivalent/kg tomatoes) when the symbiotic scenario (Scenario 1.1) is compared to the Netherlands BAU scenario (Scenario 3), when the greenhouse is heated mainly by waste heat from the pulp and paper as expected.
- The LCC indicated that when comparing all the scenarios, the average Swedish greenhouse scenario (Scenario 2) had the lowest production costs.
- The LCC also highlighted that the greenhouse building materials, machinery investment, and investment for heat generation and distribution costs accounted for the largest portions of the Frövi project CapEx (combined 57.8%). The LCC also indicated that the personnel and energy costs accounted for the largest portions of OpEx (combined 57%) for the Frövi project. Therefore, these can be understood as key leverage points for further reducing the CapEx and OpEx in similar future symbiosis projects, conceptualized in the literature.

LCA has been proven to be an effective methodology/tool to evaluate and compare the environmental impacts/benefits of IS scenarios versus BAU scenarios, aligned with previous works where it is used to evaluate the benefit of applying IS in an energy-intensive industrial park in China [17,41]. LCA and LCC are instruments proposed to overcome economic uncertainty in future investments within an IS context in the work of Mainar-Toledo et al. (2022) [17].

Examples of future explorations could be to look into how organic fertilizers' impact would decrease if waste heat from other facilities in the proximity could be used to replace heat from fossil fuels [42]. The Frövi LCA and LCC example also highlights that it is a viable option to utilize waste heat other than for replacing fossil fuels from local heating systems [26], while assessing the emission reduction potential of biofuels in Sweden.

Other interesting analyses to pursue for the case of Frövi are applications to use the low-grade waste heat for drying purposes when the waste heat from the pulp and paper is not needed, such as is the case during the summer period. Previous studies have pointed to the potential of waste heat for decreasing the environmental impacts from different production processes, where the recovery of low-grade heat could potentially improve the environmental performance of the coking industry, mainly through heat recovery for preheating coal [25]; others have suggested that using waste heat from other industrial processes for drying and wet extraction processes would improve the energy demands of the algal biofuel production facility [43].

5. Conclusions

This LCA and LCC assessment applied to a real IS demonstrator intended to assess the environmental and economic impacts of low-grade waste recycling, thus filling the gap in the literature for real case analysis of symbiosis developments. The results obtained at Frövi demonstrate the following:

1. The IS model is economically and environmentally viable, when speaking about recovering waste heat from a near industrial facility and when the greenhouse and its systems perform as expected.
2. The LCA indicated that greenhouse building materials were a key contributor to environmental impact, which highlights the importance of material choices in early project design stages.

3. The LCA also indicated that the waste heat symbiotic scenario had a significant environmental impact related to the waste heat pipe infrastructure and water pipe infrastructure when compared with the generic Sweden scenario (Scenario 2) and the generic Netherlands scenario (Scenario 3).
4. While the symbiotic scenario (Scenario 1.1) performed better than the generic Netherlands scenario (Scenario 3) in 17 out of 18 impact categories, the other scenarios (Scenarios 1.2, 1.3, and 2) indicated that there are some areas of potential improvement for future similar projects.
5. The solution established for the present study shows that the waste recovery system helps to reduce both the environmental impacts and the economic impacts of the tomatoes produced in Frövi's greenhouse.
6. Additionally, the Frövi tomato greenhouse avoids environmental and economic impacts related to the use of natural gas for heating and related to the transport from the Netherlands to Sweden required in the BAU Netherlands scenario (Scenario 3).

This research presents a novel application of LCA and LCC to an industrial symbiosis case in greenhouse agriculture, demonstrating the dual benefits of waste heat recovery in terms of both environmental and economic performance. The findings provide robust evidence that IS can be a viable strategy for decarbonizing agricultural heating systems, reducing operational costs, and improving energy efficiency. This approach serves as a replicable model for future implementations of IS in different sectors, particularly in regions aiming to optimize resource use and transition towards more circular economies.

Nevertheless, further research is needed to explore IS projects on a larger scale. Future studies should analyze and compare additional waste heat reuse scenarios, considering different heat source temperatures, geographical contexts, and agricultural applications beyond greenhouse tomato production, such as protein, shrimp, or fish farming. Additionally, validating these findings with real operational data, rather than estimated values, would strengthen the robustness of the conclusions and enhance their applicability in real-world industrial symbiosis strategies.

Despite these promising results, several barriers could limit the large-scale adoption of waste heat recovery in greenhouse agriculture. The high initial investment (CapEx) required for infrastructure, such as heat transport pipelines and greenhouse adaptation, poses a significant challenge, particularly in the absence of financial incentives. Additionally, the geographical proximity between heat suppliers and greenhouse operations is crucial, as long-distance heat transport can reduce efficiency and increase costs. Regulatory constraints also play a key role, as policies governing energy use, waste heat valorization, and agricultural production may not fully accommodate industrial symbiosis solutions.

Moreover, energy price volatility could affect the long-term economic viability of these systems, requiring further research into mechanisms that could mitigate financial risks, such as subsidies, carbon pricing incentive, or long-term energy purchase agreements. Future research should explore these aspects to better understand how policy support, business model innovation, and stakeholder collaboration can enable broader adoption of industrial symbiosis strategies for sustainable greenhouse production.

Author Contributions: Conceptualization, M.D.M.-T. and I.G.G.; methodology, M.D.M.-T. and I.G.G.; formal analysis, M.D.M.-T. and I.G.G.; investigation, M.D.M.-T., I.G.G. and J.F.; data curation, M.D.M.-T. and I.G.G.; writing—original draft preparation, M.D.M.-T., I.G.G., D.P. and T.P.; writing—review and editing, H.L., D.P. and T.P. Research supervision at WA3RM, internal review and editing, D.P. and T.P. All authors have read and agreed to the published version of the manuscript.

Funding: This research was funded by the European Commission, from the European Union's Horizon 2020 research and innovation programme, through the the project Creation of New Value

Chain Relations through Novel Approaches facilitating Long-term IS (CORALIS), under Grant Agreement number 958337.

Data Availability Statement: The datasets presented in this article are not readily available because of confidential reasons. The raw data used in this study contain sensitive CapEx and OpEx information from both the greenhouse provider and the tomato producer, who both agreed to share this information for CORALIS purposes. Requests to access the datasets should be directed to the IS developer https://wa3rm.com/.

Conflicts of Interest: Due to the participatory nature of this study, WA3RM employees and other key stakeholders were involved in the study as collaborators; in particular, employees at the IS facilitator company were involved in defining the scope of the analysis, collecting data, reviewing the results, and contributing to writing this article to submit it for publication. Employees at WA3RM were not involved in the study design, analysis, or interpretation of data. Following the previous statement, the authors declare no conflicts of interest.

Abbreviations

The following abbreviations are used in this manuscript:

BAU	Business-As-Usual
CapEx	Capital Expenditures
CHP	Combined Heat and Power Plant
EOFP	Ozone formation, terrestrial ecosystems
FEP	Freshwater eutrophication
FETP	Freshwater ecotoxicity
FFP	Fossil resource scarcity
FU	Functional unit
GLO	Global
GWP	Global warming
HOFP	Ozone formation, human health
HTPc	Human carcinogenic toxicity
HTPnc	Human non-carcinogenic toxicity
IRP	Ionizing radiation
IS	IS
LCA	Life Cycle Assessment
LCC	Life Cycle Cost
LCI	Life Cycle Inventory
LOP	Land use
MEP	Marine eutrophication
ODP	Stratospheric ozone depletion
OpEx	Operational Expenditure
PMFP	Fine particulate matter formation
SE	Swedish
SOP	Mineral resource scarcity
TAP	Terrestrial acidification
TETP	Terrestrial ecotoxicity
RER	Rest of Europe and other regions
WCP	Water consumption

Appendix A

Table A1. Inventory related to structural materials, heating system, lighting system, irrigation system, electricity generators, and construction waste (input).

Type of Material	Amount	Unit
Structural materials	6,886,735.90	kkg
Heating system	453,727.46	kg
Lighting system	130,297.00	kkg
Irrigation system	18,113.10	kg
Electricity generators	3580.00	kg
Construction waste	141,040.00	kg
Land preparation and construction energy	570,262.00	L
consumption	281,875.00	kWh
Transportation of building materials	3,741,284.00	ton-km
Tomato yield data	85.00	kg/m^2/year

Table A2. Inventory related to tomato production data (input).

Type of Material	Amount	Unit
Fertilizers and chemicals	534,815.00	kg/year
Water use	220,000,000.00	L/year
Energy use	37,200,000.00	kWh/year
Tomato plants	175,000.00	number of plants
Substrate use	7,000,000.00	m^3/year
Other supportive equipment	17,000.00	kg/year
Other main inputs	100,000,000.00	MJ/year
CO_2 for plant growth	1,700,000.00	kg/year
Collection process and packaging process	82,600.00	kg/year

Table A3. Inventory related to waste generation and treatment, estimations (input).

Type of Material	Amount	Unit
Liquid waste	121,000.00	L/year
Solid waste	509,875.00	kg/year
Volumetric waste	450.00	m^3/year

References

1. Ghanem, L.; Gürdil, G.A.K.; Demirel, B.; Eissa, M.O.S. Estimating Energy Needs for Climate-Controlled Greenhouses in Syria with a Software Tool. *Black Sea J. Eng. Sci.* **2024**, *7*, 1187–1193. [CrossRef]
2. Patel, V.K.; Singh, L.P.; Sharma, D.; Singh, K.; Sudan, S.; Koul, V.K. Trends in Greenhouse Production Technology with Special Reference to Protected Cultivation of Horticultural Crops. *Int. J. Multidiscip. Res.* **2024**, *6*. [CrossRef]
3. Nikolić, D.; Jovanović, S.; Jurišević, N.; Nikolić, N.; Radulović, J.; Radović, M.V.; Grujić, I. Sustainable Design in Agriculture—Energy Optimization of Solar Greenhouses with Renewable Energy Technologies. *Energies* **2025**, *18*, 416. [CrossRef]
4. Syrotiuk, V.; Syrotiuk, S.; Ptashnyk, V. Energy supply for the processes of greenhouse business with application of renewable sources of energy. *Bull. Lviv Natl. Agrar. Univ. Agroeng. Res.* **2019**, 56–60. [CrossRef]
5. Mahmood, F.; Al-Ansari, T. Renewable Energy Driven Sustainable Greenhouse: An Overview. In *Encyclopedia of Sustainable Technologies*; Elsevier: Amsterdam, The Netherlands, 2023; pp. 652–665. [CrossRef]
6. Paris, B.; Vandorou, F.; Balafoutis, A.T.; Vaiopoulos, K.; Kyriakarakos, G.; Manolakos, D.; Papadakis, G. Energy Use in Greenhouses in the EU: A Review Recommending Energy Efficiency Measures and Renewable Energy Sources Adoption. *Appl. Sci.* **2022**, *12*, 5150. [CrossRef]
7. Frans, S.; Juha-Matti, K. Reduction of the Climate Impact of Finnish Greenhouse Vegetables Achieved by Energy Acquisitions between 2004 and 2017. *J. Hortic. Sci. Res.* **2021**, *4*, 018. [CrossRef]
8. Iddio, E.; Wang, L.; Thomas, Y.; McMorrow, G.; Denzer, A. Energy efficient operation and modeling for greenhouses: A literature review. *Renew. Sustain. Energy Rev.* **2020**, *117*, 109480. [CrossRef]
9. Zhou, D.; Meinke, H.; Wilson, M.; Marcelis, L.F.M.; Heuvelink, E. Towards delivering on the sustainable development goals in greenhouse production systems. *Resour. Conserv. Recycl.* **2021**, *169*, 105379. [CrossRef]
10. Cuce, E.; Harjunowibowo, D.; Cuce, P.M. Renewable and sustainable energy saving strategies for greenhouse systems: A comprehensive review. *Renew. Sustain. Energy Rev.* **2016**, *64*, 34–59. [CrossRef]

11. Ammenberg, J.; Baas, L.; Eklund, M.; Feiz, R.; Helgstrand, A.; Marshall, R. Improving the CO_2 performance of cement, part III: The relevance of industrial symbiosis and how to measure its impact. *J. Clean. Prod.* **2014**. Available online: https://www.academia.edu/12877589/Improving_the_CO2_performance_of_cement_part_III_the_relevance_of_industrial_symbiosis_and_how_to_measure_its_impact (accessed on 17 March 2025).
12. Xue, X.; Wang, S.; Chun, T.; Xin, H.; Xue, R.; Tian, X.; Zhang, R. An integrated framework for industrial symbiosis performance evaluation in an energy-intensive industrial park in China. *Environ. Sci. Pollut. Res.* **2023**, *30*, 42056–42074. [CrossRef]
13. European Commission. European Platform on LCA | EPLCA. Available online: https://eplca.jrc.ec.europa.eu/ (accessed on 17 March 2025).
14. European Commission. Life-Cycle Costing. Available online: https://green-business.ec.europa.eu/green-public-procurement/life-cycle-costing_en (accessed on 17 March 2025).
15. Duflou, J.R.; Sutherland, J.W.; Dornfeld, D.; Herrmann, C.; Jeswiet, J.; Kara, S.; Hauschild, M.Z.; Kellens, K. Towards energy and resource efficient manufacturing: A processes and systems approach. *CIRP Ann.* **2012**, *61*, 587–609. [CrossRef]
16. Yan, X.; Wang, L.; Fang, M.; Hu, J. How Can Industrial Parks Achieve Carbon Neutrality? Literature Review and Research Prospect Based on the CiteSpace Knowledge Map. *Sustainability* **2023**, *15*, 372. [CrossRef]
17. Toledo, M.; Castan, M.; Millán, G.; Rodin, V.; Kollmann, A.; Peccianti, F.; Annunziata, E.; Rizzi, F.; Frey, M.; Iannone, F.; et al. Accelerating sustainable and economic development via industrial energy cooperation and shared services—A case study for three European countries. *Renew. Sustain. Energy Rev.* **2022**, *153*, 111737. [CrossRef]
18. Park, H.-S.; Behera, S.K. Methodological aspects of applying eco-efficiency indicators to industrial symbiosis networks. *J. Clean. Prod.* **2014**, *64*, 478–485. [CrossRef]
19. Briassoulis, D.; Pikasi, A.; Hiskakis, M.; Arias, A.; Moreira, M.T.; Ioannidou, S.M.; Ladakis, D.; Koutinas, A. Life-cycle sustainability assessment for the production of bio-based polymers and their post-consumer materials recirculation through industrial symbiosis. *Curr. Opin. Green Sustain. Chem.* **2023**, *41*, 100818. [CrossRef]
20. Subramanian, K.; Chopra, S.S.; Ashton, W.S. Capital-based life cycle sustainability assessment: Evaluation of potential industrial symbiosis synergies. *J. Ind. Ecol.* **2021**, *25*, 1161–1176. [CrossRef]
21. Park, J.; Duque-Hernández, J.; Díaz-Posada, N. Facilitating Business Collaborations for Industrial Symbiosis: The Pilot Experience of the Sustainable Industrial Network Program in Colombia. *Sustainability* **2018**, *10*, 3637. [CrossRef]
22. Fric, U.; Levnajić, Z.; Modic, D.; Rončević, B. Industrial symbiosis networks supporting circularity: Understanding complexity, cyclicality and resilience. *Environ. Technol. Innov.* **2025**, *37*, 104026. [CrossRef]
23. Rezaei, F.; Burg, V.; Solgi, H.; Hellweg, S.; Roshandel, R. Spatiotemporal assessment of the cumulative exergy demand of agricultural greenhouse production with industrial symbiosis. *Ecol. Indic.* **2024**, *169*, 112904. [CrossRef]
24. Xue, R.; Wang, S.; Gao, G.; Liu, D.; Long, W.; Zhang, R. Evaluation of symbiotic technology-based energy conservation and emission reduction benefits in iron and steel industry: Case study of Henan, China. *J. Clean. Prod.* **2022**, *338*, 130616. [CrossRef]
25. Wu, J.; Pu, G.; Guo, Y.; Lv, J.; Shang, J. Retrospective and prospective assessment of exergy, life cycle carbon emissions, and water footprint for coking network evolution in China. *Appl. Energy* **2018**, *218*, 479–493. [CrossRef]
26. Martin, M.; Wetterlund, E.; Hackl, R.; Holmgren, K.M.; Peck, P. Assessing the aggregated environmental benefits from by-product and utility synergies in the Swedish biofuel industry. *Biofuels* **2020**, *11*, 683–698. [CrossRef]
27. Martin, M.; Weidner, T.; Gullström, C. Estimating the Potential of Building Integration and Regional Synergies to Improve the Environmental Performance of Urban Vertical Farming. *Front. Sustain. Food Syst.* **2022**, *6*, 849304. [CrossRef]
28. Parker, T.; Kiessling, A. Low-grade heat recycling for system synergies between waste heat and food production, a case study at the European Spallation Source. *Energy Sci. Eng.* **2016**, *4*, 153–165. [CrossRef]
29. Teng, Y.; Pan, W. Estimating and minimizing embodied carbon of prefabricated high-rise residential buildings considering parameter, scenario and model uncertainties. *Build. Environ.* **2020**, *180*, 106951. [CrossRef]
30. Muñoz, E.; Curaqueo, G.; Cea, M.; Vera, L.; Navia, R. Environmental hotspots in the life cycle of a biochar-soil system. *J. Clean. Prod.* **2017**, *158*, 1–7. [CrossRef]
31. *ISO–ISO 14040:2006*; Environmental Management—Life Cycle Assessment—Principles and Framework. ISO: Geneva, Switzerland, 2006. Available online: https://www.iso.org/standard/37456.html (accessed on 15 January 2023).
32. *ISO–ISO 14044:2006*; Environmental Management—Life Cycle Assessment—Requirements and Guidelines. ISO: Geneva, Switzerland, 2006. Available online: https://www.iso.org/standard/38498.html (accessed on 15 January 2023).
33. Song, X.; Montelius, M.; Carlsson, C. Life Cycle Assessment of Per- and Polyfluoroalkyl Substances (PFAS) Remediation Technologies: A Literature Review. *Environments* **2024**, *11*, 203. [CrossRef]
34. Khankahdani, Z.H.; Ghazimoradi, M.M.; Abdollahi, M. Life Cycle Assessment. In *Encyclopedia of Toxicology*, 4th ed.; Academic Press: Cambridge, MA, USA, 2023; Volume 5, pp. 829–836. [CrossRef]
35. Nubi, O.; Murphy, R.; Morse, S. Life Cycle Sustainability Assessment of Waste to Energy Systems in the Developing World: A Review. *Environments* **2014**, *11*, 123. [CrossRef]

36. Bartzas, G.; Zaharaki, D.; Komnitsas, K. Life cycle assessment of open field and greenhouse cultivation of lettuce and barley. *In Process. Agric.* **2015**, *2*, 191–207. [CrossRef]
37. Tsoy, N.; Prado, V.; Wypkema, A.; Quist, J.; Mourad, M. Anticipatory Life Cycle Assessment of sol-gel derived anti-reflective coating for greenhouse glass. *J. Clean. Prod.* **2019**, *221*, 365–376. [CrossRef]
38. EPD. Arable Crops. Product Category Rules According to ISO 14025. 2020-11-16. Available online: www.environdec.com (accessed on 27 February 2025).
39. Huijbregts, M.A.J.; Steinmann, Z.J.N.; Elshout, P.M.F.; Stam, G.; Verones, F.; Vieira, M.; Zijp, M.; Hollander, A.; van Zelm, R. ReCiPe2016: A harmonised life cycle impact assessment method at midpoint and endpoint level. *Int. J. Life Cycle Assess.* **2017**, *22*, 138–147. [CrossRef]
40. Peters, M.S Plant Design and Economics for Chemical Engineers. McGraw-Hill (New York) 1920–2011: Free Download, Borrow, and Streaming: Internet Archive. Available online: https://archive.org/details/plantdesignecono0000pete (accessed on 27 February 2025).
41. Wang, S.; Lu, C.; Gao, Y.; Wang, K.; Zhang, R. Zhang, Life cycle assessment of reduction of environmental impacts via industrial symbiosis in an energy-intensive industrial park in China. *J. Clean. Prod.* **2019**, *241*, 118358. [CrossRef]
42. Santolin, J.; Vlaeminck, S.E.; Appiah-Twum, H.; Van Winckel, T.; Spiller, M. Consequential LCA of NPK fertilizers from microbial, animal, plant, and mineral origin highlights resource constraints and environmental impacts. *J. Clean. Prod.* **2024**, *457*, 142312. [CrossRef]
43. Chowdhury, R.; Caetano, N.; Franchetti, M.J.; Hariprasad, K. Life Cycle Based GHG Emissions from Algae Based Bioenergy with a Special Emphasis on Climate Change Indicators and Their Uses in Dynamic LCA: A Review. *Sustainability* **2023**, *15*, 1767. [CrossRef]

energies

MDPI

Article

Analysis of Precision Regulation Pathways for Thermal Substation Supply–Demand Balance

Jiaxiang Yin, Pengpeng Zhao and Jinda Wang *

School of Energy and Environment Engineering, Hebei University of Technology, Tianjin 300401, China
* Correspondence: wangjindahit@outlook.com; Tel.: +86-155-1389-9539

Abstract: Under the dual imperatives of air pollution control and energy conservation, this study proposes an enhanced optimization framework for combined heat and power (CHP) district heating systems based on bypass thermal storage (BTS). In contrast to conventional centralized tank-based approaches, this method leverages the dynamic hydraulic characteristics of secondary network bypass pipelines to achieve direct sensible heat storage in circulating water, significantly improving system flexibility and energy efficiency. The core innovation lies in addressing the critical yet under-explored issue of control valve dynamic response, which profoundly impacts system operational stability and economic performance. A quality regulation strategy is systematically implemented to stabilize circulation flow rates through temperature modulation by establishing a supply–demand equilibrium model under bypass conditions. To overcome the limitations of traditional feedback control in handling hydraulic transients and heat transfer dynamics in the plate heat exchanger, a Model Predictive Control (MPC) framework is developed, integrating a data-driven valve impedance-opening degree correlation model. This model is rigorously validated against four flow characteristics (linear, equal percentage, quick-opening, and parabolic) and critical impedance parameters (maximum/minimum controllable impedance). This study provides theoretical foundations and technical guidance for optimizing secondary network heating systems, enhancing overall system performance and stability, and promoting energy-efficient development in the heating sector.

Keywords: bypass thermal storage; control valve dynamics; model predictive control; valve opening characteristics; network optimization

1. Introduction

With the intensification of global environmental issues, atmospheric pollution mitigation and energy conservation have become core priorities for the international community. In the energy sector, combined heat and power (CHP) systems have been widely adopted in large-scale district heating projects due to their high energy efficiency [1,2]. Current research on improving the flexibility of heating systems predominantly focuses on external thermal storage devices [3]. For example, Benalcazar et al. developed a Mixed-Integer Linear Programming (MILP) model to jointly optimize thermal storage tank capacity and annual operational scheduling in CHP systems [4]. Wang et al. quantified the flexibility enhancement of CHP systems through a dynamic model integrating heat pumps, analyzing the decoupling effects of electro-thermal strategies [5]. Although these standalone thermal storage methods partially resolve the spatiotemporal conflicts between heat plant operations and heating demands, they involve high hardware investment costs and spatial limitations. The proposed bypass thermal storage (BTS) technology offers a novel pathway for optimizing heating systems. Wang et al. maximized thermal storage benefits by

 https://doi.org/10.3390/en18112691

reconfiguring the topology of district heating networks (DHNs), and implementing bypass branches on the primary side of heat exchange stations [6]. This approach directly utilizes the sensible heat of circulating water in pipelines to store surplus thermal energy from power plants, bypassing traditional reliance on standalone tanks. Kouhia et al. further quantified the technical potential of DHNs as short-term thermal storage media, confirming the economic feasibility of BTS in enhancing CHP system flexibility [7]. Compared to conventional solutions, BTS technology actively designs bypass pipelines to store excess heat in the circulating water of heating networks, achieving intraday load shifting without requiring large-scale storage equipment. This reduces the required capacity of peak-shaving heat sources, lowers system construction costs, and improves energy efficiency by shortening heat source operation time. Additionally, BTS reconstructs the operational paradigm of heat exchange stations through hydraulic decoupling: traditional stations suffer from mutual interference due to hydraulic coupling, while bypass pipelines establish pressure buffer interfaces, enabling independent control of each station and enhancing supply–demand balance accuracy [8].

As critical infrastructure, the thermal storage capacity of DHNs depends on parameters such as pipeline water volume and baseline supply/return temperatures [9]. These networks exhibit higher flexibility and environmental adaptability than traditional thermal storage tanks. In large-scale systems, indirect connections typically divide DHNs into primary networks (heat source to exchange stations) and secondary networks (exchange stations to end-users). Secondary networks directly serve users, and their operational characteristics and control strategies significantly impact heating quality and energy efficiency. Although primary network bypass technology improves flexibility through hydraulic decoupling, the high initial investment limits its widespread adoption. Therefore, this study focuses on secondary networks and their key control technologies.

In practical operations, achieving a supply–demand balance at heat exchange stations is essential for maintaining heating quality [10]. However, due to inherent time delays, conventional feedback control struggles to adapt to rapid load variations and thermal source fluctuations. This study proposes a Model Predictive Control (MPC) strategy. MPC enables proactive adjustments by constructing a thermodynamic model to predict disturbance impacts and optimizing valve opening commands using real-time pipeline temperature and user load data. Recent studies have explored similar approaches: Knudsen et al. implemented an MPC-based coordinated thermal storage/discharge control strategy, reducing pressure fluctuations to $\pm 3\%$ and response times to 15 s in a Danish heat exchange station [11]. X et al. developed a lightweight MPC algorithm (delay < 50 MS) integrated with a digital twin platform, improving dynamic response speeds by 40% [12]. Existing research primarily optimizes control methods and operational stability, whereas this study emphasizes valve control performance to enhance supply–demand balance accuracy. Notably, mechanical limitations (e.g., dead zones, hysteresis) cause random fluctuations in the openings of heat exchange control valves and bypass control valves around their setpoints. These fluctuations induce flow deviations and pressure differential variations across bypass control valves, ultimately affecting secondary network supply temperatures and disrupting thermal balance [13].

This study investigates two valves regulating bypass flow and heat exchange flow (specifically the heat exchange control valve and bypass control valve, hereafter collectively termed "control valves") to analyze their flow characteristics and critical parameters (e.g., maximum/minimum controllable impedance) [14] on the supply–demand balance of thermal substations, with numerical quantification of these effects. The findings will provide theoretical and technical guidance for optimizing secondary network control strategies, advancing the heating industry toward efficient, low-carbon, and stable operations.

2. Technical Methodology

As shown in Figure 1, this figure illustrates the logical flow of the parameter calculation for control valves, with the calculations performed using MATLAB R2022a software. The process is divided into theoretical parameter calculation and actual parameter calculation, followed by an evaluation of the valve control accuracy. Below is a detailed logical description:

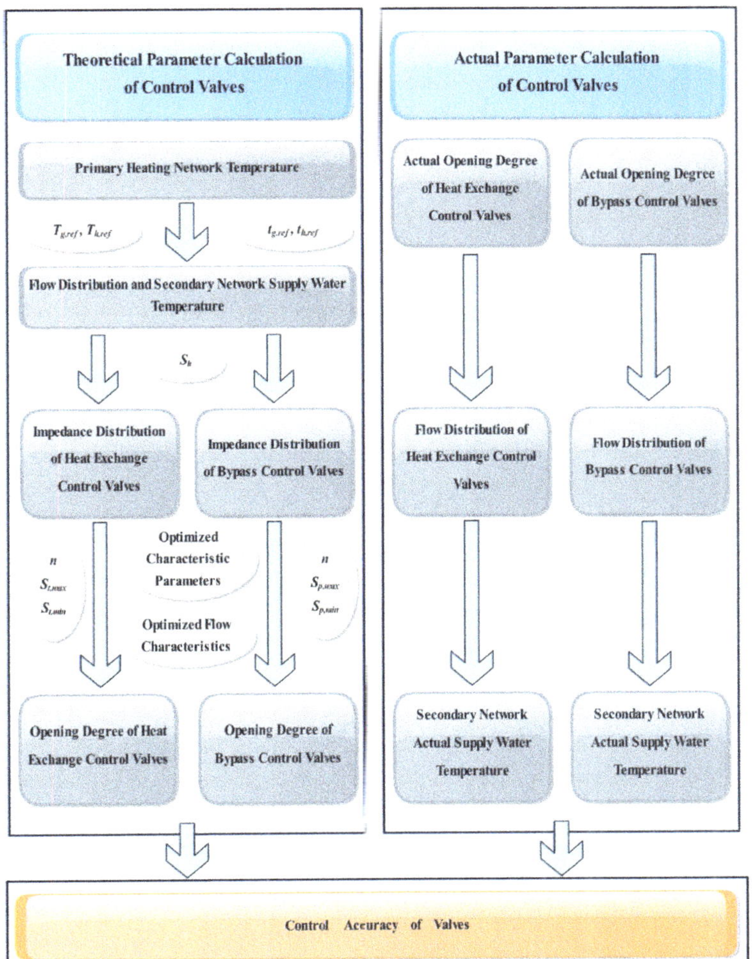

Figure 1. The logical diagram of this article.

2.1. Theoretical Parameter Calculation Section

The theoretical parameter calculation is performed under ideal conditions, assuming that the valves have no dead zone and all parameters can be precisely controlled. This calculation provides benchmark data for evaluating the performance of the valves and the balance state of the network under ideal conditions. Through Formulas (1)–(4), (7), (9), and (12), the theoretical flow distribution, impedance distribution, valve opening degree, and secondary network supply water temperature are calculated, providing a reference for subsequent practical applications.

2.2. Actual Parameter Calculation Section

In practical applications, the valves have a dead zone, causing the actual opening degree to fluctuate near the set value. This directly affects the flow distribution and the secondary network's supply water temperature. Therefore, the actual parameter calculation, based on the theoretical parameters, takes into account the influence of the dead zone. Through Formulas (13) and (14), the flow distribution is recalibrated, and through Formula (15), the actual supply water temperature of the secondary network is calculated. This process is closer to the real operating environment and can reflect the dynamic balance state of the actual network.

2.3. Comparison and Evaluation

By comparing the results of the theoretical parameter calculation and the actual parameter calculation, the impact of the valve dead zone on the network balance can be quantified [15]. Comparing the flow distribution calculated theoretically and actually, the range of flow fluctuations caused by the dead zone can be assessed. Through Formula (12), the deviation of the secondary network supply water temperature between the theoretical and actual calculations can be evaluated. Based on the comparison results of the theoretical parameter calculation and the actual parameter calculation, the control accuracy of the control valves can be comprehensively evaluated [16]. Through quantitative analysis, the optimal working conditions and applicable range of the valves can be identified, providing a scientific basis for the selection and design of the valves, ensuring that they meet the required performance standards in practical applications. This evaluation process not only helps to improve the control accuracy of the system but also provides valuable data support for subsequent optimization design.

3. Thermal Storage Scheme for Secondary Network Bypass Pipelines

Supply–Demand Balance Regulation of Thermal Substations under Secondary Network Bypass Conditions.

As illustrated in Figure 2, the circulation flow rate on the primary side of the plate heat exchanger (PHE) remains constant, a configuration designed to ensure system stability and operational reliability. Maintaining a steady primary-side flow rate is critical for preserving thermal equilibrium and operational efficiency, thereby mitigating instability and potential equipment damage caused by flow fluctuations [17]. For secondary network regulation, constant flow control (i.e., mass regulation) is implemented, ensuring hydraulic stability by minimizing pressure oscillations and hydraulic imbalances induced by flow variations. This approach guarantees a consistent heating service quality for end-users. Under this control strategy, the secondary network supply temperature exhibits a convergence trend toward the set point, while the return water temperature remains nearly constant, reflecting the thermal inertia of buildings [18]. Furthermore, the PHE's equivalent impedance in the secondary network is constant, enabling precise prediction and assessment of its hydraulic impacts during system design and commissioning [19].

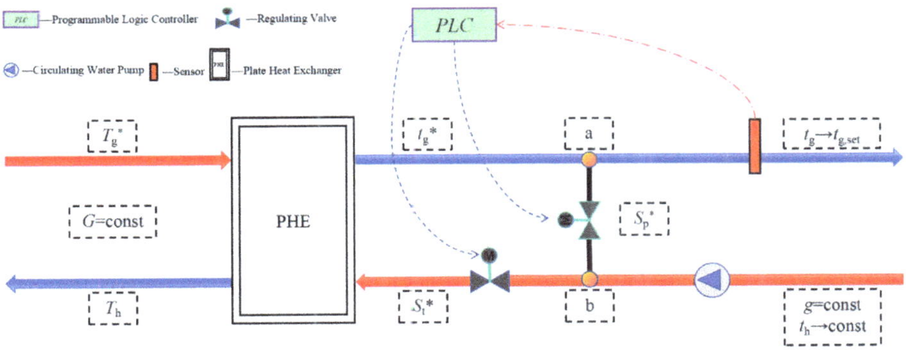

Figure 2. Schematic diagram of supply–demand balance regulation for thermal substations under secondary network bypass conditions.

4. Research on Thermal-Hydraulic Parameters and Control Valve Characteristics of PHE

4.1. Thermal-Hydraulic Parameters of PHE

When circulation pumps operate at a fixed frequency, the equivalent impedance between points a and b remains constant, ensuring a stable secondary network circulation flow rate. During pump frequency adjustments, thermal substations dynamically adjust the impedance of the heat exchange control valve and bypass control valve to maintain constant heat transfer power in the plate heat exchanger. Since the heat supply on the demand side remains constant, the set heat supply equals the heat transfer capacity flowing through the heat exchange pipes. The heat flow ratio and heat exchange output temperature are calculated using the Equation (1):

$$\begin{cases} c_p g\left(t_{g,ref} - t_{h,ref}\right) = c_p \beta g\left(t_g{}^* - t_{h,ref}\right) \\ g^B LMTD_{ref} = (\beta g)^B LMTD* \\ T_g{}^* - T_h{}^* = T_{g,ref} - T_{h,ref} = \Delta T_{ref} \end{cases} \tag{1}$$

Symbol definitions in Equation (1): c_p is specific isobaric heat capacity of water, J/(kg·°C); g is mass flow rate of heated water through the heat exchange, kg/s; $t_{g,ref}$ is baseline supply temperature of the secondary network, °C; $t_{h,ref}$ is baseline return temperature of the secondary network, °C; β is heat exchange flow ratio on the secondary side of the plate heat exchanger (dimensionless); $t_g{}^*$ is heat exchange output temperature, °C; B is empirical coefficient (constant value B = 0.3); *LMTD* is logarithmic mean temperature difference, °C; $T_{g,ref}$ is baseline supply temperature of the primary network, °C; $T_{h,ref}$ is baseline return temperature of the primary network, °C; $T_g{}^*$ is actual supply temperature of the primary network, °C; $T_h{}^*$ is actual return temperature of the primary network, °C.

The logarithmic mean temperature difference (LMTD) is calculated using Equation (2)

$$\begin{cases} LMTD_{ref} = \dfrac{\left(T_{g,ref} - t_{g,ref}\right) - \left(T_{h,ref} - t_{h,ref}\right)}{\ln \frac{T_{g,ref} - t_{g,ref}}{T_{h,ref} - t_{h,ref}}} \\ LMTD^* = \dfrac{\left(T_g{}^* - t_g{}^*\right) - \left(T_h{}^* - t_{h,ref}\right)}{\ln \frac{T_g{}^* - t_g{}^*}{T_h{}^* - t_{h,ref}}} \end{cases} \tag{2}$$

Based on Equations (1) and (2), the quantitative expressions for the relative heat transfer coefficient and the adjusted logarithmic mean temperature difference are formulated as Equation (3).

$$\begin{cases} K^* = \beta^B \\ LMTD^* = \frac{(T_g{}^* - t_g{}') - (T_h{}^* - t_{h,ref})}{\ln \frac{T_g{}^* - t_g{}'}{T_h{}^* - t_{h,ref}}} \end{cases} \tag{3}$$

4.2. Control Valve Impedance

In the calculation of heating network impedance, the equivalent impedance between points a and b remains constant. This impedance value corresponds to the original pipeline (i.e., the equivalent impedance of the secondary network without the addition of a bypass pipe). After topological reconstruction, the new equivalent impedance consists of two parts: one is the equivalent impedance of the heat regulation valve and plate heat exchanger in series in the secondary network; the other is the impedance of the bypass control valve. These two parts of the impedance are ultimately combined in parallel to form the reconstructed equivalent impedance.

Impedance Calculation for Heat Exchange and Bypass Control Valves under Different Primary Network Supply Temperatures.

$$\begin{cases} (S_h + S_t{}^*)^{-0.5} + (S_p{}^*)^{-0.5} = (S_h + S_t)^{-0.5} \\ \frac{\beta}{1-\beta} = \left(\frac{S_p{}^*}{0.5 + S_t{}^*}\right)^{0.5} \end{cases} \tag{4}$$

Symbol definitions in Equation (4): S_h is the equivalent impedance of plate heat exchanger in secondary network; $S_t{}^*$ is impedance of the heat exchange control valve; $S_p{}^*$ is impedance of bypass control valve.

4.3. Valve Opening for Different Flow Characteristics

Control valves play a critical control function in heating systems, where their operation fundamentally relies on precise valve opening adjustments. In practical heating scenarios, the impedance characteristics of control valves significantly influence the fluid flow dynamics within the network [20]. Changes in valve opening directly alter impedance, thereby affecting flow distribution and heat transfer efficiency across the system. To achieve efficient and accurate regulation, it is essential to establish a quantitative relationship between valve impedance and opening.

Control valves exhibit distinct flow characteristic curves, such as linear flow characteristic, equal percentage flow characteristic, quick-opening flow characteristic, and parabolic flow characteristic [21]. By analyzing and comparing valves with different flow characteristics, combined with the specific parameters of the heating system (e.g., temperature range, pressure differentials, and load variations), the most suitable flow characteristic can be selected to optimize system performance [22]. Control valves are defined by two independent parameters: maximum controllable impedance and minimum controllable impedance. Systematic optimization of these parameters ensures the valve meets operational requirements. In real-world applications, deviations between actual valve opening and theoretical valve opening are inevitable. By comparing these values under various operating conditions, error data can be collected and analyzed to identify error distribution patterns and primary influencing factors [23].

4.3.1. Flow Characteristics of Heat Exchange Control Valves

$$\beta = \frac{\overline{Q}(\xi)}{\sqrt{(1 - S_v)\left[\overline{Q}(\xi)\right]^2 + S_v}} \quad (5)$$

Symbol definitions in Equation (5): $\overline{Q}(\xi)$ is functional expression of relative flow vs. relative opening; R is an adjustable ratio. ξ is relative opening; S_v is valve authority.

$$\frac{d(\overline{Q})}{d(\xi)} = k\overline{Q}^n \quad (6)$$

By substituting Equation (6) into Equation (5), the quantitative relationship between the opening and impedance of the heat exchange control valve is derived, as shown in Equation (7).

$$\begin{cases} \xi_t{}^* = \frac{\left[\frac{S_v}{\beta^{-2}+S_v-1)}\right]^{\frac{1-n}{2}} - R^{n-1}}{1-R^{n-1}} & (n \neq 1) \\ \xi_t{}^* = \frac{1}{2}\log_{R_t}\left[\frac{S_v}{S_v+\beta^{-2}-1}\right] + 1 & (n = 1) \end{cases} \quad (7)$$

4.3.2. Flow Characteristics of Bypass Control Valves

$$\frac{Q}{Q_{max}} = \sqrt{\frac{S_{p,\min}}{S_p{}^*}} \quad (8)$$

Symbol definitions in Equation (8): $S_{p,min}$ is full-open impedance of bypass control valve.

By substituting Equation (8) into Equation (6), the quantitative relationship between the valve opening and impedance of the bypass control valve is derived, as shown in Equation (9).

$$\begin{cases} \xi_p{}^* = \frac{(S_{p,\min}/S_p{}^*)^{\frac{1-n}{2}} - R^{n-1}}{1-R^{n-1}} & (n \neq 1) \\ \xi_p{}^* = \left[2 - \frac{\lg(S_p{}^*) - \lg(S_{p,\min})}{\lg(R_p)}\right]/2 & (n = 1) \end{cases} \quad (9)$$

4.4. *Characteristic Parameters*

The independent parameters defining the regulation characteristics of a control valve are the maximum controllable impedance and minimum controllable impedance. The impact of each parameter on the regulation performance is analyzed. As shown in Equation (11), an increase in the equivalent impedance of the plate heat exchanger in the secondary network reduces the valve authority, leading to a significant distortion in the valve's flow characteristics. In practical applications, the valve authority should not be less than 0.3 [24].

$$R = \frac{Q_{\max}}{Q_{\min}} = \sqrt{\frac{\Delta P/S_{\min}}{\Delta P/S_{\max}}} = \sqrt{\frac{S_{\max}}{S_{\min}}} \quad (10)$$

Symbol definitions in Equation (10): Q_{max} is maximum flow rate through the control valve; Q_{min} is minimum flow rate through the control valve.

The relationship between the valve authority and impedance of the control valve is expressed in the Equation (11).

$$Sv = \frac{S_{\min}}{S_{\min} + S_h} \tag{11}$$

Symbol definitions in Equation (11): S_{min} is minimum controllable impedance of the control valve; S_h is equivalent impedance of the plate heat exchanger in the secondary network.

5. Quantitative Analysis of Valve Control Accuracy

Control Accuracy Calculation

Variations in the actual valve opening alter the flow rate and pressure differential across the bypass control valve, thereby affecting the supply–demand balance of the heat exchange station [25]. This metric quantifies the fluctuation amplitude of the secondary network temperature caused by the valve dead zone, with its calculation formula provided in Equation (12).

$$VCA = \left[(t'_g - t_g{}^*)^2 + (t''_g - t_g{}^*)^2\right]/2 \tag{12}$$

Symbol definitions in Equation (12): *VCA*—Valve Control Accuracy; $t_g{}'$ and $t_g{}''$ are secondary network supply temperature corresponding to the actual valve opening;

Flow Distribution Formula for Heat Exchange Control Valve Opening in Equation (13).

$$\begin{cases} \beta = \dfrac{\left[(1-R^{n-1})\xi+R^{n-1}\right]^{\frac{1}{1-n}}}{\sqrt{(1-S_v)\left[(1-R^{n-1})\xi+R^{n-1}\right]^{\frac{2}{1-n}}+S_v}} & (n \neq 1) \\ \beta = \dfrac{R^{(\xi-1)}}{\sqrt{(1-S_v)R^{2(\xi-1)}+S_v}} & (n = 1) \end{cases} \tag{13}$$

The relationship between valve opening and flow distribution is expressed as Equation (14):

$$\begin{cases} \beta = \left[\dfrac{1}{(S_{\text{p,min}})^{0.5} \quad \left[(1-R^{\text{n}-1})\xi_p{}^*+R^{\text{n}-1}\right]^{\frac{1}{n-1}} \quad -1} + 1\right]^{-1} & (n \neq 1) \\ \beta = \left[\dfrac{1}{(S_{\text{p,min}})^{0.5} R^{1-\xi \text{p}^*} -1} + 1\right]^{-1} & (n = 1) \end{cases} \tag{14}$$

Based on Equations (1) and (3), the secondary network supply temperature is derived as follows Equation (15):

$$t_g{}^* = 40 + 10/\beta \tag{15}$$

6. Results and Discussion

6.1 Calculation of Related Parameters

The heat output temperature, heat flow ratio, relative heat transfer coefficient, and adjusted logarithmic mean temperature difference (LMTD), calculated using Equations (1) and (3), vary with the primary network heating temperature as shown in Figure 3.

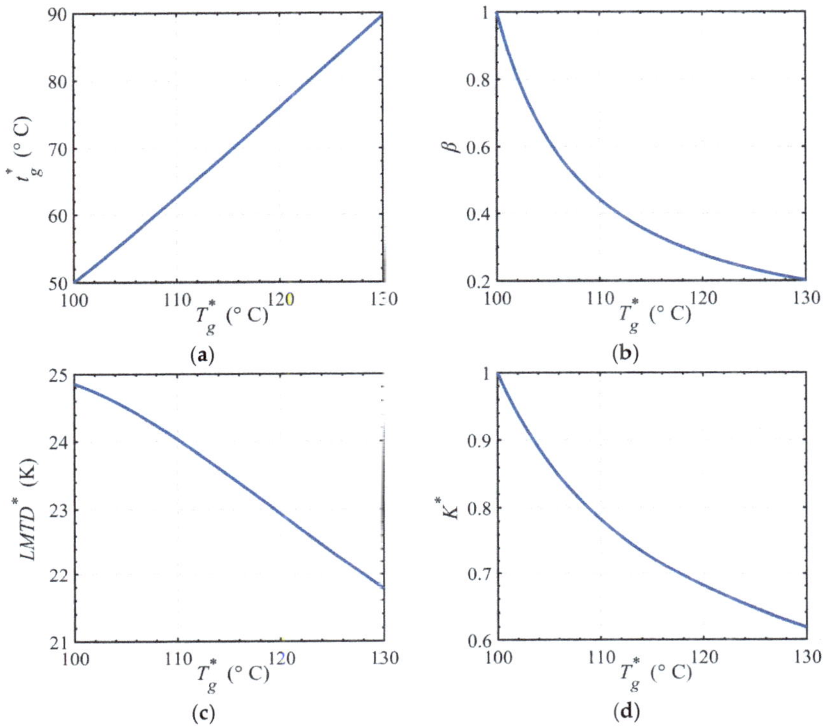

Figure 3. Temperature-dependent curves of related parameters. (**a**) Heat output temperature vs. temperature curves; (**b**) Heat flow ratio vs. temperature curves; (**c**) Relative heat transfer coefficient vs. temperature curves; (**d**) Logarithmic mean temperature difference (LMTD) vs. temperature curves.

As derived from Equation (4), the impedance of heat exchange and bypass control valves varies with the primary network supply temperature, as shown in Figure 4.

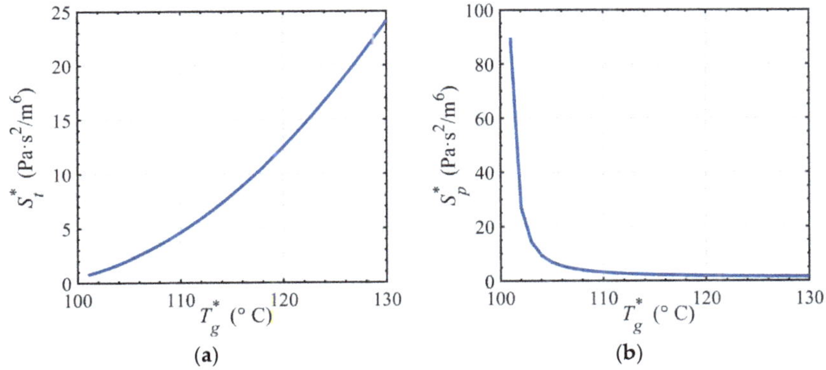

Figure 4. Temperature-dependent impedance curves of control valves. (**a**) Impedance vs. temperature curves for heat exchange control valves; (**b**) Impedance vs. temperature curves for bypass control valves.

6.2. Heat Exchange Control Valve

In practical applications, the valve authority (Sv) should not be less than 0.3. According to Equation (11), the minimum controllable impedance of the heat exchange control valve must exceed 0.21. The parameter summary of the heat exchange control valve is provided in Table 1.

Table 1. Summary of parameters for impedance control range of heat exchange control valves.

Impedance Range	Minimum Controllable Impedance	Maximum Controllable Impedance
0.75~24.13	0.21~0.75	30.29~∞

6.2.1. Impact of Minimum Controllable Impedance

For control valves with equal percentage flow characteristics, the maximum controllable impedance is fixed at 30.29. By varying the minimum controllable impedance, the relationship between valve opening and temperature is analyzed, with results illustrated in Figure 5.

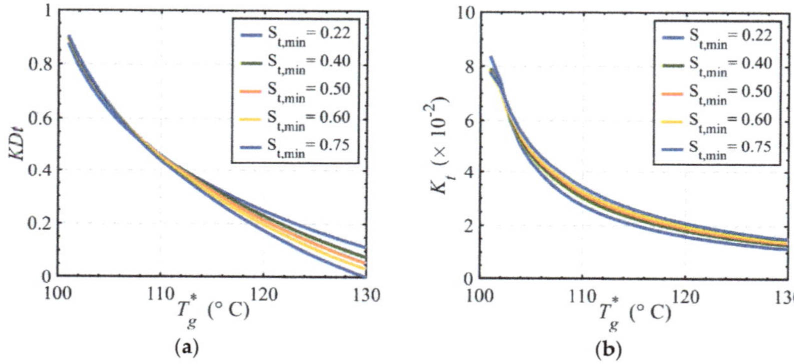

Figure 5. Control valve opening-temperature curves with varying minimum controllable impedance. (**a**) Valve opening vs. temperature curves; (**b**) Slope of valve opening curves.

As shown in Figure 5a, increasing the minimum controllable impedance shifts the entire curve downward in the coordinate system, expanding the valve opening range. For example, at a minimum controllable impedance of 0.75, the valve opening range reaches 0.9001. When the minimum controllable impedance decreases to 0.22, the opening range contracts to 0.7605. This demonstrates a positive correlation between the minimum controllable impedance and the valve opening range within a specific interval: larger minimum controllable impedance increases the operational range of the control valve. A higher minimum controllable impedance indicates that the valve must overcome greater resistance during opening, enabling a broader regulation capability to adapt to varying system demands. Curve (b) illustrates the sensitivity of the valve opening to temperature variations. Across most temperature regions, a higher minimum controllable impedance corresponds to a steeper slope in the curve. This indicates enhanced sensitivity: even minor temperature fluctuations trigger significant adjustments in the valve opening.

6.2.2. Impact of Maximum Controllable Impedance

For control valves with equal percentage flow characteristics, the minimum controllable impedance is fixed at 0.75, while the maximum controllable impedance varies starting from 30.29. The relationship between valve opening and temperature under different maximum impedance values is analyzed, with results illustrated in Figure 6.

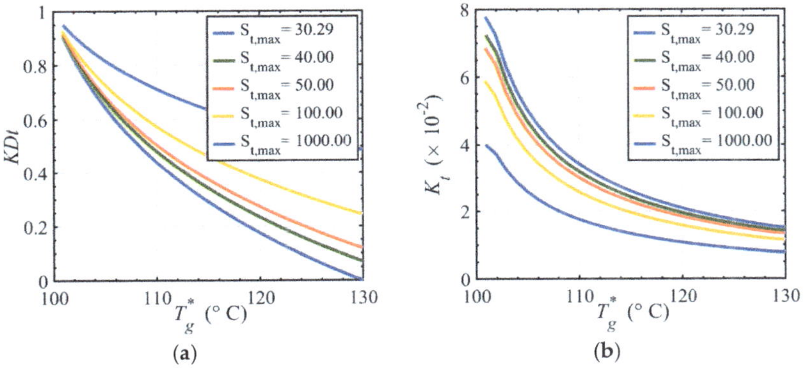

Figure 6. Control valve opening-temperature curves with varying maximum controllable impedance. (**a**) Valve opening vs. temperature curves; (**b**) Slope of valve opening curves.

As shown in Figure 6a, smaller maximum controllable impedance shifts the overall curve downward in the coordinate system, resulting in a larger valve opening range. For example, at a maximum controllable impedance of 30.29, the valve opening range reaches 0.9057. When the maximum controllable impedance increases to 1000.0, the valve opening range significantly contracts to 0.4656. This indicates a negative correlation between the maximum controllable impedance and the valve opening range within a specific interval: lower maximum controllable impedance expands the operational range of the control valve. The maximum controllable impedance reflects the flow resistance imposed by the fully open valve on fluid dynamics. A smaller maximum controllable impedance implies reduced flow resistance, enabling the valve to operate over a wider opening range to accommodate diverse heating demands. Curve (b) demonstrates that under the same primary network supply water temperature, a smaller maximum controllable impedance corresponds to a steeper slope, indicating higher sensitivity of the control valve opening to temperature variations. This implies that even slight temperature changes will trigger substantial adjustments in valve opening.

6.2.3. Precision Comparison of Heat Exchange Control Valves with Different Flow Characteristics and Metrics

The flow characteristics and control precision of heat exchange control valves play a critical role in ensuring stable system operation and efficient heat supply. Based on previous research, this study selects a set of characteristic parameters with a maximum controllable impedance of 30.29 and a minimum controllable impedance of 0.75. Under these conditions, the valve opening vs. temperature curves for four flow characteristics (linear, equal percentage, quick-opening, and parabolic) are plotted and analyzed, as illustrated in Figure 7.

For heat-exchange control valves in secondary network bypass configurations, the quick-opening characteristic control valve (Curve a) exhibits distinct segmented behavior. The average slope of its characteristic curve reaches 0.1115 in the low-temperature range (100–105 °C), indicating extreme sensitivity to temperature variations and exceptionally high control accuracy. However, in the mid-to-high temperature range (105–130 °C), the average slope drops significantly to 0.0093, leading to severe degradation of control precision. Curve (b) characterizes the fluctuation of heat exchange control valve errors induced by dead zone effects under varying temperatures. Specifically, a higher vertical coordinate value indicates stronger operational volatility and greater deviations from the preset opening at the corresponding temperature. For valves with quick-opening characteristics, the mean slope of the accuracy curve in the low-temperature region (100–105 °C) is 0.6229, indicating relatively minor fluctuations and errors between the actual and preset valve openings within this range. In contrast, the medium-to-high-temperature region (105–130 °C) exhibits a significantly steeper mean slope of 3.6162, corresponding to amplified operational volatility. Therefore, considering its performance across temperature ranges, the quick-opening valve is optimal for scenarios where primary network supply temperatures fluctuate primarily in the low-temperature zone, leveraging its high-precision regulation in this regime while avoiding precision losses at elevated temperatures.

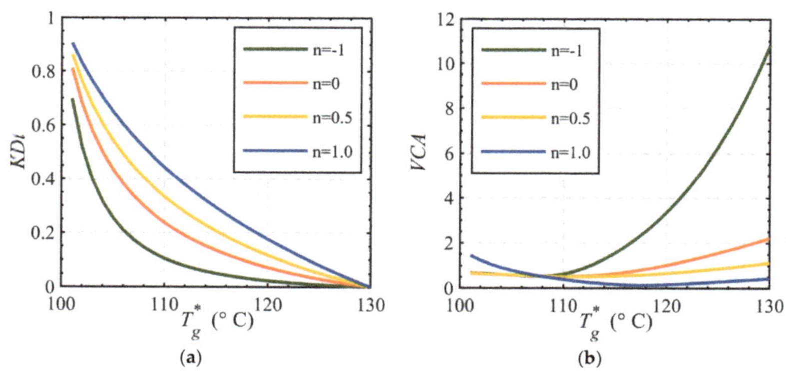

Figure 7. Valve opening vs. temperature curves for control valves with four flow characteristics. (**a**) Valve opening vs. temperature curves; (**b**) Control accuracy vs. temperature curves.

Replacing the valve with an equal percentage characteristic valve results in a more balanced slope distribution. For equal percentage valves, higher full-open impedance (up to an upper bound) increases the average slope across the entire temperature range, enhancing control accuracy. The equal percentage valve achieves an average slope of 3.17 over 100–130 °C, delivering robust overall control precision. The accuracy curve of the equal percentage-type control valve displays a mean slope of 0.4349 across the full temperature range. This value suggests that within this temperature interval, the actual opening of the valve exhibits minor fluctuations and reduced deviations compared to the preset opening, signifying lower errors. This demonstrates the broad applicability of equal percentage valves in heating systems, providing stable and reliable regulation under diverse thermal conditions.

6.3. *Bypass Control Valves*

The parameter summary of bypass control valves is presented in Table 2.

Table 2. Summary of impedance control ranges for bypass control valves.

Impedance Range	Minimum Controllable Impedance Range	Maximum Controllable Impedance Range
1.57~89.68	0~1.57	89.68~+∞

6.3.1. Impact of Minimum Controllable Impedance

Using control valves with linear flow characteristics as the research subject, the valve opening vs. temperature curves under different minimum controllable impedance values are illustrated in Figure 8, while the maximum controllable impedance is fixed at 89.68.

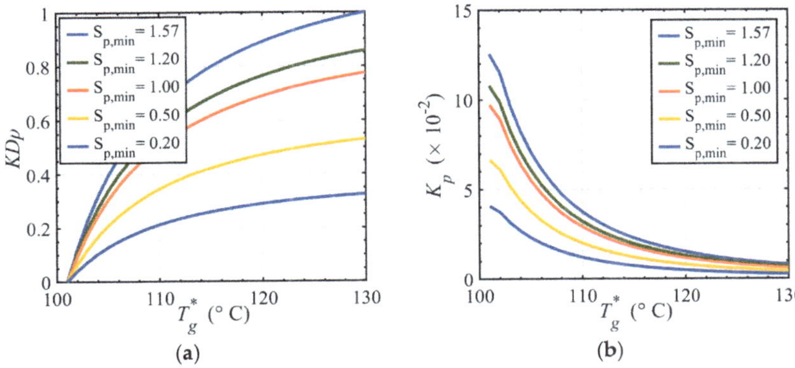

Figure 8. Control valve opening-temperature curves with varying minimum controllable impedance. (**a**) Valve opening vs. temperature; (**b**) Slope of valve opening curve.

As shown in Curve (a) of Figure 8, increasing the minimum controllable impedance shifts the entire curve upward within the coordinate system. For instance, at a minimum controllable impedance of 1.57, the valve opening range reaches 1.0 (full scale). When the minimum controllable impedance decreases to 0.2, the valve opening range contracts to 0.3252. This demonstrates a positive correlation between the minimum controllable impedance and the valve opening range within a specific interval: larger minimum controllable impedance expands the operational range of the control valve.

6.3.2. Impact of Maximum Controllable Impedance

For control valves with linear flow characteristics (fixed minimum controllable impedance at 1.57), the valve opening vs. temperature curves under different maximum controllable impedance values are illustrated in Figure 9.

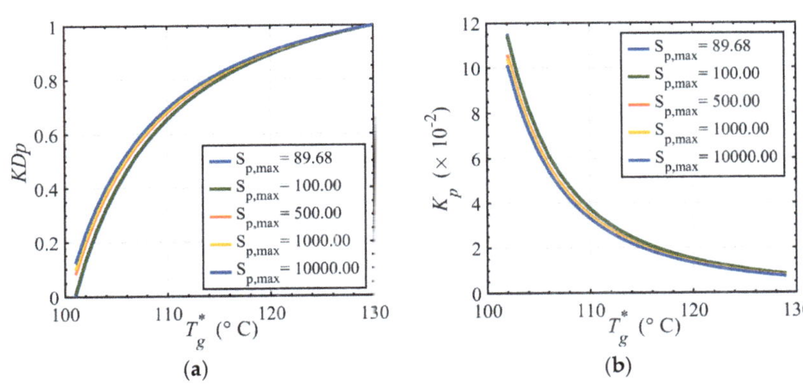

Figure 9. Control valve opening-temperature curves with varying maximum controllable impedance. (**a**) Valve opening vs. temperature; (**b**) Slope of valve opening curve.

As shown in Figure 9a, smaller maximum controllable impedance shifts the overall curve downward in the coordinate system, resulting in a larger valve opening range. A maximum controllable impedance of 89.68 corresponds to a valve opening range of 1.0 (full scale). When the maximum controllable impedance increases to 10,000, the valve opening range decreases to 0.8837. This demonstrates that the maximum controllable impedance regulates the effective operating range of the valve. Within a specific range, a lower maximum controllable impedance expands the valve opening range.

6.3.3. Precision Comparison of Bypass Control Valves with Different Flow Characteristics and Metrics

In conclusion, the bypass control valve is selected with a maximum controllable impedance of 89.68 and a minimum controllable impedance of 1.57. The valve opening vs. temperature curves for four flow characteristics (linear, equal percentage, quick-opening, and parabolic) are illustrated in Figure 10.

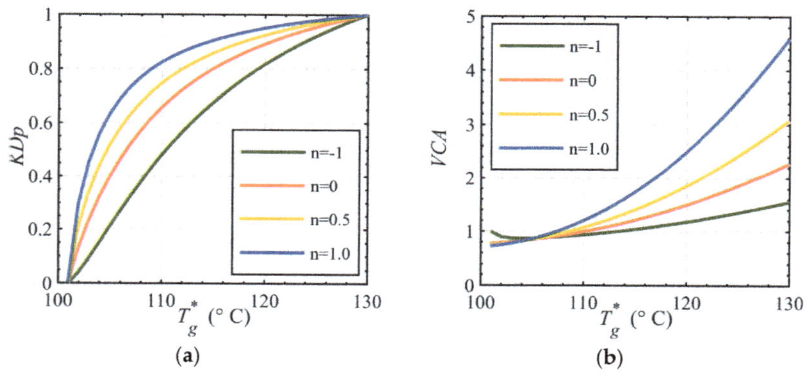

Figure 10. Valve opening vs. temperature curves for control valves with four flow characteristics. (**a**) Valve opening vs. temperature; (**b**) Control accuracy vs. temperature.

For bypass control valves with equal percentage flow characteristics in secondary network bypass systems, Curve (a) shows an average slope of 9.85 in the low-temperature range (100–106 °C), achieving exceptional control accuracy. In the mid-to-high temperature range (106–130 °C), the average slope decreases to 1.68, resulting in significant degradation of control precision. The accuracy curve of the equal percentage flow characteristic control valve exhibits a mean slope of 0.8090 in the low-temperature region (100–106 °C), indicating minor fluctuations and reduced deviations between the actual and preset valve openings within this range. In contrast, the mean slope increases significantly to 2.4378 in the medium-to-high-temperature region (106–130 °C), corresponding to amplified fluctuations and diminished control precision. Consequently, when evaluating the performance of the equal percentage flow characteristic bypass control valve across different temperature intervals, it is better suited for scenarios where the primary network supply temperature predominantly fluctuates within the low-temperature range. Under such conditions, the valve leverages its high-precision regulation advantage in the low-temperature region while mitigating the inherent limitations of reduced accuracy in high-temperature operations.

When replacing the bypass control valve with a quick-opening flow characteristic valve, Curve (a) demonstrates more balanced slope distribution across the entire temperature range. For quick-opening heat exchange control valves, higher full-open impedance (up to an upper bound) correlates with increased average slope values, enhancing control accuracy. Specifically, the average slope across the full temperature range reaches 3.49, indicating robust overall control performance. The accuracy curve of the quick-opening characteristic control valve exhibits a mean slope of 1.1114 across the full temperature range, indicating minor fluctuations and reduced deviations between the actual opening and preset opening within this interval. In summary, the equal percentage characteristic valve demonstrates broad applicability in heating systems, delivering stable and reliable regulation performance under varying temperature conditions.

7. Optimization Verification Based on Real Heating Network

7.1. Basic Information of the Heating System

The centralized heating system is located in Zhangjiakou City, Hebei Province. The heating period lasts 141 days (from 1 November to 31 March of the following year). The design heat load was calculated using the climate correction method, with baseline parameters under −13.6 °C temperature conditions. The heating radius is approximately 11.4 km, covering a total heating area of 11.5 million m², and the total design heat load reaches 350 MW. The system comprises one heat source (an extraction condensing turbine cogeneration unit) and 51 heating stations. All heating stations have undergone network topology reconstruction, with bypass branches added to the original pipeline structure.

7.2. Model Control Strategy

This study adopts a Model Predictive Control (MPC) strategy, which integrates historical meteorological data and real-time forecasting information to achieve a 6-h-ahead precise prediction of heat load demand. Based on these predictions, the system dynamically generates reference temperatures as control input parameters. Utilizing a Programmable Logic Controller (PLC), the system flexibly adjusts the opening degrees of heat exchange control valves and bypass control valves, thereby enabling refined control of the district heating network.

7.3. Verification Method

During the 2024–2025 heating season, our team selected 16 heating stations with similar operating conditions from the Zhangjiakou Flexible Heating Network Project. Each station was configured with different control valve types as outlined in Table 3.

Table 3. The valve configuration of the heat station.

Group	Heating Station	Heat Exchange Control Valve	Bypass Control Valve
1	Ideal City—North High Area	Quick-opening Type	Quick-opening Type
2	Ideal City—South High Area	Quick-opening Type	Linear Type
3	Ningxi Home—South High Area	Quick-opening Type	Parabolic Type
4	Fuyuanli High Area	Quick-opening Type	Equal Percentage Type
5	Tianxiu Garden No. 2 Station	Linear Type	Quick-opening Type
6	Jingyang Heat Station	Linear Type	Linear Type
7	Yiqingyuan Heat Station	Linear Type	Parabolic Type
8	Fuguiyuan Heat Station	Linear Type	Equal Percentage Type
9	Xinmao Home	Parabolic Type	Quick-opening Type
10	Zhangheng New City	Parabolic Type	Linear Type
11	Tejia Residential Building	Parabolic Type	Parabolic Type
12	Industrial Road Residential Area	Parabolic Type	Equal Percentage Type
13	Zhongxing Residential Building	Equal Percentage Type	Quick-opening Type
14	Guangming Village	Equal Percentage Type	Linear Type
15	Emerald Town	Equal Percentage Type	Parabolic Type
16	Qiantun Heat Station	Equal Percentage Type	Equal Percentage Type

Based on the actual measurement data of each heat station, the stability of the secondary network supply temperature when the primary network supply temperature changes between 70 and 100 °C is observed. If the average deviation between the secondary network supply temperature and the set supply temperature is too large, then the valve configuration is unreasonable; otherwise, the configuration is reasonable.

7.4. Engineering Data Analysis

The average temperature deviation of the secondary network supply for each heat station is shown in Table 4:

Table 4. The heat exchange flow fluctuates.

Heating Substation	1	2	3	4	5	6	7	8
average	3.8	3.9	4.0	4.0	3.0	3.1	3.2	3.3
Heating Substation	**9**	**10**	**11**	**12**	**13**	**14**	**15**	**16**
average	2.7	2.9	3.1	3.3	2.5	2.7	2.9	3.1

From the figure above, it can be seen that the 13th heating substation has the smallest average temperature deviation, only 2.5 °C. The experimental validation section constructed 16 combination scenarios through pairing four characteristic types of heat exchange control valves with four characteristic types of bypass control valves. From 55 heating stations, 16 typical stations with complete operational data were selected for comparative analysis. Based on the research conclusion—minimum temperature deviation fluctuation occurs when using equal percentage heat exchange control valves paired with quick-opening bypass control valves—the following two contrast groups were specifically chosen:

Group 1 (Numbers 13, 14, 15, 16): The heat exchange control valves are fixed as equal percentage type, with bypass control valve configurations covering various characteristics such as linear and parabolic, to validate the stability advantages of equal percentage heat exchange control valves in diverse bypass scenarios.

Group 2 (Numbers 1, 5, 9, 13): The bypass control valves are fixed as quick-opening type, with heat exchange control valves covering various characteristics such as linear and parabolic, to quantify the enhancement effect of quick-opening bypass control valves on the system's disturbance rejection capability.

The control valve configurations of each heating substation are shown in Table 5.

Table 5. Valve configuration of the selected heat exchange stations.

Group	Heating Station	Heat Exchange Control Valve	Bypass Control Valve
1	Ideal City—North High Area	Quick-opening Type	Quick-opening Type
5	Tianxiu Garden No. 2 Station	Linear Type	Quick-opening Type
9	Xinmao Home	Parabolic Type	Quick-opening Type
13	Zhongxing Residential Building	Equal Percentage Type	Quick-opening Type
14	Guangming Village	Equal Percentage Type	Linear Type
15	Emerald Town	Equal Percentage Type	Parabolic Type
16	Qiantun Heat Station	Equal Percentage Type	Equal Percentage Type

As shown in Figure 11a, where all heat exchange control valves are equal percentage type control valves, it can be seen that when the bypass control valves are quick-opening type control valves, the fluctuation of the secondary network supply temperature is lower. In Figure 11b, where all bypass control valves are quick-opening type control valves, it is observed that when the heat exchange control valves are equal percentage type control valves, the fluctuation of the secondary network supply temperature is also lower. The above observed data can intuitively validate the correctness of the conclusion of this paper.

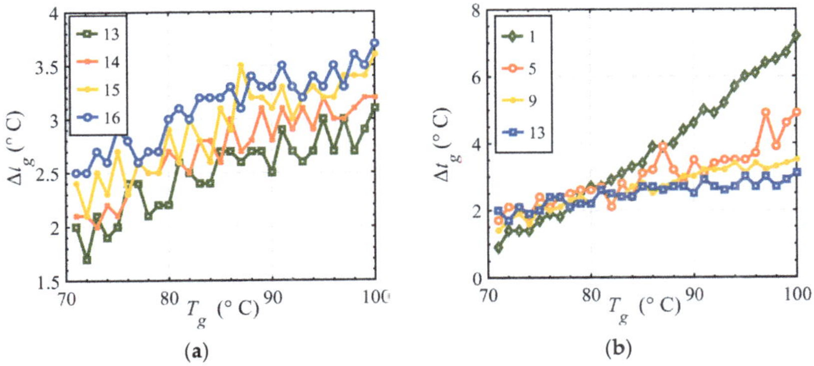

Figure 11. (**a**) The comparison for Heat Substations Nos 13, 14, 15, and 16. (**b**) The comparison for Heat Substations Nos 1, 5, 9, and 13.

8. Conclusions

This study focuses on a heating network system based on BTS technology, establishing a theoretical framework for valve selection and control strategies with engineering significance through quantitative analysis of dynamic regulation performance under specific operating conditions. By overcoming the limitations of traditional single-performance-index evaluation, the research reveals the intrinsic mechanism between valve control impedance characteristics and supply–demand balance from a system coupling perspective, providing new cognitive dimensions for heating system optimization.

8.1. System Impact Mechanism of Control Impedance Characteristics

This study demonstrates that when the controllable impedance range of control valves covers system operational requirements, a combination of higher minimum control impedance and lower maximum control impedance significantly enhances the regulation stability of heat exchange stations. This is manifested in smoother valve opening adjustments, stabilized dynamic responses in heat exchange processes, and effective suppression of overload phenomena and oscillation effects during regulation. Notably, the minimum control impedance is constrained by a critical threshold of valve authority. When valve authority falls below the theoretical optimal range, nonlinear distortion occurs in valve operating characteristics, leading to systemic attenuation of regulation accuracy.

8.2. Valve Type Selection

Comparative analysis demonstrates that selecting equal percentage valves for heat exchange regulation and quick-opening valves for bypass regulation achieves optimal control performance with minimal variation, enabling control valves to deliver their maximum regulatory efficiency. Experimental verification in practical engineering applications confirms that this configuration attains the most stable supply–demand equilibrium in thermal stations. Critical implementation requires explicit consideration of the operational temperature range in the primary network water supply. For bypass regulation valves, quick-opening valves exhibit enhanced control capability in low-temperature regimes (100–106 °C), while equal percentage valves demonstrate superior control performance in high-temperature ranges (106–130 °C).

8.3. Practical Implications

The research outcomes hold significant real-world applicability. For instance, urban heating network decarbonization retrofits can implement closed-loop control strategies integrating "peak-shifting heat storage—dynamic allocation—energy efficiency optimization", directly improving energy utilization rates and heating service quality. Additionally, bypass pipeline heat storage technology enhances the peak-shaving capacity of combined heat and power (CHP) units, addressing the core challenge of renewable energy (e.g., wind and solar power) intermittency and volatility, thereby promoting renewable energy integration.

8.4. *Research Boundaries and Knowledge Expansion Pathways*

Current research is constrained by regional engineering data limitations (Zhangjiakou flexible heating network), leaving theoretical gaps in:

1. Universal applicability validation under extreme cold/humid–hot climates;
2. Nonlinear dynamic modeling of multi-valve coupling in complex networks;
3. Long-term material aging effects on valve characteristics.

Future research will expand along three dimensions:

1. Cross-scale data validation: Establish multi-climate zone data collection networks covering frigid regions (e.g., Shenyang) and hot summer/cold winter zones (e.g., Shanghai), developing GIS-based model parameter correction mechanisms.
2. Complex system modeling: Introduce fluid network theory and Nash equilibrium models to construct impedance distribution optimization models considering multi-valve coupling effects, achieving optimal energy flow allocation through equilibrium solutions.
3. Long-term performance evolution: Implement a 5-year tracking observation program for the Zhangjiakou project, developing time-varying prediction models for valve characteristic degradation and optimizing lifecycle-based control strategies.

Author Contributions: J.Y.: Writing—original draft, Conceptualization, Data Curation, Visualization. J.W.: Supervision, Writing—review and editing, Software, Funding Acquisition. P.Z.: Conceptualization, Data Curation, Visualization. All authors have read and agreed to the published version of the manuscript.

Funding: This study was supported by the National Natural Science Foundation of China (No. E2208104) and the Hebei Natural Science Foundation (No. E2024202065).

Data Availability Statement: The original contributions presented in this study are included in this article, further inquiries can be directed to the corresponding author.

Acknowledgments: We would like to thank the Hebei Gongda Keya Group Co., Ltd. for providing the necessary data and field study platform. Jinda Wang would like to thank the support from Initial Research Funding of Hebei University of Technology.

Conflicts of Interest: The authors declare that the research was conducted in the absence of any commercial or financial relationships that could be construed as conflicts of interest.

Abbreviations

The following abbreviations are used in this manuscript:

CHP	Combined Heat and Power
BTS	Bypass Thermal Storage
LMTD	Logarithmic Mean Temperature Difference
PHE	Plate Heat Exchanger
MPC	Model Predictive Control
VCA	Valve Control Accuracy

Nomenclature

The following nomenclature is used in this manuscript:

ymbol	Significance Represented	Symbol	Significance Represented
$_{g,ref}$	reference supply water temperature of the primary network, K	$S_p{}^*$	Impedance distribution of the secondary side bypass control valves of the heat substation, Pa·s^2/m^6
$_{h,ref}$	reference return water temperature of the primary temperature, K	LMTD$_{ref}$	reference LMTD of PHE, K
$_{g,ref}$	reference supply water temperature of the secondary network, K	LMTD *	LMTD of PHE after throttling regulation, K
$_{h,ref}$	reference return water temperature of secondary network, K	ξ	relative opening of the control valve
$_p$	constant-pressure specific heat of the heat medium,	$\xi_t{}^*$	relative opening of the heat exchange control valve
	Characteristic curve exponent	$\xi_p{}^*$	relative opening of the bypass control valve
	proportion of the bypass control valve cost in the total cost of the bypass branch	$t_g{}'$, $t_g{}''$	secondary-side supply temperature corresponding to the actual valve opening
$T_g{}^*$, $T_h{}^*$	design supply and return water temperatures of the primary network, K	R	rangeability
$_v$	Valve authority	R_t	rangeability of heat exchange control valves
$_g{}^*$, $t_h{}^*$	design supply and return water temperatures of the secondary network, K	R_p	rangeability of bypass control valves
3	empirical parameters to characterize the convective heat transfer intensity of radiators	$S_{p,min}$	minimum impedance of bypass regulating on the secondary side of the heat substation, Pa·s^2/m^6
K	overall heat transfer coefficient of PHE regulation, W/(m^2·K)	$S_{p,max}$	maximum impedance of bypass regulating on the secondary side of the heat substation, Pa·s^2/m^6
S_h	equivalent impedance of plate heat exchanger in the secondary network, Pa·s^2/m^6	$S_{t,min}$	minimum impedance of heat-exchange regulating on the secondary side of the heat substation, Pa·s^2/m^6
$S_t{}^*$	impedance of the secondary side heat exchange regulating of the heat substation, Pa·s^2/m^6	$S_{t,max}$	maximum impedance of heat-exchange regulating on the secondary side of the heat substation, Pa·s^2/m^6
KDt	relative opening of the valve	Kt	slope of the primary network supply water temperature curve about valve opening

References

1. Barco-Burgos, J.; Bruno, J.C.; Eicker, U.; Saldaña-Robles, A.L.; Alcántar-Camarena, V. Review on the integration of high-temperature heat pumps in district heating and cooling networks. *Energy* **2022**, *239*, 122378. [CrossRef]
2. Ma, M.; Tang, X.; Shi, C.; Wang, M.; Li, X.; Luo, P.; Zhang, B. Roadmap towards clean and low-carbon heating to 2060: The case of northern urban region in China. *Energy* **2023**, *284*, 129181. [CrossRef]
3. Wang, H.; Yin, W.; Abdollahi, E.; Lahdelma, R.; Jiao, W. Modelling and optimization of CHP-based district heating system with renewable energy production and energy storage. *Appl. Energy* **2015**, *159*, 401–421. [CrossRef]
4. Benalcazar, P. Optimal sizing of thermal energy storage systems for CHP plants considering specific investment costs: A case study. *Energy* **2021**, *234*, 121323. [CrossRef]
5. Wang, L.; Zhang, S.; Fu, Y.; Liu, M.; Liu, J.; Yan, J. Heat–power decoupling for the CHP unit by utilizing heat storage in the district heating system integrated with heat pumps: Dynamic modeling and performance analysis. *Energy* **2024**, *306*, 132485. [CrossRef]
6. Wang, J.; Pan, B.; Rong, L.; Sun, C.; Qi, C. Topology reconstruction of the district heating network for maximizing comprehensive benefits of thermal storage. *Energy* **2024**, *313*, 134065. [CrossRef]
7. Kouhia, M.; Laukkanen, T.; Holmberg, H.; Ahtila, P. District heat network as a short-term energy storage. *Energy* **2019**, *177*, 293–303. [CrossRef]
8. Jiang, Y.; Wan, C.; Botterud, A.; Song, Y.; Xia, S. Exploiting Flexibility of District Heating Networks in Combined Heat and Power Dispatch. *IEEE Trans. Sustain. Energy* **2020**, *11*, 2174–2188. [CrossRef]
9. Vivian, J.; Quaggiotto, D.; Zarrella, A. Increasing the energy flexibility of existing district heating networks through flow rate variations. *Appl. Energy* **2020**, *275*, 115411. [CrossRef]
10. Deng, W.; Yao, J.; Ma, D. Robust adaptive precision motion control of hydraulic actuators with valve dead-zone compensation. *ISA Trans.* **2017**, *70*, 269–278. [CrossRef]

11. Knudsen, B.R.; Rohde, D.; Kauko, H. Thermal energy storage sizing for industrial waste-heat utilization in district heating: A model predictive control approach. *Energy* **2021**, *234*, 121200. [CrossRef]
12. Zheng, X.; Shi, Z.; Wang, Y.; Zhang, H.; Tang, Z. Digital twin modeling for district heating network based on hydraulic resistance identification and heat load prediction. *Energy* **2024**, *288*, 129726. [CrossRef]
13. Xu, B.; Su, Q.; Zhang, J.; Lu, Z. Analysis and compensation for the cascade dead-zones in the proportional control valve. *ISA Trans.* **2017**, *66*, 393–403. [CrossRef] [PubMed]
14. Xu, X.; Wang, Y.; Fang, L.; Wang, Z.; Li, Y. Numerical study on cavity-vortex evolution in cavitation flow in nuclear control valve with different openings and structural parameters. *Ann. Nucl. Energy* **2025**, *211*, 110880. [CrossRef]
15. Lee, J.; Moon, G.; Lee, J.; Jun, C.; Choi, J. Evaluation of Methods for Estimating Long-Term Flow Fluctuations Using Frequency Characteristics from Wavelet Analysis. *Water* **2023**, *15*, 2968. [CrossRef]
16. Damarla, S.K.; Sun, X.; Xu, F.W.; Shah, A.; Huang, B.A.; IEEE. A Sigmoid Function based Method for Detection of Stiction in Control Valves. In Proceedings of the 7th IEEE International Symposium on Advanced Control of Industrial Processes (AdCONIP), Univ British Columbia, Vancouver, BC, Canada, 7–9 August 2022; pp. 210–215.
17. Xuan, W.C.; Xu, C.T.; Qian, C.Y.; Wang, J.X.; Jiang, Z.K.; Ma, R.X.; Yu, B.B.; Shi, J.Y.; Chen, J.P. Numerical simulation and thermodynamic test analysis of plate heat exchanger based on topology optimization. *Appl. Therm. Eng.* **2024**, *255*, 17. [CrossRef]
18. Soret, G.M.; Vacca, P.; Tignard, J.; Hidalgo, J.P.; Maluk, C.; Aitchison, M.; Torero, J.L. Thermal inertia as an integrative parameter for building performance. *J. Build. Eng.* **2021**, *33*, 101623. [CrossRef]
19. Bordin, C.; Gordini, A.; Vigo, D. An optimization approach for district heating strategic network design. *Eur. J. Oper. Res.* **2016**, *252*, 296–307. [CrossRef]
20. Li, S.X.; Yu, M.Y.; Wu, H.L.; Hu, Y.G.; Ma, T.Q.; Liu, B.C. Optimization of V-shaped regulating ball valve profile based on proxy optimization algorithm. *Eng. Comput.* **2023**, *40*, 1723–1748. [CrossRef]
21. Fan, Y.; Wang, J.; Lei, Y.; Huang, J.; Peng, L.; Xu, X. Study on flow field and flow control characteristics of pressure independent control valve. *Fluid Mach.* **2022**, *50*, 68–74.
22. Blizard, A.; Stockar, S. A Graph-Based Technique for the Automated Control-Oriented Modeling of District Heating Networks. *J. Dyn. Syst. Meas. Control-Trans. ASME* **2024**, *146*, 12. [CrossRef]
23. Daneshwar, M.A.; Noh, N.M. Identification of a process with control valve stiction using a fuzzy system: A data-driven approach. *J. Process Control* **2014**, *24*, 249–260. [CrossRef]
24. Muniak, D.P. Control Valve with a Constant Inner Authority Value. *J. Therm. Sci.* **2018**, *27*, 487–495. [CrossRef]
25. Nguyen, Q.K.; Jung, K.H.; Lee, G.N.; Park, S.B.; Kim, J.M.; Suh, S.B.; Lee, J. Experimental study on pressure characteristics and flow coefficient of butterfly valve. *Int. J. Nav. Archit. Ocean. Eng.* **2023**, *15*, 100495. [CrossRef]

energies

MDPI

Review

Application Advances and Prospects of Ejector Technologies in the Field of Rail Transit Driven by Energy Conservation and Energy Transition

Yiqiao Li [1,*], Hao Huang [1], Shengqiang Shen [2,*], Yali Guo [2], Yong Yang [2] and Siyuan Liu [1]

1 Zhan Tianyou College, Dalian Jiaotong University, Dalian 116028, China; cqhuanghao@outlook.com (H.H.)
2 School of Energy and Power Engineering, Dalian University of Technology, Dalian 116024, China; ylguo@dlut.edu.cn (Y.G.)
* Correspondence: liyiqiao@djtu.edu.cn (Y.L.); zzbshen@dlut.edu.cn (S.S.)

Abstract: Rail transit as a high-energy consumption field urgently requires the adoption of clean energy innovations to reduce energy consumption and accelerate the transition to new energy applications. As an energy-saving fluid machinery, the ejector exhibits significant application potential and academic value within this field. This paper reviewed the recent advances, technical challenges, research hotspots, and future development directions of ejector applications in rail transit, aiming to address gaps in existing reviews. (1) In waste heat recovery, exhaust heat is utilized for propulsion in vehicle ejector refrigeration air conditioning systems, resulting in energy consumption being reduced by 12~17%. (2) In vehicle pneumatic pressure reduction systems, the throttle valve is replaced with an ejector, leading to an output power increase of more than 13% and providing support for zero-emission new energy vehicle applications. (3) In hydrogen supply systems, hydrogen recirculation efficiency exceeding 68.5% is achieved in fuel cells using multi-nozzle ejector technology. (4) Ejector-based active flow control enables precise $\pm$ 20 N dynamic pantograph lift adjustment at 300 km/h. However, current research still faces challenges including the tendency toward subcritical mode in fixed geometry ejectors under variable operating conditions, scarcity of application data for global warming potential refrigerants, insufficient stability of hydrogen recycling under wide power output ranges, and thermodynamic irreversibility causing turbulence loss. To address these issues, future efforts should focus on developing dynamic intelligent control technology based on machine learning, designing adjustable nozzles and other structural innovations, optimizing multi-system efficiency through hybrid architectures, and investigating global warming potential refrigerants. These strategies will facilitate the evolution of ejector technology toward greater intelligence and efficiency, thereby supporting the green transformation and energy conservation objectives of rail transit.

Keywords: ejector; engine waste heat recovery; compressed air energy storage; hydrogen supply system; pantograph lift control; energy-saving technology

1. Introduction

Under the background of accelerating the transformation of the global energy structure, carbon emissions in the rail transit field account for more than 20% of total emissions [1]. Breakthroughs in energy efficiency enhancement and new energy substitution are demanded through technological innovation within the rail transit field. As a field characterized by high energy consumption, improving energy efficiency in the rail transit

https://doi.org/10.3390/en18153951

field holds considerable significance for global low carbon development. Among emerging technologies, ejector technology has attracted increasing attention for its potential to drive the green transformation of the rail transit field due to its energy-saving advantage of zero mechanical energy losses.

Ejectors are composed of nozzles, suction chambers, mixing chambers, and diffusers, as illustrated in Figure 1. Its working principle is based on the conversion of pressure potential and thermal energy of a high-temperature, high-pressure motive fluid into kinetic energy as it passes through the nozzle. The resulting high-velocity, low-pressure jet enters the suction chamber, creating a local pressure lower than that of the suction fluid, which is thereby drawn into the system. Due to the strong shear effect caused by the pressure difference between the motive fluid and the suction fluid, the suction fluid continuously transfers mass and energy with the motive fluid. Consequently, the mixing degree of the two fluids tends to be uniform in the mixing chamber. As the discharge fluid passes through the diffuser, there is an increase in the flow cross-sectional area, which results in a decrease in fluid velocity and an increase in pressure. Meanwhile, part of the kinetic energy is converted back into pressure potential energy and heat energy. Finally, the medium temperature and medium pressure of the discharge fluid are obtained at the outlet of the diffuser. Throughout this process, the pressurization of the suction fluid is realized without directly consuming mechanical energy. Owing to the inherent advantage of zero additional mechanical energy consumption, ejector technology is utilized not only in the rail transit field but also extensively across diverse domains including desalination [2], refrigeration cycles [3], petroleum refining [4], aerospace [5], nuclear power plant safety engineering [6], and waste heat recovery [7]. Significant values have been consistently demonstrated in energy conservation, pressure regulation, energy recovery, vacuum generation, and precise control.

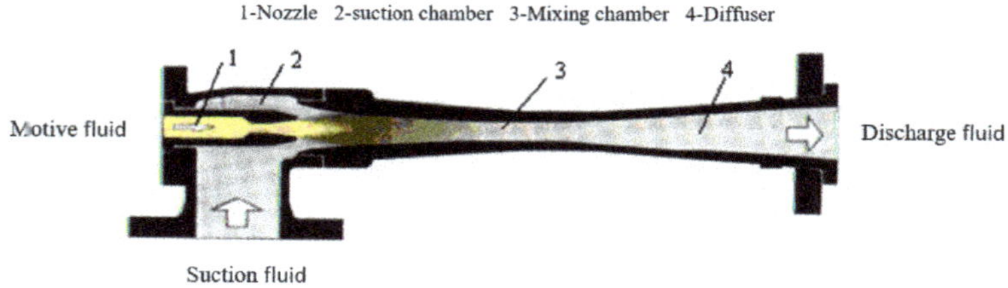

Figure 1. Schematic diagram of typical ejector structure.

The ejector features a compact structure, stable operation, and low maintenance cost. Furthermore, it aligns with the core requirements of new energy technologies having low consumption and high efficiency to achieve performance improvements through structural innovation rather than additional energy input. In the rail transit field, ejector technology has been primarily applied in systems such as engine waste heat recovery, fuel cell hydrogen supply system optimization, compressed air-powered pressure pneumatic reduction, pantograph lift control, and energy-saving vacuum toilets. Research targeting these systems not only directly affects the energy efficiency level of the relevant systems in the field of rail transit but also fosters technological innovation by promoting energy-saving transformation in associated industrial processes and facilitating the integration of renewable energy systems. As a result, ejector technology provides essential technical support for building a synergistic emission reduction framework that spans both the rail transit sector and associated industries.

Despite the potential performance advantages offered by ejector systems, their implementation in rail transit is hindered by several significant constraints. The fixed geometry design frequently transitions into subcritical mode under variable operating conditions, which complicates maintenance and leads to reduced entrainment rates and pressure recovery coefficients [8]. Based on the application of a proton exchange membrane fuel cell (PEMFC), fluctuations in anodic pressure exceed ±0.15 bar during load drift. This situation underscores the need for dynamic control strategies to mitigate mechanical stress [9]. Further challenges include the degradation of high-speed mixing chambers and the obstruction of vacuum toilet ejectors, which require special materials or frequent maintenance in harsh environments [10]. Preliminary studies on multi-nozzle ejectors indicate that geometric optimization has the potential to enhance operational stability. However, a trade-off exists between adaptability and system complexity [11].

Due to the technical maturity and widespread application of ejector technology in the fields of seawater desalination [2] and refrigeration [3], the current reviews of ejector technology mainly focus on these two fields. However, the review of the latest developments of ejector technology within the rail transit field remains limited owing to the relatively emerging nature of the demand for this technology in the field. Consequently, the research progress pertaining to engine waste heat recovery, fuel cell hydrogen supply system optimization, and compressed air-powered pressure reduction technology in the rail transit field was reviewed in this paper. In addition, the synergy potential, remaining shortcomings, and future research directions of these technologies in improving the energy efficiency of rail transit systems and new energy applications were discussed. Under the background of global energy conservation and energy transition, these unique discussion priorities will help readers clearly understand the research hotspots, the latest progress, and the existing shortcomings of ejector technology in the field of rail transit, thus addressing the major gaps in the current review.

This review organized the research progress on ejector technology aimed at rail transit energy conservation. The review commences with an introduction of ejector energy-saving principles and their application value in Section 1. Theoretical fundamentals and intelligent control strategies are discussed in Section 2. Applications in engine waste heat recovery are reviewed in Section 3, focusing on the analysis of working fluids, transcritical systems, and structural optimization. Section 4 examins ejector applications within hydrogen supply systems for fuel cells, addressing geometric parameter optimization and multi-nozzle innovations. Compressed-air vehicle pressure reduction systems are discussed in Section 5, including thermodynamic optimization and system integration. Finally, Section 6 summarizes other applications such as pantograph control, vacuum toilets, ventilation systems, and cleaning devices. This review addresses the advancements and challenges of ejector technology in supporting the rail transit energy transition.

2. Performance and Intelligent Control of Ejector Technology

Based on the multifaceted application value and energy-saving potential of ejectors, the theoretical principles governing ejector operation are detailed in this section, which critically examines emerging intelligent control strategies.

2.1. Performance of Ejectors

The operation and performance of an ejector are known to affect the performance of an ejector cycle. The important parameters related to the performance of the ejector mainly include the critical discharge pressure, the entrainment ratio (ER), the pressure ratio, and the efficiency of the ejector.

The ejector performance undergoes a sharp decrease when the discharge pressure exceeds a certain value, which is identified as the critical discharge pressure of the ejector.

The entrainment ratio ER refers to the ratio of entrained fluid flow rate $\dot{m}_s$ to the motive fluid flow rate $\dot{m}_p$, indicating the effectiveness of fluid entrainment and mixing within the ejector system, which is defined as follows [12]:

$$ER = \frac{\dot{m}_s}{\dot{m}_p} \tag{1}$$

The pressure ratio N represents the ratio of pressure increase in the suction fluid to the pressure drop in the motive fluid by the ejector system, which can be defined as follows [12]:

$$N = \frac{P_d - P_s}{P_p - P_d} \tag{2}$$

where P_d, P_s, and P_p denote the diffuser discharge pressure, suction fluid pressure, and motive fluid pressure, respectively.

The efficiency of ejector η, in essence, refers to the relationship between the power generated by the ejector (output power $\dot{E}_{out}$) and the power consumed by the ejector (input power $\dot{E}_{in}$). Based on the one-dimensional theory assuming that mixing is completed in the constant area mixing chamber and that the spacing between the nozzle exit and the mixing chamber entrance is zero, the efficiency of the ejector can be defined as follows [13,14]:

$$\eta = \frac{\dot{E}_{out}}{\dot{E}_{in}} = ER \cdot N \tag{3}$$

where ER and N are the entrainment ratio and pressure ratio.

The ejector performance has been demonstrated to be significantly influenced by geometric parameters and operating conditions. In studies of public transportation air conditioning systems, Saban Unal et al. [15] demonstrated that two-phase ejectors could improve the coefficient of performance (COP) by approximately 15% when evaporation and condensation temperatures are properly optimized. Through simulations using R134a refrigerant, Baek et al. [16] identified that the mixing chamber diameter and length are the dominant factors affecting the ejection ratio. Meanwhile, Bai et al. [17] observed that improper geometric parameters, including an excessively small throat diameter or overly large mixing chamber, could lead to pressure rise failure or refrigerant ejection failure. The complexity of two-phase flow introduces additional challenges, as shown by Chen et al. [18], whose experiments revealed approximately 20% error in numerical simulations under critical conditions. Yang et al. [19] further noted that neglecting non-equilibrium steam condensation phenomena could lead to an 11.71% overestimation of the ejection ratio. Experimental studies by Li et al. [20] demonstrated that a nozzle expansion angle of 2.0° optimized the main flow expansion state and maximized the ejection ratio. In seawater desalination ejector refrigeration systems, R114 refrigerant was found to perform best due to its low saturation pressure and high water production rate (Petrovic et al. [21]). Optimization strategies include improving mixing chamber design through constant rate momentum change or constant rate of kinetic energy change theory (Kumar et al. [22]), implementing multi-stage ejectors such as two-stage structures to recover redundant momentum and enhance the ejection ratio (Yadav et al. [23]), and controlling wall heat conduction by using low thermal conductivity materials to reduce non-isothermal losses (Haida et al. [24]).

2.2. Intelligent Control Strategies for Improving Performance

Traditional control strategies of ejectors frequently struggle to achieve rapid and optimal adaptation across wide operational ranges, which directly affects system efficiency and stability. Based on the comprehensive review by Al-Doori et al. [25], artificial intelligence (AI) and machine learning (ML) methodologies have been extensively integrated into ejector refrigeration systems to address the limitations of traditional experimental and computational approaches, primarily focusing on performance prediction, geometric optimization, and adaptive control. Artificial Neural Networks (ANNs) combined with Genetic Algorithms (GAs) enable multi-parameter optimization of ejector geometry (e.g., nozzle throat diameter, area ratio (AR), nozzle exit position (NXP)), achieving significant performance enhancements. Liu et al. [26] demonstrated a 35.39% efficiency increase and 8% average performance improvement by eliminating vortices and optimizing secondary flow dynamics using ANN-GA frameworks. Similarly, Zhang et al. [27] utilized ANNs to predict ejector ER and critical back pressures with 68% higher experimental agreement and <15% prediction error under varying operating conditions. This approach surpassed conventional thermodynamic models by effectively capturing the nonlinear relationships between geometric parameters and system outputs. Gaussian Process Regression combined with computational fluid dynamics (CFD) data, as implemented by Ringstad [28], provides robust design optimization for CO_2 ejectors. This methodology allows for the efficient exploration of design spaces and the identification of optimal geometries that enhance pressure recovery ratios by up to 76% in variable–geometry designs. Furthermore, hybrid ML strategies such as particle swarm optimization combined with GA have been shown to minimize energy consumption in hybrid ejector air conditioning systems, resulting in 17% energy savings and peak load efficiency improvements of 23% at 23 °C ambient temperatures (Wang et al. [29]). ANN-based controllers optimize spindle positions to boost ER by 41% across diverse pressure conditions (Li et al. [30]). Extreme learning machines accelerate off-design performance predictions for multi-evaporator systems, reducing computational costs while achieving $R^2 > 0.96$ for critical parameters like COP (Bencharif et al. [31]). Despite these advances, challenges remain in data quality requirements, model interpretability, and integration with high-fidelity CFD simulations for multiphase flows, necessitating future research into hybrid physics-informed ML architectures and standardized validation protocols.

Advanced flow diagnostics and control in ejector systems must be strengthened to provide critical empirical data for ML model training and real-time performance monitoring. Tunable diode laser absorption spectroscopy enables non-invasive line-of-sight measurements of temperature and species concentration downstream of supersonic nozzles, with Al-Manea et al. [32] reporting spatial thermal capturing errors below 5% at distances 15–30 mm from jet centerlines. Velocity fields and turbulent structures within mixing chambers are quantified through particle image velocimetry (PIV), which reveals boundary layer separations and shock wave interactions that reduce ejector efficiency; Dvořák and Kotek [33] correlated PIV-derived velocity contours with pressure recovery losses, identifying optimal mixing chambers for minimizing entropy generation. Density gradients and shock trains are visualized using Schlieren imaging, with Chen et al. [34] linking nozzle exit flow patterns to ejector efficiency reductions under over-expanded conditions. Surface temperature distributions associated with nucleation events during non-equilibrium condensation are captured by infrared thermography, though its accuracy is constrained by material emissivity and requires carbon coatings for error mitigation (Pradere et al. [35]). The sensor data obtained from these techniques is integrated into ML controllers for adaptive ejector operation. For example, pressure and temperature sensors guide ANN-regulated spindle adjustments to maintain optimal NXP, preventing boundary layer separation under variable condenser pressures (Han et al. [36]). However, sensor deployment faces inherent limitations, including optical interference in radial ejectors, thermal shock vulnerability of acrylic test sections, and particle seeding artifacts in PIV measurements that perturb flow dynamics. Consequently, while sensors enable critical validation of CFD and ML models, their role remains supplementary to computational frameworks for holistic system control.

ML algorithms can be trained using extensive datasets generated from high-precision CFD simulations and experimental tests that cover diverse operating scenarios. The data capture complex frequently involves relationships between real-time input parameters, such as pressure, temperature, flow rate, and the position of adjustable components, including needle valves or swirl actuators. Additionally, these inputs are related to critical output performance indicators, which include the ER, discharge pressure, and flow stability. The integration of trained models (such as ANNs or reinforcement learning agents) with extensive sensor networks that provide real-time data streams facilitates the development of intelligent predictive control systems. These systems possess considerable potential to address recognized specific applications. This approach enables the dynamic optimization of adjustable ejector geometries, such as needle valve placement within the throat area, nozzle spacing in multi-nozzle designs, or swirl strength, to maximum energy efficiency, thermal efficiency, and stability under transient conditions. Adverse flow transitions, including subcritical operation or backflow, can be anticipated and forestalled before performance decreases. Complex multi-nozzle ejector operations can be coordinated, and ejector control can be integrated into broader system management strategies. This intelligent control mode represents a key future direction for enhancing ejector adaptability, stability, and overall efficiency, which is fundamentally important for fully realizing the potential of ejector technology in promoting energy savings, improving operational reliability, and advancing the green transformation of rail transit systems.

3. Application of Ejectors in Engine Waste Heat Recovery Technology

A critical challenge in the field of rail transit energy saving is that fuel consumption is relatively low with respect to the energy conversion ratio. The effective thermal efficiency of the engine is typically around 30%, meaning that approximately 70% of the energy from fuel combustion is lost. Specifically, this energy loss is distributed as follows: around 10% through radiation, 30% through exhaust gases, and another 30% through the refrigeration system (as shown in Figure 2) [37]. In total, nearly 70% of the energy is dissipated as waste heat [38]. The combustion waste heat temperature of the engine ranges from 474 K to 874 K, which is typically characterized by high temperature, high flow rate, and high pressure [39]. This waste heat not only exacerbates air pollution but also causes overheating of the water tank, which results in the dynamic performance of the vehicle being impaired. Consequently, reducing the generation of engine waste heat and improving its utilization are vital strategies for reducing air pollution and enhancing vehicle performance stability.

3.1. Vehicle Ejector Refrigeration Air Conditioning and the Technical Advantages

The utilization rate of primary energy can be improved with important economic and social benefits by replacing the traditional compressed refrigeration vehicle air conditioning with ejector refrigeration technology, which utilizes vehicle waste heat energy. The ejector refrigeration system compared with the vehicle compression refrigeration system has the advantages of no moving parts, a simplified structure, a small footprint, durability, and lower cost.

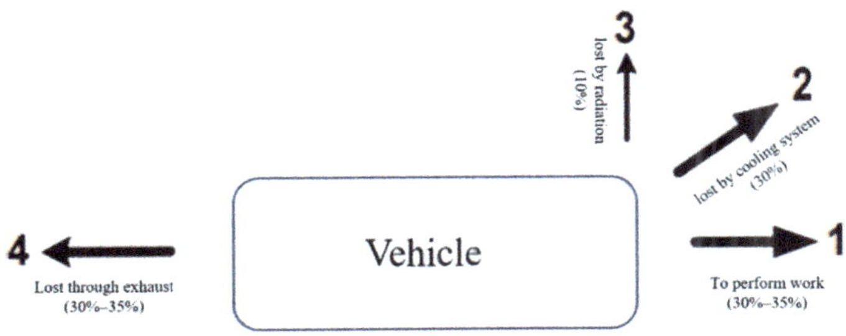

Figure 2. Classical energy usage of the engine.

The ejector cycle is driven by vehicle exhaust waste heat to generate a refrigeration effect, as shown in Figure 3. The high-pressure working medium is heated and vaporized (9→10) in the generator using waste heat, forming high-pressure steam that enters the ejector nozzle. Within the nozzle, expansion and acceleration occur as pressure potential energy is converted into kinetic energy. A low-pressure zone formed at the nozzle outlet results in the low-pressure gaseous secondary flow from the evaporator. Momentum exchange and thorough mixing of both fluid streams occur in the mixing chamber. The mixed fluid subsequently decelerates and increases pressure in the diffuser section, recovering part of the pressure potential energy before entering the condenser (6→7). The condensed liquid working medium is divided into two paths. One path is pressurized by the pump and returns to the generator (7→9), completing the ejection cycle. The other path undergoes pressure reduction through the throttle valve and enters the evaporator (8→5) for endothermic refrigeration. By replacing the traditional throttle valve with an ejector, expansion energy is recovered while the enthalpy is significantly reduced, leading to lower compressor power consumption. Simultaneously, the evaporator functions as the condenser for the compression sub-cycle (1→2→3→4), where heat exchange reduces the condensation load of the compression cycle and enables synergistic operation of the dual-cycle synergy.

Compared to traditional throttle valves, ejectors demonstrate a higher COP, reducing air conditioning system energy consumption by 12% to 17% [40].

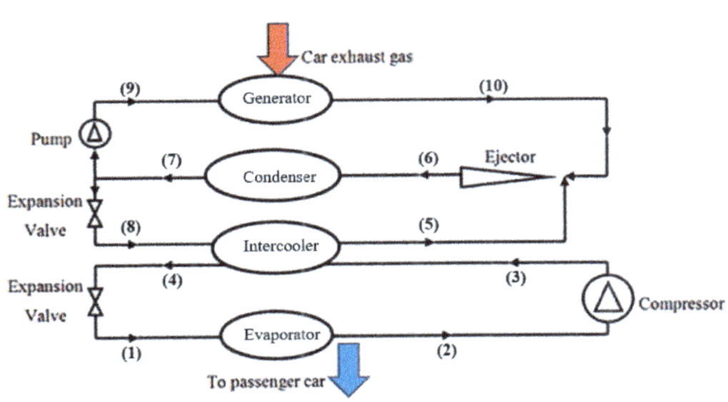

Figure 3. Vehicle ejector refrigeration air conditioning system [41].

3.2. Research Progress of Ejector Refrigeration Air Conditioning System

3.2.1. Different Working Medium Ejector Refrigeration for Vehicles

In ejector refrigeration air conditioning systems, Freon-based working medium such as R11, R12, and R22 are characterized by high thermal stability, efficient refrigeration performance, and a mature technical foundation, while their high ozone depletion potential (ODP) and global warming potential (GWP) are restricted by international agreements. Although ammonia and water are both environmentally friendly, each presents drawbacks—ammonia is toxic, and water's evaporation temperature is limited, resulting in the failure to meet low temperature requirements. Working mediums including CO_2, hydrocarbons, HFOs, and R32 possess zero ODP and low GWP, which are environmentally friendly and offer the potential for improved energy efficiency and adaptability to new technologies such as supercritical systems.

R134a is one of the most commonly utilized refrigerants in vehicle ejector refrigeration and air conditioning systems, where its effectiveness has been demonstrated through multiple performance improvements. When implemented in ejector systems, R134a improves refrigeration capacity and COP while simultaneously reducing system weight, refrigerant charge, and heat exchanger size across various conditions. Takeuchi et al. [42] reported enhancements in refrigeration capacity ranging from 25% to 45%, and increases in the COP between 45% and 65% for transit refrigeration systems operating at a cold storage temperature of 256 K. Additionally, they noted a 40% reduction in system weight and a 68% decrease in refrigerant charge. Oshitani et al. [43] observed a 15% COP increase in double evaporator systems operating at 319 K. In exhaust heat-driven systems exposed to temperatures at 803~1273 K, Zegenhagen and Ziegler [44] documented refrigeration capacity gains of 2.3–5.3 kW. Zhu and Elbel [45] investigated adjustable swirl nozzles under fluctuating conditions (as shown in Figure 4). Ünal et al. [46] achieved a 4% reduction in condenser size and a 55% reduction in evaporator size for bus air conditioning systems operating at 328 K condensation and 279 K evaporation temperatures. Galindo et al. [8] studied diesel intake refrigeration. Critical constraints were also highlighted, including a significant reduction in thermodynamic efficiency under off-design conditions with fixed ejector geometry. These efficiency losses are characterized by COP decreases from 0.099 to 0.151, primarily caused by the ejector transitioning from the efficient critical mode (double-choking) into the degraded subcritical mode (single-choking) [8], and the necessity for control mechanisms such as adjustable swirl nozzles to manage flow during dynamic operation [44]. These studies confirm the significant potential of R134a-based systems for COP enhancement, capacity reduction, and lightweight design optimization. However, operational flexibility under dynamic conditions presents challenges, particularly due to the fundamental limitations associated with momentum transfer inefficiencies and shock-induced losses associated with two-phase flow in ejectors [47].

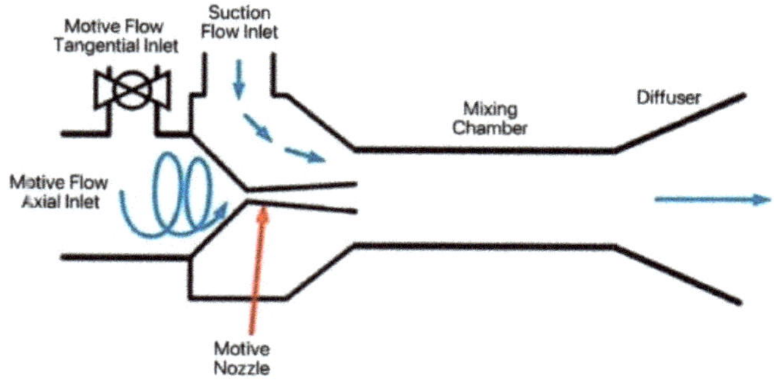

Figure 4. Variable ejector with adjustable needle in motive nozzle [45].

Alternative working mediums, including water and refrigerant mixtures, have been investigated for specific vehicle ejector refrigeration applications to utilize their distinct advantages. Jaruwongwittaya et al. [48] studied a two-stage ejector cycle using water (condensation and evaporation temperatures of 328 K and 279 K, respectively), where optimized ER yielded a COP of 0.29~0.89. This configuration demonstrated superior adaptability to high condensation temperature relative to single-stage systems, with potential engine load reductions exceeding 30%. Keeratiyadathanapat et al. [41] proposed a hybrid compressed–ejector system using R134a for compression and R141b for ejection, partially powered by waste heat. This approach increased system COP by 10~20% and decreased energy consumption by approximately 20% compared to traditional compression cycles. Through CFD corrections, the mathematical model's prediction error was reduced from 15.5% to 5.5%.

Low GWP refrigerants such as R1234yf have been investigated as environmentally friendly alternatives in ejector systems. Lawrence and Elbel [49] showed that R1234yf offers a 1–3% higher theoretical COP gain than R134a due to its higher isentropic expansion potential. Experimental results from their R134a-based ejector system demonstrated up to 10% higher COP than a conventional throttling cycle, with the enhanced performance attributed to marginally more favorable cycle configurations and operating conditions for ejector operation. While directly replacing R134a with R1234yf can decrease system capacity and COP, Sukri et al. [50] noted that adding an ejector significantly offsets these limitations by decreasing compressor power requirements and increasing COP by as much as 15%. The ER of R1234yf is significantly enhanced through geometric optimization of ejector components, specifically through adjustments to the nozzle exit diameter, mixing chamber diameter, and NXP. Galindo et al. [51] achieved an ER of 0.139 by optimizing the nozzle exit diameter (3 mm), mixing chamber diameter (3.6 mm), and NXP (5.5 mm) under adverse operating pressures specific to internal combustion engine waste heat recovery. In contrast, Suresh et al. [40] found that increasing the ejector AR to 5.33 (by reducing the throat diameter from 3.2 mm to 2.5 mm) boosts the ER of R1234yf and R1243zf by 82% under critical operating conditions, emphasizing the strong dependence of ejector performance to geometric scaling.

To broaden the technology approach, absorption and mixed working mediums have also been explored for vehicle waste heat-driven refrigeration fields. Venkataraman et al. [52] reviewed the application of fuel cell exhaust-driven absorption refrigeration systems in transit applications. They stated that a solid oxide fuel cell (SOFC) as the auxiliary power unit can provide stable waste heat to drive the absorption refrigeration cycle, thereby avoiding the performance reduction typically associated with coupling to variable engine exhaust. The temperature of the SOFC exhaust gas heat source is suitable for driving the refrigeration system. Among available working fluid pairs, ammonia–water (NH_3–H_2O) is appropriate for refrigerated transport at 254 K, while lithium bromide–water is more suitable for automotive air conditioning applications. Currently, there are no mature vehicle absorption refrigeration systems that have been commercialized. It is worth noting that the mixed working medium exhibits advantages in complex cycles. Pan et al. [53] proposed a coupled system integrating a supercritical CO_2 Brayton cycle with ejector expansion refrigeration for engine waste heat recovery, utilizing a mixed working medium including R32 and CO_2. They adopted a numerical simulation approach to investigate the refrigeration performance and key parameter influences of these refrigerants. The results show that when the mass ratio of R32 to CO_2 is 0.9/0.1, the refrigeration capacity of the system reaches a value of 225.5 kW, and the integrated COP is measured at 2.05. The ejector ER is observed to increase to 1.39. When the turbine inlet pressure is 25 MPa and the temperature is 594 K, the waste heat recovery efficiency is optimal under these conditions. These findings demonstrate that tailored working medium combinations can enhance system adaptability and performance, providing flexible solutions for various heat source conditions and temperature ranges in rail and automotive applications. As summarized in Table 1, comparative studies of key performance indicators highlight the limitations of current ejector-based refrigeration systems under varying refrigerants.

Table 1. Comparative summary of key performance indicators for ejector-based refrigeration systems in vehicle engine waste heat recovery.

Refrigerant	COP Gain/COP Value	ER	Energy Saving Effect
R134a	10–20% [40]	-	AC energy consumption reduced by 12–17% [40]
R1234yf	Theoretical COP 1–3% > R134a [50]	0.139 (optimized) [51]	Compressor power consumption reduced 15% [50]
Water	COP: 0.29–0.89 [48]	Optimized two-stage ER [48]	Engine load reduced >30% [41], fuel consumption and greenhouse gas emissions reduced
$R32/CO_2$ Mixture	Integrated COP: 2.05 [53]	1.39 [53]	Refrigeration capacity: 225.5 kW [53]
CO_2 Transcritical	System COP increased 12.8% [54] Efficiency improved 16–35% [55]	-	Energy savings potential: 22.16% annually [55]

The reviewed studies demonstrate significant potential for R134a-based ejector systems to improve refrigeration capacity (25~45%), COP (45~65%), and lightweight design (40% weight reduction) [42], but off-design performance remains a critical constraint, with COP dropping sharply under fixed geometries [8]. Adjustable nozzles [45] and multiple working medium systems (e.g., R134a/R141b) [41] mitigate dynamic inefficiencies, yet operational flexibility requires further optimization. Low GWP alternatives including R1234yf show promise (1~3% COP gain [49]), but geometric tuning (e.g., AR optimization [40]) is essential to compensate for capacity losses. Absorption systems (e.g., NH_3–H_2O [52]) and mixed-media cycles (e.g., $R32/CO_2$ [53]) offer solutions for waste heat recovery but face commercialization barriers.

3.2.2. Transcritical System of Vehicle Ejector

A transcritical system refers to a system in which the refrigerant undergoes a supercritical state in the cycle. Its operating pressure exceeds the critical point. Unlike conventional subcritical systems that release heat through condensation, transcritical systems discharge heat through gas cooling, making them particularly suitable for high-temperature refrigeration applications. At present, CO_2 serves as the mainstream working medium in the transcritical system. Other new working mediums, such as HFOs, are mostly adopted in subcritical systems due to critical temperature constraints or safety limitations. However, technology development may enable further applying of low-GWP working medium within the transcritical field.

In 2019, Ipakchi et al. [56] proposed a CO_2-based transcritical cycle architecture integrating ejectors, turbines, and compressors for waste heat recovery in combined refrigeration and power supply systems. Using numerical simulation, they studied the thermal economy and the mechanistic impact of discharge pressure on system performance. The results show that a 10 bar increase in ejector discharge pressure raises energy efficiency by 16.4% while decreasing exergy efficiency by 9.2%. After optimization, the energy efficiency of the system reaches 27.42%, with an exergy efficiency of 24.21%. The corresponding net output power is 7.55 kW, the net present value is 0.3419 M$, and the payback period is 4.5 years. These results verify the economic feasibility of CO_2 as a working medium in vehicle waste heat recovery applications. In 2020, Chen et al. [54] conducted experiments evaluating the thermodynamic performance of a transcritical CO_2 dual-rotor intercooling refrigeration system under high-temperature environments. The results show that under the condition of 319 K, the refrigeration capacity increases by 19.8% and the COP by

12.8% compared with the basic cycle, while a significant reduction in compressor exhaust temperature occurs. This confirms that the CO_2 working medium breaks through the traditional performance limit through the intermediate refrigeration technology under the high-temperature condition of mobile vehicle air conditioning. In 2022, Song et al. [55] proposed a comprehensive analytical framework for transcritical CO_2 refrigeration and heat pump technology, demonstrating its efficacy in new energy vehicle air conditioning, building heating, hot water supply, and ice and snow venues. They quantify an annual energy-saving potential at 22.16% and show that introducing expansion work recovery technologies (e.g., ejectors and expanders) can improve energy efficiency by 16% to 35%. In 2023, Wu et al. [57] developed a CO_2 transcritical power cycle coupled with ejector refrigeration for cascade utilization of engine exhaust gas and cylinder liner water waste heat. Using numerical simulation and multi-objective optimization, they studied the thermo-economic performance of this integrated system. The result shows that the ejector refrigeration subsystem driven by cylinder liner water reduces the CO_2 condensation temperature to 301.67 K, achieving a 13.01% increase in net power and a 6.91% reduction in electricity cost. The exergy loss rate decreases from 24.94% to 17.62% (ambient temperature 284~304 K), and a 100% water utilization rate of the cylinder liner is reached alongside stable transcritical operation.

Transcritical CO_2 systems demonstrate strong potential in high-temperature refrigeration and waste heat recovery, offering energy efficiency improvements (16–35%) [55] and significant COP gains (12.8%) [54] through advanced architectures like dual-rotor intercooling and ejector integration. However, trade-offs exist: increasing ejector discharge pressure boosts energy efficiency (16.4%) but reduces exergy efficiency (9.2%) [56], highlighting the need for balanced optimization. While CO_2 excels in low-GWP applications and offers stability in transcritical mode [57], its economic feasibility (a 4.5-year payback [56]) and performance under dynamic conditions require further validation. Emerging low-GWP refrigerants (e.g., HFOs) remain limited to subcritical systems.

3.2.3. Structure and Parameter Optimization of Vehicle Ejector

Structural and parameter optimization of ejectors involves the precise design and adjustment of their key geometric components. These key dimensions include the nozzle throat diameter, nozzle outlet position, mixing chamber diameter and length, diffusion angle, and so on. The optimal parameter combination is typically determined by a combination of theoretical analysis, numerical simulation, and experimental verification to ensure efficient fluid mixing, supercharging, and energy transfer, thereby improving the overall performance, operational stability, and adaptability to specific operating conditions of ejectors. Structural and parameter optimization of ejectors aims to improve core performance indicators in vehicle ejector refrigeration systems. These indicators include the ER, critical pressure, COP, and operating stability in the vehicle ejector refrigeration system. Moreover, vibration noise is reduced and system reliability is enhanced under space-constrained operating conditions. System reliability and response speed are also increased during frequent start and stop conditions. Therefore, the optimization of ejector structure and parameters realizes a substantial improvement in the overall performance of the vehicle ejector refrigeration system

Innovative design and structural optimization of vehicle ejectors play a critical role in enhancing the system energy recovery and operating condition adaptability. Takeuchi et al. [42] proposed a novel two-phase ejector system for transport refrigeration applications to address expansion energy loss in refrigeration cycles. Using experimental and numerical simulation approaches, they investigated the energy recovery mechanisms. Their results demonstrated that the adoption of a two-stage expansion nozzle significantly reduced vortex-induced energy losses in the mixing section, leading to a 25~45% increase in cooling capacity and a 45~65% improvement in the system's COP under identical component conditions. Similarly, to minimize the evaporation temperature difference loss, Oshitani et al. [43] proposed an ejector-based system for dual evaporator refrigeration in vehicle air conditioning to address COP loss caused by evaporation temperature differences. They adopted an experimental approach combined with numerical simulation to elucidate the functional mechanisms of this system under operational constraints. The results show that a dual-stage variable ejector structure improves cabin air conditioning performance by 15% and reduces refrigeration cooling time by 20%. After optimization of the hybrid segment diameter and length parameters, the air conditioning performance of the system is enhanced by 15% at 319 K. In terms of adaptability, Zhu and Elbel [45] developed an adjustable dual-inlet swirl nozzle mechanism for throttling control in vehicle air conditioning ejectors, addressing sensitivity to fluctuating operating conditions. They adopted an experimental approach to study the flow regulation characteristics and operational stability of this control system under transient conditions. The results show that the mass flow rate of the R134a refrigerant can be reduced by 36% by adjusting the swirling strength with fixed nozzle geometry. The pressure at the nozzle inlet increases by 244 kPa for every 0.24 increase in the swirl strength when the inlet pressure is in the range of 826~1034 kPa. The control effect of this mechanism is remarkable when the pressure at the nozzle outlet is less than 500 kPa. A computational framework integrating CFD ejector characterization and one-dimensional thermodynamic modeling was developed by Galindo et al. [8] for multi-objective optimization of geometric parameters and operating conditions in internal combustion engine intake refrigeration cycles. They studied the system COP and cooling capacity enhancement of this integrated ejector refrigeration system using a numerical simulation approach combined with GAs. The results show that the intake air temperature can be reduced to 274 K and the refrigeration capacity can reach 3.66 kW, improving the energy recovery rate and adaptability to multiple operating conditions after the structure optimization under the designed working conditions. COP and refrigeration capacity can be improved more effectively through multi-objective optimization.

Regarding the optimization of vehicle ejectors' specific structures, Lawrence and Elbel [49] proposed a double evaporator ejection cycle framework for vehicle air conditioning systems to address low-pressure refrigerant expansion loss. They studied the flow loss and frictional loss optimization under operational constraints using an experimental approach combined with numerical simulation. The results indicate that appropriate matching between the ejector throat diameter and the mixing segment diameter can improve the ejection performance. The 5 mm mixing segment reduces the flow loss by more than 40% compared to the 3 mm mixing segment. Frictional losses are significantly decreased by reducing the nozzle expansion angle from 4.0° to 2.3°. The optimization results in a peak ejection efficiency of 0.15. When the condensation temperature is 319 K and the evaporation temperature is 283.9~286 K, the COP of the system is increased by 10% at most. Galindo et al. [51] developed a parametric optimization methodology for ejectors in internal combustion engine waste heat recovery systems using the R1234yf refrigerant. They studied structural parameter effects on system ER and efficiency using numerical simulation. By adjusting the nozzle outlet diameter (D_e3), the mixing chamber diameter (D_e4), and the nozzle outlet position (L_e2), the optimal combination was determined to be D_e3 = 3.0 mm, D_e4 = 3.6 mm, L_e2 = 5.5 mm under the condition of the throat diameter fixed at 1.8 mm. The ER reaches the maximum value of 0.139. The results show that the deviation of 0.1 mm in the diameter of the mixing chamber leads to a 13.3% decrease in the ER. The ER decreased by 13.2% when the nozzle outlet diameter deviated from the optimal value. The compact design meets the space constraints. The critical or subcritical mode mathematical model can effectively predict the ER of variable operating conditions. Suresh et al. [40] proposed a parametric optimization framework for ejector geometric parameters (AR and NXP) in waste heat-driven air conditioning systems using low-GWP refrigerants. They analyzed the ER and ejection efficiency characteristics under varied thermodynamic conditions using numerical simulation methods. The results show that increasing the AR can significantly improve the ER. The ER of R1234yf and R1243zf reaches 0.51 at an AR of 5.33, which is 82% higher than the initial AR. The ER of R134a and R440a increase about 2.3 times under the same conditions. NXP has a significant effect on ER. The optimal NXP is 10 mm, where the ER of R1234yf is the highest. The structure optimization shows that 10 mm NXP maximizes the ejection efficiency by forming a stable supersonic secondary flow throat. The design is suitable for compact waste heat recovery systems.

The structural optimization of vehicle ejectors demonstrates substantial performance improvements, with two-stage nozzle designs enhancing cooling capacity by 25–45% and COP by 45–65% [42], while dual evaporator systems achieve 15% COP gains [43]. However, geometric sensitivity remains a challenge—minor deviations (0.1 mm in mixing chamber diameter) can reduce ER by 13.3% [51], highlighting manufacturing precision requirements. (AR) increases boost ER (82% for R1234yf at AR = 5.33 [40]); optimal NXP (e.g., 10 mm [40]) must balance efficiency with compactness. Although ejector innovations (e.g., swirl control [45], throat diameter matching [49]) mitigate losses, real-world applicability under transient conditions requires further validation.

In summary, the engine waste heat provides a feasible driving heat source for the vehicle ejector refrigeration air conditioning system. The system features no moving parts, a compact structure, and utilizes waste heat instead of mechanical compression. This scheme can reduce the energy consumption of air conditioning by 12~17% [40] and demonstrates nice performance in low-GWP working medium applications. However, the existing research still faces key challenges. Firstly, the fixed geometry ejector makes it easy to enter the subcritical mode under variable engine conditions. Secondly, real-time control strategies for adapting to heat source fluctuations during vehicle operation remain underdeveloped. Thirdly, experimental and operational data on low-GWP refrigerants in transcritical ejector systems remain limited. Future research requires attention in the following directions: developing an adaptive ejector structure responsive to operating conditions. The thermodynamic behavior of a low-GWP working medium in the transcritical cycle should be explored. This will improve the practicality and environmental benefits of vehicle waste heat recovery.

4. Application of Ejector in Hydrogen Supply System for Fuel Cell Engine

The hydrogen supply system of a fuel cell engine is significant for the hydrogen power device, which is typically composed of a hydrogen storage device, a pressure-reducing valve, a hydrogen ejector, a circulating pump, and a safety control unit. High-pressure hydrogen is stored in a dedicated tank and is subjected to multi-stage pressure reduction to enable precise control of hydrogen flow and pressure. As illustrated in Figure 5, the system ensures coordinated hydrogen supply and tail gas recirculation according to the dynamic demand of the fuel cell stack. The system is widely adopted in the fields of new energy vehicles, rail transit, ship power, distributed generation, and industrial reserve power. Key advantages include zero emissions, high energy conversion efficiency, and enhanced environmental compatibility. Furthermore, ejector-based hydrogen recirculation systems are particularly well-suited for long-endurance operations, low-temperature environments, and high-load conditions, making them ideal for clean energy applications in demanding operational scenarios.

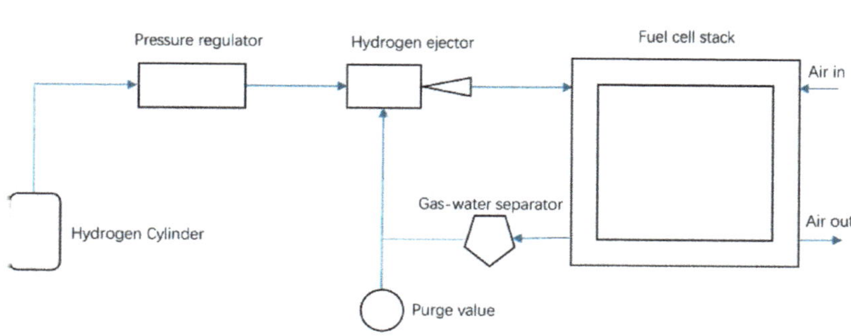

Figure 5. Hydrogen supply system for fuel cell engine [58].

The hydrogen ejector is the core component in a fuel cell engine's hydrogen supply system, which can deliver hydrogen with stable flow and pressure. Hydrogen ejectors demonstrate fast response, high control precision, and high utilization rate of hydrogen. They can quickly respond to the pressure and flow demand of the fuel cell system. This technology not only improves engine performance but also significantly reduces emissions. Therefore, hydrogen ejectors are important for realizing zero-emission vehicles [59]. The hydrogen ejector has been adopted as a component in the recycling system of fuel cell vehicles, including the Hyundai Nexo fuel cell vehicle (FCV) and Honda Clarity FCV [60]. In addition, a new concept is proposed: the ejector is integrated into the hydrogen refueling process of the rail transit hydrogen refueling station. The integrated scheme can improve the effective utilization of high-pressure hydrogen [61]. The application of hydrogen ejectors represents a transformation in the vehicle industry towards cleaner and more efficient transit solutions.

4.1. Progress on Hydrogen Supply Systems for Fuel Cell Engines

4.1.1. Progress on Early-Stage Hydrogen Supply Systems for Fuel Cell Engines

The significant potential of ejector technology in improving hydrogen fuel mixing and pressure control was demonstrated in earlier related studies. In 2004, Marsano et al. [62] employed numerical simulation to design a hybrid ejector with both a constant cross-section and constant pressure for the anode recirculation system of solid oxide fuel cells. The results show that for the constant pressure ejector at an operating pressure of 380 kPa, the suction flow mass flow rate at the ejector inlet increases by 23% compared to the constant section ejector. The Mach number of the mixing chamber is reduced to 0.4. The system reforming efficiency increases by 15% relative to the baseline design. The pressure recovery coefficient of the diffusion section is 0.92. Carbon deposition is successfully inhibited in the range of 60~100% load. This early study reveals the critical role of ejectors in fuel reforming systems and lays a theoretical foundation for subsequent on-board applications. A dual ejection hydrogen fuel engine system was proposed by Kim et al. [63] in 2005 to address high-load tempering, direct ejection mixing non-uniformity, and high-pressure hydrogen supply sealing challenges. The experimental results show that the thermal efficiency of the pure direct ejection scheme is reduced by 22% due to short mixing time, hydrogen leakage, and ejection loss. The dual ejection system enables a smooth transition in the range of 59~74% load. A thermal efficiency improvement of up to 22% over the pure direct ejection scheme is demonstrated in the low-load area. However, the system still faces efficiency degradation in the high-load direct ejection area due to inadequate mixing. In 2006, Karnik et al. [64] proposed an ejector-based dual-pressure control strategy for anode hydrogen recirculation in PEMFC. The results show that the full-order observer can be constructed only by measuring the anode pressure. The critical role of ejectors in improving fuel mixing uniformity and suppressing carbon deposition within hydrogen fuel cell recirculation systems is collectively emphasized by these foundational studies.

The potential of ejector technology for resolving cost and stability challenges in fuel cell hydrogen supply systems was demonstrated in earlier related studies. In 2007, Wee [65] proposed a critical assessment framework aimed at addressing the commercialization bottlenecks in PEMFC hydrogen supply systems. Key constraints associated with conventional hydrogen production were identified in this framework, including CO_2 emissions and anode catalyst poisoning risks, as well as deficiencies in high-purity hydrogen storage and transit infrastructure. He evaluated practical solutions to advance large-scale PEMFC deployment while mitigating environmental and operational challenges. The experimental data show that the metal hydride hydrogen storage system achieves a driving range of 1.35 km/g in the light vehicle. However, this system is characterized by low volumetric energy density and poor portability. The study highlights that the stability and cost control of the hydrogen supply system represent fundamental obstacles impeding the application of PEMFC in the transit field. To achieve reduced costs for the hydrogen supply system, Ahluwalia and Wang [66] proposed the utilization of an ejector for hydrogen recirculation in 2008, which has the potential to reduce parasitic power consumption by 30% in vehicle fuel cell systems. At the same time, the system cost is reduced to 108 USD/kW. The thermal management system reduces the heat dissipation area by 40% compared to the traditional design. This achievement marks the engineering validation of the ejector technology in the vehicle fuel cell system, lowering costs while boosting efficiency. To improve hydrogen supply system stability, Zhu and Li [67] adopted the numerical simulation method to investigate the ejection characteristics of the convergent nozzle ejector in the anode recirculation line of the PEMFC under both critical and subcritical operating conditions. The results show that the ER exceeds 2.0 when the primary flow pressure exceeds 5 bar. The pressure recovery coefficient of the mixing chamber can be increased to 0.95. Additionally, the boundary layer thickness of the secondary flow velocity is reduced by 38%.

Early studies of ejectors for fuel cell hydrogen supply systems found the ER in critical mode to be insufficient, with deviations in critical pressure ratio threshold prediction and low-pressure recovery coefficient in the mixing chamber. Furthermore, the recirculation ratio proved unstable under transient conditions while secondary flow ejection efficiency remained limited. These shortcomings resulted in fluctuations in carbon deposition suppression, reduced precision in anodic humidity control, and an imbalance in steam recovery rate, highlighting the pressing need for dynamic characteristic modeling and the development of multivariable cooperative control strategies to ensure reliable, high-performance operation in hydrogen supply systems.

4.1.2. Progress on Recent Hydrogen Supply Systems for Fuel Cell Engines

With the development of a multi-stage ejector structure and cooperative control strategy, the research of hydrogen supply systems for fuel cell engines has made progress in the limited working range of the traditional single ejector, pressure fluctuation, insufficient hydrogen recirculation stability, and system efficiency attenuation under the condition of wide power output in recent years.

In 2020, Kuo et al. [68] conducted both experimental and numerical investigations to evaluate the performance of a passive venturi ejector aimed at enhancing hydrogen cycle stability in PEMFC. The results show that the system's startup time is shortened to 1000 s, the hydrogen utilization rate is increased to 98.5%, and the net efficiency of the system is improved by 19%. Voltage stability is maintained by the ejector even at power ratings as low as 3.8%. Nitrogen penetration is reduced by 42% and steam-carrying efficiency is increased by 28%. In the same year, to solve the problem of insufficient power adaptability of the hydrogen cycle ejector in the PEMFC system, hydrogen ER improvement under low power conditions was studied by Han et al. [11] through nozzle switching. The impact of secondary flow temperature, pressure, and humidity on entrainment performance of a multi-nozzle ejector was investigated through experiments combined with numerical simulation. The hydrogen supply range is extended to 0.27~1.6 g/s through design optimization involving alterations to the mixing chamber diameter and the inclination angle of the symmetrical nozzle throat. Numerical simulation and experimental verification demonstrate that the multi-nozzle structure significantly improves the hydrogen ER under low power conditions by alternating between the central nozzle and the symmetrical nozzle. It is found that the secondary flow temperature, pressure, and humidity have a significant impact on the entrainment performance, which provides an essential parameter for the hydrogen supply optimization of a wide power PEMFC system. In 2021, Zhao et al. [69] proposed a comparative assessment framework for PEMFC hydrogen supply systems. This framework analyzes cycle configurations, equipment characteristics, and control strategies to evaluate the performance differences and operational viability between ejectors and hydrogen compressors. They identified critical advantages, limitations, and control challenges across four distinct hydrogen supply architectures, with a focused analysis on the trade-offs between ejector and compressor systems. The results indicate that compressor-based recirculation systems offer a wider adjustment range but rely on external energy input. Ejector-based systems feature a simpler structure yet require geometric optimization to expand their operational range.

To resolve the narrow working range and pressure fluctuations inherent in single ejector configurations for PEMFC hydrogen supply, Chen et al. [70] in 2022 proposed a dual ejector system that optimizes geometric parameters for distinct low-power and high-power operational intervals. This system was investigated through numerical simulation combined with experimental validation for optimal design and dynamic control. Experimental results show that the dual-easer system achieves hydrogen recycling across the full power range of 70 kW PEMFC. Optimized control reduces the anode pressure fluctuation to less than 44 kPa during switching, effectively reducing the impact on battery life. In 2023, a nested double-nozzle ejector design integrating a bypass channel was proposed by Chen et al. [71] to address the limitations of narrow working ranges associated with single nozzle ejectors in PEMFC hydrogen supply systems. This design optimizes seven geometric parameters (e.g., nozzle diameter, axial spacing) to extend the operational range. A numerical simulation approach combined with experimental methods was employed to validate the system's hydrogen entrainment capability and anode pressure fluctuation performance enhancement under variable fuel cell power conditions. Experiments show that this design achieves full coverage of power output ranging from 9 to 100% in a 150 kW PEMFC system. The required hydrogen ER is satisfied while anode pressure fluctuation is significantly reduced compared to the conventional dual ejector system. The nested nozzle design avoids pressure mutation during mode switching by sharing the mixing cavity, thereby improving the system's stability and efficiency.

4.2. Parameter Optimization and Structural Innovation of Hydrogen Ejector

4.2.1. Optimization of Geometrical Parameter and Structure of Hydrogen Ejectors

Optimizing the geometric parameters of hydrogen ejectors plays a crucial role in enhancing the overall efficiency and stability of fuel cell hydrogen recirculation systems. Dadvar and Afshari [72] adopted the numerical simulation method to study the design parameters of the ejector of the anode recirculation system of the PEMFC. The results show that when the ratio of the nozzle throat diameter to the mixing chamber diameter is 3.81, the ER is increased to 0.8. The critical current density is reduced to 0.45 A/cm^2, and the power coverage is extended to 35~100 kW. The hydrogen circulation efficiency improves by 22.7% compared with traditional compressor-based systems. Pressure fluctuations at the anode inlet are reduced by 38%. The total efficiency of the system is 54.3% and the platinum catalyst loading is reduced to 0.3 mg/cm^2. This parameter optimization study proves that geometric size matching can significantly improve the cooperation efficiency of the ejector and the stack. In studying the influence of ejector geometric parameters on the recycling ratio for the optimization problem of hydrogen recycling efficiency in an 80 kW PEMFC system, Kuo et al. [73] adopted a numerical simulation method. The results show that a nozzle throat diameter to mixing chamber diameter ratio of 2.2 leads to a stoichiometric hydrogen ratio reaching its optimal value of 1.5, which is 11% higher than the benchmark configuration. This study validates the effectiveness of specific geometric parameter matching for optimizing hydrogen cycle efficiency. Yin et al. [74] adopted a numerical simulation method to investigate the optimization strategy of a double-ring cavity ejector structure, aimed at meeting the wide operating condition for hydrogen cycling demand in a PEMFC. The results indicate that optimizing the nozzle design can effectively eliminate the vortex phenomenon at the outlet, resulting in an 18.8% increase in ER at idle speed. When the diameter of the mixing chamber is 6.5 mm, the system achieves optimal hydrogen ER, system recirculation ratio, and hydrogen concentration distribution uniformity in the current range of 31.1~559.8 A. The synergistic effect of the two nozzles improves the uniformity of hydrogen concentration distribution by 15% in the high-pressure area. Adopting a numerical simulation method, Antetomaso et al. [75] studied the geometric parameter optimization strategy based on nozzle diameter scaling for the scalability of hydrogen cycle ejectors in a PEMFC system. The results show that reducing the nozzle diameter from 1.05 mm in the 5000 W system to 0.1 mm in the 300 W system ensures the ER remains in a stable range of 1.2~2.3. The gradient of pressure drop in the mixing chamber was reduced by 42%.

As to the study of the multi-parameter coupling optimization, Ding et al. [76] investigated the influence law of mixing chamber diameter (D_m) and nozzle expansion angle (θ) on non-equilibrium condensation strength and ejection performance of hydrogen circulation ejectors for PEMFC system through numerical simulation. The results indicate that non-equilibrium condensation reduces ejector efficiency. However, this negative effect diminishes with the increase of D_m and θ. There is a sensitive change zone of condensation intensity with D_m = 2.65 mm and a significant weakening threshold with θ = 11.0°. An average increase of 16.8% and a maximum increase of 22.8% in the effective working range are achieved with the optimized combination of D_m = 2.40 mm and θ = 11.0°. Similarly, Arabbeiki et al. [77] adopted the numerical simulation method to investigate the influence mechanism of the five-dimensional geometric parameters of the hydrogen circulation ejector for PEMFC. These parameters include nozzle throat diameter, outlet position, and mixing chamber length, all of which significantly affect the ER and the choked flow pattern. The results indicate that the diameter of the nozzle throat and the outlet position exert considerable influence on the ejection performance. The ER is increased by 20% after optimization. Shock wave formation is suppressed to enhance the stability of the flow field. The critical flow state of Mach 1.0 is maintained under design and off-design conditions. The applicability of the ejector is broadened in the wide power range of 80–101 kW.

The reviewed studies demonstrate that geometric optimization of hydrogen ejectors significantly enhances PEMFC system performance, with ER improvements up to 20% [77] and hydrogen circulation efficiency gains of 22.7% [72]. However, several critical challenges emerge. While diameter ratios (2.2–3.81) effectively optimize stoichiometric ratios [72,73], their applicability across power ranges (300 W–100 kW) requires careful scaling [75]. Non-equilibrium condensation effects, though mitigated by parameter optimization (16.8% average improvement [76]), remain a fundamental limitation. The most promising approaches combine multi-parameter optimization [76,77] with innovative designs like double-ring cavities [74], yet real-world validation under transient conditions is lacking.

4.2.2. Innovation of Adjustable Structure of Hydrogen Ejector

The innovation of the hydrogen ejector's adjustable structure recommends dynamic adjustment components such as variable nozzles, movable throats, or adjustable mixing chambers. The innovation enables key geometric parameters, including flow area, expansion ratio, or mixing efficiency, to be adjusted in real time according to actual operating conditions when combined with intelligent control technology. Therefore, the dynamic optimization of the internal flow characteristics and energy conversion process is achieved. This significantly enhances the ejector's adaptability, operational efficiency, and flexibility under variable operating conditions.

In 2012, Brunner et al. [78] combined numerical simulation and experimental methods to study the design, hydrogen recirculation flow rate, and stoichiometric ratio performance of an adjustable ejector for the hydrogen recycling requirements of PEMFC. The geometric parameters of the nozzle were optimized by CFD, and the electronic control needle valve was designed to adjust the nozzle opening to realize the dynamic matching of flow. The experimental results show that within a hydrogen flow range of 0.04~0.14 mol/s, controlling the needle valve displacement maintains a secondary flow pressure increase of 12~34 kPa. This pressure rise enables a consistent ER exceeding 60% across the operating range, satisfying Ballard's minimum stoichiometric requirement of 1.6 for hydrogen recirculation. Employing experiments and numerical simulations, Jenssen et al. [79] in 2017 developed a variable geometry ejector for cascading fuel cell stacks. The results show that adopting a 50:50 anode cascade ratio reduces the global hydrogen excess rate to 1.25 and sets the throat diameter of the ejector to 4.6 mm. At the same time, this adoption extends the minimum operating power of the system to 3.8% of the rated power and achieves 99.2% voltage stability under load. In addition, the anodic voltage drop is reduced by 28% compared with the traditional design. The ejection efficiency is increased by 19% in the 10~100 kW power range. The introduction of a variable geometry ejector breaks the power adaptability limitation of the traditional fixed structure for the system. In 2022, Zhao et al. [69] proposed a comprehensive analytical framework for PEMFC gas supply systems, highlighting ejector optimization mechanisms within hydrogen supply architectures to extend operational flexibility through variable geometry nozzle adjustments. They adopted a review methodology to evaluate the structural characteristics of adjustable ejectors, identifying technological advancements that enhance working ranges through dynamic nozzle opening control. The numerical simulation results show that the optimized nozzle AR significantly improves the ejection performance, with ER reaching 62.5%~80%. In 2025, Seth et al. [80] adopted numerical simulation to study the performance of variable geometry ejectors based on moving needle valves in meeting the wide power range hydrogen cycling requirements of PEMFC. The results show that by adjusting the nozzle radius, the system's working range can be extended from 100 kW with fixed ejectors to a range of 17~100 kW. The application of mobile needle valve technology indicates the maturity of the technology direction for the next-generation adaptive ejector.

Variable geometry ejectors demonstrate significant advantages in PEMFC hydrogen recirculation, with ER reaching 60–80% [69,78] and system power coverage expanding by 83% [80]. However, their practical implementation faces challenges. While adjustable needle valves enable dynamic flow matching (12–34 kPa pressure increase [78]), mechanical complexity and control precision remain concerns. The 50:50 cascade configuration [79] improves voltage stability (99.2%) and reduces hydrogen excess rate (1.25), but scalability to larger systems requires validation. Although variable geometry designs extend operating ranges (17–100 kW [80]), their reliability under long-term cycling and transient conditions is unproven.

4.2.3. Multi-Nozzle Cooperative Structure of Hydrogen Ejector

The multi-nozzle cooperative structure in hydrogen ejectors involves the integration of multiple nozzles within a single ejector unit. Through the optimization of the spatial layout of nozzles, ejection angle, and flow distribution parameters, this design leverages flow field coupling mechanisms and dynamic control strategies to achieve synergistic flow behavior and enhanced mixing among nozzles. This technology aims to enhance the suction capability, improve the efficiency of energy conversion, and adapt to the needs of complex operating conditions.

The full power range coverage of the ejector is enabled by the breakthrough in the multi-nozzle structure. In 2020, Xue et al. [58] adopted the method of numerical simulation combined with experimental verification to design a multi-nozzle ejector for the PEMFC hydrogen circulation system. They investigated the effect of the simultaneous operation of multiple nozzles working together on both the circulation ratio and eddy current suppression. The results indicate that the cycle ratio of the dual-nozzle operation reaches its maximum at a main pressure of 6.5 bar. This operational mode spans the 35~100 kW power range, effectively suppresses mixing section eddy current, and maintains a stable anode inlet pressure. The total power generation is increased by 17.5 times compared with the single nozzle mode. The proposed cooperative working mode utilizing multiple nozzles marks a new phase in actively controlling the flow field within the ejector. Given the hydrogen cycling demand of PEMFC under wide operating conditions, Han et al. [11] in 2020 adopted numerical simulation methods to study the entrainment performance of multi-nozzle ejectors, which consist of a central nozzle and symmetrical double nozzles (as shown in Figure 6). The results show that the working range of the multi-nozzle ejector is extended to 0.27~1.6 g /s after optimizing the mixing chamber diameter and the symmetrical nozzle throat inclination angle. This configuration achieves a 45% increase in working range compared with the traditional single nozzle. The ER increases to more than 0.2 in the low-power section. Furthermore, the CFD simulation results show that the kinetic energy loss can be effectively suppressed by optimizing the collision position of the symmetric nozzle. In 2023, Feng et al. [9] adopted numerical simulation methods to study the design of multi-nozzle cooperative ejectors for the hydrogen recycling requirements of PEMFC under wide operating conditions. The results show that a nozzle throat diameter to mixing chamber diameter ratio of 3.2 and an ejection coefficient of 1.2 enable coverage of the power range of 35–100 kW. At this time, the anodic pressure fluctuation is less than $\pm$0.15 bar, which is 42% lower than the single nozzle scheme. The optimized multi-stage ejection structure improves the hydrogen recycling efficiency to 68.5%, and the net efficiency of the system is 58.6%. The platinum catalyst loading is reduced to 0.25 mg/cm^2, and the heat dissipation area of the thermal management system is reduced by 45% compared with the traditional design.

Regarding the dual-nozzle, in 2021, Song et al. [81] adopted experiments combined with numerical simulation methods to design a dual-nozzle ejector aimed at addressing the hydrogen cycling problem of PEMFC under wide power conditions. The results indicate that the dual-nozzle mode can cover the power range of 17~85 kW under the hydrogen supply pressure of 250~700 kPa. The ER is increased to 2.1. In comparison to the single nozzle mode, the pressure drop is reduced by 42% at low load conditions, while the mixing uniformity is increased by 37% at high load conditions. Additionally, the structure also reduces the diameter of the nozzle throat to 1.38 mm, and the pressure recovery coefficient within the diffusion section reaches 0.92. The steam-carbon ratio is stable above 2.4. Furthermore, the nitrogen permeability on the anode side is reduced by 53%. Under the low-temperature starting condition, Chen et al. [71] adopted numerical simulation combined with an experimental method to study the parameter optimization of nested dual-nozzle ejectors for the wide operating condition hydrogen cycling requirements of PEMFC. The results show that seven geometric parameters are optimized, including the diameters of the large and small nozzles and axial spacing, which the nested nozzle ejector enables a power range of 9% to 100%. The ER increases to 1.5~2.5, significantly broader than the narrow operating range achievable with conventional single nozzle ejectors. The CFD simulation shows that the pressure gradient of the mixing chamber is reduced by 19%. The nested structure reduces the pressure fluctuation at the anode inlet by 89% compared to the conventional dual ejector system. The design of nested double nozzles marks the innovation of the ejector structure into the stage of three-dimensional flow field fine control.

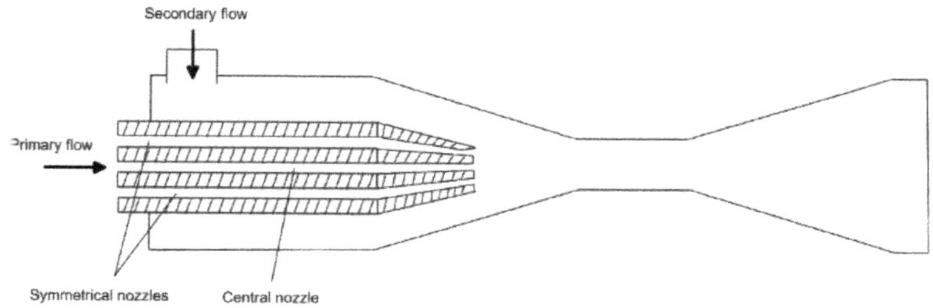

Figure 6. Schematic diagram of multi-nozzle ejector [11].

The development of multi-nozzle ejectors represents a significant advancement in PEMFC hydrogen recirculation systems, demonstrating remarkable performance improvements. The dual-nozzle configuration achieves a 45% wider operating range [11] and increases ER to 2.1 [81] while reducing pressure drops by 42% at low loads. However, these designs introduce new challenges: (1) The complex flow dynamics in multi-nozzle systems require precise geometric optimization, as seen in the 19% pressure gradient reduction achieved through seven-parameter optimization [71]. (2) While anode pressure fluctuations are reduced by 42–89% [9,71], the manufacturing tolerances for such systems become more stringent. (3) The benefits of increased system efficiency (up to 68.5% [9]) must be balanced against the added complexity of multi-nozzle control strategies.

In summary, the hydrogen supply system of fuel cell engines significantly improves the hydrogen cycle efficiency and operating condition adaptability through technological innovation and structural optimization of hydrogen ejectors. The cooperative design of multiple nozzles and the innovation of an adjustable structure extend the power coverage to the whole working range. The dynamic intelligent control technology reduces the anode pressure fluctuation to less than ±0.15 bar. The net efficiency of the system is improved to 58.6% [9]. The technology faces challenges, including insufficient hydrogen recycling stability across wide power output ranges, limited efficiency in suppressing transient pressure fluctuations, efficiency attenuation from reduced mixing uniformity at high loads, and performance fluctuations during low-temperature startup ejection. Future research needs to focus on the development of multi-scale dynamic optimization algorithms, the application of new erosion-resistant materials, and the construction of a cooperative regulation model for gas–liquid–solid multiphase flow. Overcoming the bottlenecks of dynamic pressure instability and multi-nozzle flow field interference is crucial to advancing hydrogen supply systems toward high efficiency and long lifespan.

5. Application of Ejector in New Pressure Reduction System for Compressed Air-Powered Vehicles

Novel compressed air-powered vehicle (APV) pressure reduction systems intelligently regulate compressed air pressure, enabling efficient and stable energy conversion between high-pressure storage tanks and power units. Core components include multi-stage reduced pressure valves, pressure sensors, and electronic control units, which dynamically optimize air pressure output and recover braking residual pressure. Ambient air is compressed to high pressure by external air compressors and stored in onboard gas storage tanks, as depicted in Figure 7. During vehicle operation, the high-pressure air is reduced to motive pressure by the pressure-reducing valve before entering the compressed air engine, where it expands to perform work that drives the vehicle. Throttling energy losses are reduced through the application of multi-stage expansion technology and power recovery-type pressure-reducing valves, thereby enhancing overall energy utilization. This system is primarily applied in zero-emission compressed APV, including urban short-distance logistics vehicles, park shuttles, and industrial special vehicles. This technology is particularly suited to low-temperature or closed environments, such as cold chain transit and mine tunnel operations, achieving both environmental protection and endurance improvement. In the new compressed air power vehicle systems, APV represents a new energy vehicle utilizing high-pressure air as the power source. These vehicles could perform truly zero pollution through compressed air-powered engines. The decompression method for high-pressure gas is regarded as a key technology in compressed APV. Due to the current high energy consumption associated with high-pressure gas decompression, a novel pneumatic vehicle decompression system has been developed, employing an air ejector to replace the throttle valve. This system increases output power by more than 13% [82].

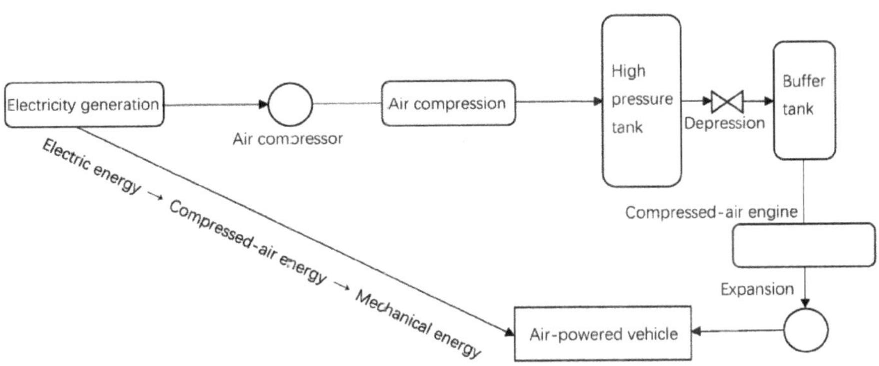

Figure 7. Main components of APV [83].

5.1. Progress in Air-Powered Vehicles and Compressed Air Energy Storage

Research on APV and compressed air energy storage has advanced from fundamental theory verification to complex system optimization. In compressed APV power systems, approximately 50% of energy loss occurs during the decompression process [83], highlighting the critical importance of optimizing pneumatic engine operation to improve overall energy efficiency. This finding reveals how traditional pressure-reducing valves constrain pneumatic system efficiency, making pressure regulation technology innovation a research hotspot. Although the technical method has gradually clarified, breakthroughs in system energy conversion efficiency remain challenging.

Many scholars investigate the technical characteristics and environmental economy of APV through reviews and numerical simulations. The results indicate that, despite advantages such as zero emissions and low maintenance costs, the technology is limited by bottlenecks, including low energy density, thermodynamic loss during expansion, and gas storage pressure attenuation. These bottlenecks lead to shorter driving ranges and lower overall energy efficiency. Meanwhile, hybrid integration, multi-level expansion optimization, and energy recovery technology are proposed as methods to enhance practicality, with the short-distance low-speed scenario regarded as the main application direction at this stage.

Verma [84] proposed a critical feasibility framework for compressed APV technologies, which integrates technical principles, development status analysis, and system viability assessments. This framework aims to evaluate structural configurations and environmental and economic advantages, as well as identify critical bottlenecks in the transition towards sustainable mobility. Verma assessed deployment barriers and scalability potential of air-powered propulsion systems in modern transport infrastructure. The results show that compressed air technology presents several benefits, including zero pollution, low maintenance costs, and convenient fuel preparation. However, low energy density is observed, and gas storage pressure significantly decreases with use. Furthermore, the expansion process causes a sharp temperature drop, leading to icing problems. The practical application process is seriously restricted by these factors. To quantify energy efficiency, environmental impact, and economic viability across the operational and manufacturing phases of compressed APV, Papson et al. [85] developed a comprehensive lifecycle assessment framework. They adopted a numerical simulation methodology to evaluate fuel economy and endurance performance under the European urban driving cycle (UDC), with parallel analysis of carbon emissions throughout the vehicle lifecycle. The results indicate that the comprehensive efficiency of the APV pump to the wheel is merely 14.7%, which is limited by the low energy density of compressed air and the thermodynamic loss of the expansion process. The UDC range is determined to be only 29 miles, with full cycle carbon emission of 626 g CO_2 per mile and a fuel cost of USD 0.21 per mile. These indicators are significantly worse compared to those of gasoline and electric cars. Although this technology demonstrates zero exhaust emission characteristics, it is primarily suitable for low-speed, short-distance applications. Shi et al. [83] conducted a review to assess the technical status and formulate improvement strategies for APV, addressing low energy density, limited thermodynamic efficiency, and insufficient driving range. The influence of compressed air energy density, multi-stage expansion efficiency, and energy recovery technology on pump-to-wheel efficiency has been analyzed through theoretical modeling and comparison with existing data. Structure optimization increases APV energy conversion efficiency to 19.8%, while maintaining zero exhaust emissions. However, the vehicle range remains limited to just 29 miles, and the energy density of compressed air is significantly lower than that of lithium-ion batteries and conventional fuel systems. Therefore, the energy loss and system integration problems in the compression/expansion process need to be further studied.

Regarding hybrid APV, Marvania and Subudhi [86] proposed a comprehensive technical synthesis for compressed air engines, which integrates historical development, working mechanisms, thermodynamic efficiency modeling, and aerodynamic hybrid propulsion schemes. They evaluated critical feasibility constraints (including low energy density, rapid tank pressure decay, and irreversible expansion losses) that limit the practical implementation of pure compressed air power systems. The experimental results show that a 10 L gas storage tank supports only 2 km of endurance. However, the aerodynamic hybrid system can reduce fuel consumption by 8%~59% through braking energy recovery in cooperation with an internal combustion engine. The multi-stage expansion strategy, exhaust gas recirculation, and gas storage pressure constant technology should be optimized to improve future practicality. Therefore, it is suggested that this technology be prioritized for short-distance low-speed scenarios. To address energy efficiency and commercialization bottlenecks for compressed air hybrid power systems, Wasbari et al. [87] proposed a comprehensive optimization framework that integrates regenerative braking, multi-stage compression/expansion, and energy recovery techniques across series, parallel, and hybrid architectures. They evaluated technological feasibility approaches to address operational constraints and enable sustainable commercialization. The results showed that series, parallel, and hybrid architectures significantly improve fuel economy in the UDC. However, the commercialization of these systems is limited by the energy density of compressed air and the safety concerns of high-pressure storage tanks. Therefore, endurance improvements require lightweight materials, heat loss utilization, and system integration optimization.

The application of ejector technology to compressed air energy storage system pressure regulation is gradually improving problems, including irreversible thermodynamic loss during high-pressure gas multi-stage decompression, limited throttle efficiency of traditional pressure-reducing valves, and insufficient energy conversion rate. This decompression method of using ejectors to improve the compressed air energy storage system can be used in a new compressed APV. Guo et al. [88] designed a dual ejector system framework for adiabatic compressed air energy storage (A-CAES) through comparative optimization analysis to mitigate pressure regulation losses. They adopted an experimental approach combined with numerical simulation to validate the ejector's operational efficacy in reducing throttling losses and enhancing energy recovery efficiency. The results show that replacing the traditional pressure-reducing valve with an ejector increased the system's energy conversion efficiency from 61.95% to 65.36%, representing a 3.41% improvement. This outcome verifies the ejector's effectiveness in reducing pressure losses and optimizing the energy storage system's energy conversion efficiency. Ejector technology has been introduced into the pressure control field, providing a new method for improving pneumatic system energy efficiency. Chen et al. [89] employed a novel ejector-based throttling strategy to mitigate throttling losses in A-CAES systems. The results show that optimal values exist for the ejector ER and the active flow pressure, while the system performance is less affected by ambient temperature. By classifying pressure into three levels (high/medium/low), high-pressure air drives the ejector to compress low-pressure air to medium pressure. This strategy reduces throttling exergy destruction by 23.1% and increases the system's round-trip efficiency by 2.00 percentage points (from 66.22% to 68.22%). Zhou et al. [90] employed numerical simulation to reduce throttling losses in compressed air energy storage systems by replacing throttle valves with ejectors. The results indicate that by recovering the outlet air pressure energy from the third-stage compressor, the ejector reduces the compressor's unstable speed range from 0.491 to 0.771. Simultaneously, the system's average exergy efficiency is observed to increase with increasing initial gas storage pressure, with the optimization effect being particularly significant at low initial pressure.

APV exhibits zero-emission advantages but faces critical limitations in energy density and thermodynamic efficiency, restricting its viability for mainstream mobility. While structural optimizations can improve energy conversion efficiency to 19.8% [83], the technology remains impractical for long-range applications, with a maximum UDC range of only 29 miles and high lifecycle emissions (626 g CO_2/mile) [85]. Hybrid architectures mitigate some limitations—reducing fuel consumption by 8–59% [86]—but cannot overcome the fundamental energy density gap versus batteries or fuels. Ejector-based pressure regulation shows promise in CAES systems, improving efficiency by 3.41% [88] and reducing exergy losses by 23% [89], yet these gains are insufficient to offset APV's core drawbacks. For sustainable deployment, APV systems must prioritize niche applications (e.g., low-speed, short-distance transport) while integrating hybrid solutions and advanced materials to address energy storage and expansion losses.

5.2. System Integration Optimization of Air Ejector

Compressed air potential energy historically powered certain manufacturing or transit systems as a versatile energy source. In the late 19th century, this form of energy was adapted for vehicle propulsion. During the latter half of the 20th century, considerable efforts were directed toward using compressed air for electrical energy storage [91]. Currently, advances in ejector technology are being incorporated into compressed APV.

The efficiency of air ejectors is significantly improved through thermodynamic optimization in compressed air energy storage systems, where exergy loss and heat dissipation are simultaneously reduced, enabling adaptation to high fluctuating energy demand. In 2016, Guo et al. [88] proposed a comparative dual-system framework for A-CAES to mitigate exergy loss during pressure reduction by integrating multi-level pressure regulation through ejector technology. They contrasted a traditional pressure-reducing valve operating between 6.40 and 2.50 MPa (System 1) with stepwise ejector control (System 2), developing thermodynamic optimization methods to reduce throttling loss. They adopted a combined numerical simulation and experimental methodology to quantify exergy recovery and validate thermodynamic enhancement effects in multi-stage pressure regulation processes. The results demonstrate that the ejector increases turbine inlet pressure through momentum exchange. Consequently, the power output of System 2 rose from 3.80 MW to 5.51 MW, and the overall energy conversion efficiency improved from 61.95% to 65.36%. Parameter analysis indicates that mixing a working fluid at 5.00 MPa with a suction fluid at 1.80 MPa effectively balances pressure loss and flow gain, thereby validating the proposed technique's optimization potential for boosting energy storage density. In 2023, Rabi et al. [92] analyzed the ejector thermodynamic optimization strategy to optimize compressed air energy storage systems. The results demonstrate that the incorporation of ejectors significantly enhances gas–liquid mass transfer and energy recovery, thereby effectively addressing the problems of high heat dissipation and low efficiency in traditional compressed air energy storage systems. This thermodynamic optimization strategy enables ejector technology to effectively accommodate the intermittent and highly fluctuating energy output characteristics of compressed APV. In 2024, Liu et al. [93] adopted numerical simulation to study pressure loss during throttling processes in compressed air energy storage systems and to investigate thermodynamic optimization of ejector-enhanced systems. The results indicate that the round-trip efficiency of the new system was improved by 3.07%, while exergy loss decreased by 401.9 kW compared to the traditional adiabatic system during the decompression process. Based on advanced exergy analysis, it is recommended that system improvements should prioritize optimizing the ejector structure, followed by the optimization of the turbine and compressor.

Through intelligent design and optimal configuration, the air ejector multi-system cooperative integration was achieved. Meanwhile, the deep integration of air ejector technology with other related systems, such as power, control, and sensing, is required. Efficient cooperation and information exchange among subsystems have been realized, thereby improving the overall system performance, energy efficiency, and reliability. In 2021, Sadeghi and Ahmadi [94] proposed a combined cooling, heat, and power supply system that integrates compressed air energy storage with CO_2 ejector refrigeration technology. The results indicate that the system consumes 72.02 MW of power during the charging phase while generating 136.56 MW in the discharge phase, providing a 1.96 MW cooling load and a 65.8 MW heat load. The cycle efficiency is found to be 81.15%, and the exergy efficiency is 56.57%. After optimization, the exergy efficiency increases to 68.19%.

Adopting the numerical simulation method in 2022, Cao et al. [95] studied an A-CAES system aimed at optimizing the configuration of an integrated single ejector in the final compression stage. The results show that by ejecting the third-stage compressed air as the secondary flow, the compressor backpressure fluctuation range is reduced by 39.87%. When the initial gas storage pressure is 5.0 MPa and the ER is 0.168, the system achieves a round-trip efficiency higher than that of the traditional constant voltage operation modes by 2.73%. Meanwhile, reaching the highest value of 57.94%, energy consumption is reduced by 5.36%. The findings demonstrate that the ejector significantly broadens the compressor's working range and improves energy storage efficiency. Additionally, this confirms that ejector technology contributes to improving compressor stability under variable working conditions, which is particularly important for the frequent start and stop conditions of compressed air energy storage systems. In 2023, Yang et al. [96] utilized a numerical simulation method to investigate pressure regulation in compressed air energy storage systems, specifically discussing the power generation time and total power generation of an energy storage system integrated with an ejector and burner. The results indicate that with the ejector coefficient of 0.8 and the burner heat power of 10 MW, the system power generation time is extended to 12.45 h, with total power generation reaching 140,052 kWh. Compared to the non-supplementary combustion mode, these values represent increases of 15.6 times and 17.5 times, respectively. The ejector mixes high-pressure air with low-pressure ambient air, significantly improving the air mass flow rate at the turbine inlet, while the burner provides thermal energy compensation. This integrated approach effectively overcomes the inherent limitation of short power generation duration in traditional A-CAES systems.

The integration of ejector technology in CAES demonstrates significant thermodynamic improvements yet faces critical challenges in practical implementation. While Guo et al. [88] achieved a 3.41% efficiency gain (61.95% to 65.36%) through dual-system ejector optimization, this improvement remains marginal relative to the energy demands of large-scale applications. Similarly, Liu et al. [93] reported a 3.07% round-trip efficiency increase, highlighting the limitations of incremental gains from ejector-based throttling loss mitigation. The multi-system integration approach shows promise, with Yang et al. [96] achieving a 17.5-fold power generation increase through ejector–burner hybridization, but this introduces complexity in control and thermal management. The technology's adaptability to fluctuating energy demands is validated by Rabi et al. [92], yet system-level reliability under dynamic conditions requires further validation. While ejectors enhance compressor stability (39.87% backpressure fluctuation reduction [95]), their scalability to industrial-scale CAES remains unproven.

In summary, ejector technology has been recommended for use in the pressure reduction systems of compressed air-powered vehicles, offering significant improvements in energy conversion efficiency. The multi-stage pressure regulation and residual pressure recovery strategy effectively reduces traditional throttle valve losses, improving system output power efficiency by 13% [82]. However, technological development is limited by core bottlenecks, which include turbulence loss caused by thermodynamic irreversibility, the synergistic contradiction between lightweight gas storage and thermal management, and insufficient ejector dynamic matching capability. Future efforts should aim to achieve coordinated improvements in both energy density and efficiency, address temperature drops during the expansion process, enable dynamic matching of multi-stage decompression parameters, and enhance the system's round-trip efficiency. These advancements are essential to overcoming key technical challenges and promoting the practical implementation of energy-efficient compressed air technologies.

6. Other Applications of Ejectors in the Field of Rail Transit

As rail transit develops towards high speed, intelligence, and sustainability, ejector technology innovation expands from single-function realization to multi-objective collaborative optimization, offering significant value for a wide range of related systems. Active airflow control dynamically adjusts bow mesh contact force within the pantograph pneumatic system with simultaneous aerodynamic noise suppression. The train vacuum toilet system enhances sewage collection efficiency and reduces water consumption through negative pressure suction. For the train ventilation system, multi-stage ejection combined with personalized air distribution strategies optimizes thermal environment adjustment precision and enhances energy efficiency. Within track cleaning and maintenance devices, stubborn pollutants are efficiently removed from infrastructure and vehicle surfaces through high-pressure ejector technology.

6.1. Pantograph Pneumatic System

As a critical component device in electric traction vehicles, the pantograph pneumatic system achieves precise control of pantograph lifting and contact pressure through coordinated operation of pneumatic actuators, solenoid valves, and sensors. This system dynamically adjusts the contact state between the pantograph slide and the catenary wire, with widespread application in high-speed railways, urban rail transit, and trolley buses. Rising operating speeds of high-speed trains make aerodynamic optimization of pantographs a key research direction to improve bow network flow quality.

Ikeda et al. [97] proposed an ejector-integrated pantograph design for high-speed trains, leveraging controlled airflow from the slide surface to modulate aerodynamic noise suppression and lift dynamics. They adopted a wind tunnel experimental methodology to investigate the impact of ejector flow characteristics on critical aerodynamic performance metrics under operational conditions. Unilateral ejection alters the lift coefficient by 0.04, translating to a 20 N lift force variation for a full-scale pantograph (1 m span) at 300 km/h, enabling active uplift force control. Bilateral ejection reduces narrowband noise at the vortex-shedding frequency (430 Hz) by 5 dB (from baseline levels), though inducing broadband noise above 1 kHz due to ejector flow. PIV confirmed that ejected jets suppress vortex roll-up and stabilize trailing edge shear layers, simultaneously enhancing aerodynamic stability and mitigating aeroacoustic noise. Building on prior research, subsequent scholars began exploring more efficient pneumatic control systems. To modulate aerodynamic stabilization in high-speed trains, Ikeda et al. [98] proposed a dual-source ejector-integrated pantograph lift control system leveraging external and self-supplied airflow. They adopted a combined numerical simulation and wind tunnel experimental approach to investigate the aerodynamic control effectiveness of these ejector configurations under operational flow regimes. The results indicate that the lift coefficient can be significantly altered by the internal circulation airflow within the self-supplied ejector, achieving a lift adjustment range of up to ± 20 N at 300 km/h. Numerical simulations further reveal that the trailing edge vortex system can be reconstructed and the separation flow inhibited through ejector flow. This method effectively resolves contact force instability caused by pantograph lift fluctuation, offering a practical solution for low-noise pantograph design and current collection performance optimization.

Suzuki et al. [99] proposed a self-circulating ejector system for pantograph aerodynamic regulation in high-speed trains, utilizing internal airflow circulation to enhance lift stability and aerodynamic control. They adopted a combined numerical simulation and wind tunnel experimental approach to investigate the aerodynamic–aeroacoustic performance of this integrated system under operational conditions. The results show that trailing edge ejection can significantly alter the flow structure, forming a cyclic vortex system and increasing the lift coefficient by 0.25. During experimental validation, the circular ejector slightly reduces the lift control range while effectively suppressing lift fluctuations in the high-speed regime. The practicality of this ejector system for the collaborative optimization of pantograph aerodynamic noise and lift force is confirmed. This research improves the early ejector design through structural innovation, pushing circulating airflow control technology toward engineering applications. With growing emphasis on high-speed train energy efficiency requirements, recent research has increasingly focused on aerodynamic drag optimization as a key area of improvement. Huang et al. [100] proposed an ejector-integrated aerodynamic optimization framework for high-speed train pantographs, leveraging position-, velocity-, and slot-width-adjusted flow field modulation to achieve drag reduction. They adopted a numerical simulation with an improved delay-detached eddy simulation model to analyze ejector control technology effects on flow structure and resistance characteristics under actual aerodynamic operating conditions. The results demonstrate that when the distance between the ejector slot and the pantograph leading edge is less than 0.6 times its height and the ejector velocity is below 0.6 times the train speed, the drag reduction rate continuously increases with rising ejector velocity, reaching a maximum reduction of 25.03%. Reducing the ejector slot width enhances the drag reduction effect, though ejector energy consumption requires consideration. Increasing ejector velocity contributes more significantly to drag control than expanding slot width. Overall, train energy saving is only applicable when the ejector speed remains below 0.6 times the train speed. This research provides a theoretical basis for designing ejector drag reduction systems in high-speed trains.

The integration of ejector technology in high-speed train pantographs demonstrates significant potential for aerodynamic optimization yet presents notable trade-offs in practical implementation. While Ikeda et al. [97] achieved a 5 dB reduction in narrowband noise (430 Hz) through bilateral ejection, this came at the cost of induced broadband noise above 1 kHz, highlighting the challenge of balancing noise suppression across frequency ranges. The technology shows greater promise in aerodynamic stabilization, with Ikeda et al. [98] demonstrating $\pm$20 N lift adjustment at 300 km/h, effectively addressing contact force instability. However, Huang et al. [100] revealed critical limitations in energy efficiency, where maximum drag reduction (25.03%) is only achievable when ejector velocity remains below 0.6 times train speed—a constraint that may limit operational flexibility. The self-circulating ejector design by Suzuki et al. [99] improved the lift coefficient by 0.25 but sacrificed some control range, underscoring the need for design compromises.

6.2. Train Vacuum Toilet System

As an efficient sanitary treatment device based on negative pressure suction technology, the train vacuum toilet system generates a pressure difference in the pipeline through a vacuum generator to enable rapid collection and sealed transmission of excreta. Compared to traditional gravity drainage systems, this approach offers significantly enhanced water-saving efficiency. Additionally, the use of narrower pipelines reduces both the overall weight and space requirements of the train. Owing to these advantages, vacuum toilet systems are widely applied in high-speed railways, subways, long-distance buses, and aviation.

Guo [10] experimentally combined a dynamic pressure response model to study the energy-saving optimization of a pneumatic ejector circuit, addressing the energy consumption problem of vacuum sanitation systems. A variable vacuum system incorporates a mass sensor to detect excreta volume and activate a control linkage that adjusts air supply, dynamic regulating vacuum degree based on waste amount. For ejectors, a pneumatic circuit pressure variation equation was established, confirming the feasibility of generating a small vacuum at low waste volumes. The experiments demonstrate that, compared to fixed vacuum systems, this design achieves over 50% energy savings within 0.2×10^{-3}~1.6×10^{-3} m^3 waste volumes while exceeding 99% collection rates. Research on the aerodynamic characteristics of similar vacuum sanitation systems reveals multi-dimensional propulsion characteristics. Guo et al. [101] proposed an intelligent sanitary ware system integrating liquid level/mass sensors and adjustable pneumatic ejector circuits for demand-matching vacuum discharge power, addressing energy-saving requirements in next-generation sanitation systems. A combined experimental and numerical simulation approach was adopted to develop and validate a control strategy capable of dynamically adjusting the ejector air supply duration based on real-time excreta volume. The results demonstrate that compared with traditional fixed vacuum systems during manual waste collection, the variable vacuum system saves over 30% of energy. The effectiveness of the pneumatic ejector in achieving adaptive energy adjustment within the vacuum sanitation system is verified, providing a theoretical basis for energy-saving optimization of ejector-driven vacuum systems.

To enhance aerodynamic efficiency and vacuum-generation performance in train sanitation applications, Fujino et al. [102] developed a multi-parameter diagnostic framework for optimizing vacuum ejector systems, integrating valve sizing, pipeline configurations, and gas-path layouts. They adopted a combined experimental and numerical simulation approach utilizing high-response fast-flow sensors for transient pressure/flow measurements to quantify energy efficiency dynamics under operational transient conditions. The results indicate that pressure loss can be reduced by increasing the diameter of the pressure relief valve and optimizing pipe size and path, which increases the ejector vacuum discharge rate by 18%. It is proven that pneumatic component optimization reduces system supply pressure to 0.6 MPa, achieving 26% lower compressed air consumption while maintaining ejector functionality. This provides a quantitative basis for the energy-saving design of vacuum toilets. Advancing sensor technology drives intelligent progression of vacuum regulation research. Fujino et al. [103] developed a multi-parameter impact assessment framework for aerodynamic energy consumption in train vacuum sanitation systems, quantifying the influence mechanisms of ejector operational states and sewage storage tank volumes on system efficiency. The experimental methodology implemented high-response pneumatic power meters to measure real-time pressure, flow rate, and airflow energy dissipation. By actively adjusting sewage tank volumes and controlling the start–stop sequences of the ejector, the study provided detailed insights into system energy dynamics. It is found that reducing the sewage tank volume increases the vacuum pressure to −55 kPa while shortening the ejector air supply time to 2.5 s. The results show that the optimized system maintains a sewage transmission distance of 2.24 m, reducing aerodynamic energy consumption by approximately 20% compared to the initial state. The feasibility of energy consumption optimization through ejector timing control and sewage storage tank structure improvement is verified.

The reviewed studies demonstrate significant progress in ejector-based vacuum sanitation system optimization yet reveal critical limitations in practical implementation. While Guo [10] achieved a remarkable 50% energy savings through dynamic vacuum adjustment, this performance is constrained to a narrow waste volume range (0.2×10^{-3}~1.6×10^{-3} m^3), potentially limiting broader applicability. Fujino et al. [102] showed that component optimization could reduce air consumption by 26%, but the achieved 18% discharge rate improvement may not justify the required system redesign costs in existing installations. The intelligent system by Guo et al. [101] presents a 30% energy saving advantage yet raises concerns about sensor reliability and maintenance complexity in harsh sanitation environments. Fujino's subsequent work [103] demonstrated 20% energy reduction through tank volume optimization, but the achieved −55 kPa vacuum pressure may be insufficient for certain high-demand applications.

6.3. Train Ventilation System

As a key system for maintaining carriage air quality and environmental comfort, the train ventilation system is controlled through dynamic airflow organization and pressure balance. It enables air circulation and filtration, temperature and humidity control, and targeted pollutant removal. Its core framework integrates efficient filtering modules, low-noise wind units, and an intelligent control unit. The system performs effectively in high speeds or closed environments while maintaining stable cabin air pressure and fresh air supply, and is widely applied in high-speed rail, subway, intercity train, and other rail transit fields.

Starcheous et al. [104] proposed a cascaded power converter-integrated heating system for railway transit units, utilizing hybrid-controlled multi-stage ejectors wherein the first stage mixes converter-generated high-temperature air with ambient air, and the second stage dynamically regulates recirculation rates through valve modulation to achieve compartment-specific thermodynamic precision. A combined theoretical and experimental approach was employed to develop and validate this novel system for dynamic temperature regulation across operational load ranges. The experiments show that implementing an adjustable ejector increased the maximum system pressure to 0.6 MPa while thermal efficiency improved by 40% over conventional electric heating systems. Through the secondary recovery of waste gas heat, the energy utilization rate reaches 78%. Research confirms that the ejector structure effectively solves the contradiction between high-temperature air transit and comfort control. The scheme provides an innovative solution for energy-saving heating of transit equipment. Although energy efficiency is significantly improved, the scheme's fine regulation of airflow distribution within the carriage remains to be further explored. By analyzing airflow uniformity and thermal comfort through detailed CFD modeling of passenger thermal loads and diffuser geometries to optimize airflow distribution efficacy, Iranzo et al. [105] proposed a comparative ventilation framework for railcar environmental control systems. They adopted a numerical simulation to evaluate vertical ejector versus horizontal guide diffuser airflow regulation during actual ventilation operational. Research shows that the original diffuser's vertical ejector caused the local wind speed in the passenger area to reach 3.7 m/s—exceeding the comfort standard limit of 1 m/s—while the improved design reduces the maximum wind speed to 2.4 m/s through horizontal diversion and better balances the temperature field and velocity field. The results provide an important basis for optimizing railway vehicle ventilation systems. Schmeling et al. [106] developed a multi-nozzle personalized ventilation system aimed at optimizing thermal comfort in high-temperature rail transit environments. The system features six adjustable airflow nozzles designed to modulate localized thermal sensations based on passenger feedback and equivalent temperature metrics. They adopted an experimental methodology combining equivalent temperature measurements and subjective evaluations from 40 subjects to analyze flow rate-dependent thermal environment dynamics across 294~314 K operational ranges. The results show that the maximum flow rate of the ejector could reduce the chest equivalent temperature by up to 9 K, significantly improving thermal comfort. Although the nozzle orientation remains fixed, individualized control of airflow rates led to a 20% increase in subjective comfort ratings. However, the study also identified a residual risk of thermal discomfort in the head and foot regions due to insufficient airflow distribution. This research marks a paradigm shift from global energy saving to precise and comfortable regulation of railway ventilation systems. Dynamic perception technology should be further combined in the future to achieve full-area thermal environment optimization.

The reviewed studies highlight significant advancements in ejector-based thermal regulation for rail transit but reveal critical limitations in practical implementation. While Starcheous et al. [104] achieved a 40% thermal efficiency improvement and 78% energy utilization rate, the system's reliance on 0.6 MPa pressure raises operational safety and maintenance cost concerns. Iranzo et al. [105] demonstrated that optimized diffuser designs reduced local wind speeds from 3.7 m/s to 2.4 m/s, yet this still exceeds the 1 m/s comfort standard, indicating persistent challenges in balancing ventilation efficacy with passenger comfort. Schmeling et al. [106] showed promising results with a 9 K reduction in equivalent temperature and 20% improvement in subjective comfort ratings, but the system's inability to address thermal discomfort in the head and foot regions reveals fundamental limitations in localized airflow distribution.

6.4. Rail Transit Cleaning and Maintenance Device

Rail transit cleaning and maintenance equipment is specialized for removing pollutants from rail infrastructure and vehicle surfaces, integrating high-pressure ejector cleaning technology, intelligent identification and positioning technology, and environmental protection recovery treatment technology. This equipment efficiently removes dust from track slabs and contact nets, along with stubborn stains on car bodies. Featuring modular designs with multi-modal adaptability, such systems are deployed in various forms, including track grinding vehicles, tunnel-cleaning robots, and platform maintenance units. It is widely used across high-speed railway lines, urban subways, freight railways, and cross-sea bridge track scenarios. Vasic et al. [107] adopted combined experimental and numerical simulations to investigate cleaning technologies for a low-adhesion railway track. These technologies include high-pressure water ejectors, sand ejectors, and Sandite, a mixture of quartz sand and gel. A high-pressure water ejection system employing 1000 bar pressure and a 30-degree nozzle angle effectively removes the contaminated film from the track blade. However, the residual wet film requires treatment with Sandite to increase the adhesion coefficient to 0.09. The results indicate that the cleaning efficiency of the ejector peaks at 40 mph, although leaf contamination is prone to reappearing. Sodium bicarbonate spray applied at 4.5 mph selectively strips pollutants and exhibits environmental protection characteristics. Although such combined mechanical–chemical cleaning schemes have achieved initial results, their high-speed cleaning efficiency and long-term stability still require further optimization.

In summary, the innovative application of ejectors in rail transit has significantly enhanced overall system performance across multiple subsystems. The pantograph pneumatic system utilizes self-supply gas ejection technology enabling ±20 N dynamic lift adjustments at 300 km/h [98]. A vacuum toilet system incorporating intelligent sensing and variable vacuum ejectors achieves over 50% energy savings [10]. The ventilation system, combined with multi-stage ejection and personalized nozzle design, has demonstrated a 40% improvement in thermal efficiency [104]. Track cleaning devices employing 1000 bar high-pressure ejector technology efficiently remove pollutants. However, challenges remain in balancing high-frequency noise generation with ejector energy consumption. Problems of low-load vacuum stability and scaling attenuation coexist, while carriage airflow uniformity and dynamic thermal response precision require optimization. High-speed cleaning, durability, and antifouling capabilities are limited. Future research requires overcoming bottlenecks in high-frequency noise suppression, low-load vacuum stability improvement, carriage airflow uniformity optimization, and high-speed cleaning durability enhancement to realize collaborative performance and sustainability upgrades for ejectors in rail transit systems.

7. Conclusions

This paper comprehensively reviewed the application and research progress of ejector technology in rail transit driven by energy conservation and energy transition. This review summarizes current technical achievements, identifies research shortcomings, and indicates future research hotspots.

Based on the comprehensive review, the following conclusions can be drawn:

(1) Vehicle ejector refrigeration air conditioning is driven by vehicle waste heat energy and utilizes ejector refrigeration technology to replace the traditional compressed refrigeration system in automobiles. This system can reduce the energy consumption of the air conditioning system by 12~17%. Its application has important economic and social benefits for improving the primary energy utilization rate.

(2) The hydrogen ejector serves as the core component of the fuel cell engine's hydrogen supply. It not only improves engine performance but also significantly reduces emissions, resulting in a 9% reduction in high-pressure hydrogen demand. It represents one of the key technologies for realizing zero-emission vehicles.

(3) In pneumatic vehicle pressure reduction systems, throttle valves are replaced by air ejectors, a design modification that increases system output power by more than 13%.

(4) In pantograph pneumatic systems, the lift force is dynamically adjusted by ±20N at 300km/h through self-supply gas ejection technology. Vacuum toilet systems exceed 50% by intelligent sensing and variable vacuum ejectors. Ventilation systems improve thermal efficiency by 40% through the combination of multi-stage ejection and personalized nozzle design. Track cleaning devices efficiently remove pollutants by adopting 1000 bar high-pressure ejector technology.

However, there are still many challenges affecting the current research:

(1) In rail transit engine waste heat recovery, the fixed geometry ejector readily enters subcritical mode under variable engine operating conditions, causing the cooling capacity to be suddenly reduced and the coefficient of performance to be significantly attenuated. Data scarcity exists for global warming potential working fluids applied in the transcritical cycles. The coefficient of performance of traditional compression refrigeration systems is much lower than the theoretical value for ejector refrigeration, limiting the system's actual energy efficiency improvement.

(2) In hydrogen ejector technology, insufficient stability of hydrogen recycling under wide power output causes anodic pressure fluctuations exceeding ±0.15 bar during transient conditions. Mixing uniformity decreases at high load, resulting in the system's net efficiency decaying to 58.6%. The ejection ratio fluctuates more than 7.3% when starting at low temperatures, while hydrogen recovery efficiency reaches only 97.2%.

(3) In pressure reduction systems for new compressed air-powered vehicles, turbulence loss is caused by thermodynamic irreversibility, with conventional pressure relief valves inducing 50% energy loss. Although the ejector increases power output by 13%, the system round-trip efficiency is only 57.94%. Synergistic contradictions exist between lightweight gas storage and thermal management.

(4) In other rail transit applications, a prominent contradiction exists between ejector noise reduction and energy consumption balance in the pantograph pneumatic system. Although the bilateral ejector reduces narrowband noise by 5 dB, wideband noise above 1 kHz is triggered. Insufficient vacuum stability occurs in the vacuum toilet system under low load conditions, while scaling-induced performance reduc-

tion is observed. The ventilation system's airflow distribution uniformity requires optimization. The insufficient high-speed cleaning durability of the high-pressure ejector cleaning device easily allows pollutant reproduction.

8. Future Perspectives

Future research will focus on the following topics, as summarized in Table 2:

(1) Through dynamic intelligent regulation and structural innovation, a machine learning-based ejector pressure threshold prediction algorithm should be developed to improve variable working condition adaptability. A nested double-nozzle ejector requires a different design to extend power coverage. An adjustable swirl nozzle and mobile needle valve structure should be developed to achieve dynamic flow matching. Multi-nozzle flow field interference suppression technology should be further developed.

(2) Through multi-system energy efficiency collaborative optimization, a hybrid pneumatic–electric architecture integrating residual pressure recovery and a multi-level decompression strategy should be established. Integration between fuel cell hydrogen supply systems and refueling station ejectors requires enhancement to improve hydrogen recycling efficiency. The cylinder liner water-driven ejector cooling system is to be strengthened to reduce the exhaust exergy loss rate.

(3) Through the application of transcritical cycles and environmentally friendly working medium, research on CO_2 transcritical ejector refrigeration systems can be deepened through backpressure optimization to improve round-trip efficiency. The injectivity characteristics of global warming potential working fluids require investigation under wide operating conditions to enhance the coefficient of performance. Ammonia working medium applications are to be expanded in low-temperature refrigerated transit to overcome the environmental limitations of conventional refrigerants.

(4) Through the construction of a multi-physics field simulation platform integrating computational fluid dynamics and electrochemical mechanisms (e.g., fuel cell multiphase flow coupling), an ejector-stack cooperative model should be established to overcome hydrogen recirculation instability across wide power ranges.

Table 2. Structured future research agenda for ejector technology in rail transit, grouped by application domain.

Application Domain	Specific Technology Focus	Future Research Direction
Waste heat recovery	Dynamic regulation and structural innovation	Developing machine learning-based ejector pressure threshold prediction algorithms; designing nested double-nozzle ejectors; creating adjustable swirl nozzles and mobile needle valve structures.
Hydrogen systems	Flow field optimization and system integration	Developing multi-nozzle flow field interference suppression technology; constructing multi-physics simulation platforms for ejector-stack cooperation.
Air-powered vehicle	System architecture and efficiency enhancement	Optimizing multi-level decompression strategies; designing thermally–mechanically coupled gas storage systems.

Through the above technological breakthroughs, the cross-generation upgrade of ejectors to high energy efficiency, long life, and intelligence will be promoted, and the green and low carbon transformation of rail transit will be facilitated.

Author Contributions: Conceptualization, Y.L. and H.H.; methodology, Y.L.; software, Y.L.; validation, Y.L., H.H., Y.G., Y.Y. and S.L.; formal analysis, Y.L.; investigation, Y.L.; resources, S.S.; data curation, Y.L.; writing—original draft preparation, Y.L.; writing—review and editing, H.H.; visualization, H.H.; supervision, Y.G.; project administration, S.L.; funding acquisition, Y.L. All authors have read and agreed to the published version of the manuscript.

Funding: This research was funded by the Fundamental Research Funds for the Provincial Universities of Liaoning, grant number LJ212410150010.

Conflicts of Interest: The authors declare no conflict of interest.

Nomenclature

ER	Entrainment ratio
PEMFC	Proton exchange membrane fuel cell
COP	Coefficient of performance
ML	artificial intelligence
AI	Machine learning
ANNs	Artificial neural networks
Gas	Genetic algorithms
AR	Area ratio
NXP	Nozzle exit position
CFD	Computational fluid dynamics
PIV	Particle image velocimetry
ODP	Ozone depletion potential
GWP	Global warming potential
FCV	Fuel cell vehicle
APV	Air-powered vehicle
A-CAES	Adiabatic compressed air energy storage
SOFC	Solid oxide fuel cell
$\dot{m}_s$	Ratio of entrained fluid flow rate
$\dot{m}_p$	Motive fluid flow rate
N	Pressure ratio
P_d	Diffuser discharge pressure
P_s	Suction fluid pressure
P_p	Motive fluid pressure
η	Efficiency of ejector
$\dot{E}_{out}$	Output power
$\dot{E}_{in}$	Input power

References

1. Bahman, N.; Alalaiwat, D.; Abdulmohsen, Z.; Al Khalifa, M.; Al Baharna, S.; Al-Mannai, M.A.; Younis, A. A critical review on global CO_2 emission: Where do industries stand? *Rev. Environ. Health* **2022**, *38*, 681–696. [CrossRef] [PubMed]
2. Sadeghi, M.; Yari, M.; Mahmoudi, S.; Jafari, M. Thermodynamic analysis and optimization of a novel combined power and ejector refrigeration cycle—Desalination system. *Appl. Energy* **2017**, *208*, 239–251. [CrossRef]
3. Li, D.; Groll, E.A. Transcritical CO_2 refrigeration cycle with ejector-expansion device. *Int. J. Refrig.* **2005**, *28*, 766–773. [CrossRef]
4. Khafaji, H.; Shahsavand, A.; Shooshtari, S.H.R. Simultaneous optimization of crude oil refinery vacuum distillation column and corresponding ejector system. *Energy* **2024**, *294*, 130702. [CrossRef]
5. Chen, H.; Cai, C.; Jiang, S.; Zhang, H. Numerical modeling on installed performance of turbofan engine with inlet ejector. *Aerosp. Sci. Technol.* **2021**, *112*, 106590. [CrossRef]
6. Takeya, Y.; Miwa, S.; Hibiki, T.; Mori, M. Application of steam injector to improved safety of light water reactors. *Prog. Nucl. Energy* **2015**, *78*, 80–100. [CrossRef]
7. Mohammed, R.H.; Qasem, N.A.; Zubair, S.M. Enhancing the thermal and economic performance of supercritical CO2 plant by waste heat recovery using an ejector refrigeration cycle. *Energy Convers. Manag.* **2020**, *224*, 113340. [CrossRef]
8. Galindo, J.; Dolz, V.; Tiseira, A.; Ponce-Mora, A. Thermodynamic analysis and optimization of a jet ejector refrigeration cycle used to cool down the intake air in an IC engine. *Int. J. Refrig.* **2019**, *103*, 253–263. [CrossRef]
9. Feng, J.; Han, J.; Pang, Z.; Peng, X. Designing Hydrogen Recirculation Ejectors for Proton Exchange Membrane Fuel Cell Systems. *Energies* **2023**, *16*, 1201. [CrossRef]

10. Guo, Z. An energy conservation design in vacuum sanitation systems. In Proceedings of the 2011 International Conference on Electronics, Communications and Control (ICECC), Ningbo, China, 9–11 September 2011.
11. Han, J.; Feng, J.; Hou, T.; Peng, X. Performance investigation of a multi-nozzle ejector for proton exchange membrane fuel cell system. *Int. J. Energy Res.* **2021**, *45*, 3031–3048. [CrossRef]
12. Elmore, E.; Al-Mutairi, K.; Hussain, B.; El-Gizawy, A.S. Development of Analytical Model for Predicting Dual-Phase Ejector Performance. In Proceedings of the International Mechanical Engineering Congress and Exposition, Phoenix, AZ, USA, 11–17 November 2016.
13. Winoto, S.H.; Li, H.; Shah, D.A. Efficiency of Jet Pumps. *J. Hydraul. Eng.* **2000**, *126*, 150–156. [CrossRef]
14. Karassik, I.J. *Pump Handbook*; McGraw-Hill: New York, NY, USA, 2001.
15. Ünal, Ş.; Tuncay, Y. Thermodynamic analysis of the two-phase ejector air-conditioning system for buses. *Appl. Therm. Eng.* **2015**, *79*, 108–116. [CrossRef]
16. Baek, S.; Ko, S.; Song, S.; Ryu, S. Numerical study of high-speed two-phase ejector performance with R134a refrigerant. *Int. J. Heat Mass Transf.* **2018**, *126*, 1071–1082. [CrossRef]
17. Bai, T.; Yan, G.; Yu, J. Experimental investigation on the dynamic malfunction behavior of the two-phase ejector in a modified auto-cascade freezer refrigeration system. *Energy Convers. Manag.* **2019**, *183*, 382–390. [CrossRef]
18. Chen, W.; Shi, C.; Hu, M.; Chong, D.; Wang, J.; Yan, J. Numerical and Experimental Analysis of Two Phase Flow in Ejector. *Energy Procedia* **2014**, *61*, 1298–1301. [CrossRef]
19. Yang, Y.; Zhu, X.; Yan, Y.; Ding, H.; Wen, C. Performance of supersonic steam ejectors considering the nonequilibrium condensation phenomenon for efficient energy utilisation. *Appl. Energy* **2019**, *242*, 157–167. [CrossRef]
20. Li, Y.; Deng, J.; Ma, L. Experimental study on the primary flow expansion characteristics in transcritical CO2 two-phase ejectors with different primary nozzle diverging angles. *Energy* **2019**, *186*, 115839. [CrossRef]
21. Petrovic, A.; Delibasic, B.; Filipovic, J.; Petrovic, A.; Lomovic, M. Thermoeconomic and environmental optimization of geothermal water desalination plant with ejector refrigeration system. *Energy Convers. Manag.* **2018**, *178*, 65–77. [CrossRef]
22. Kumar, V.; Singhal, G.; Subbarao, P. Realization of novel constant rate of kinetic energy change (CRKEC) supersonic ejector. *Energy* **2018**, *164*, 694–706. [CrossRef]
23. Yadav, S.; Pandey, K.; Kumar, V.; Gupta, R. Computational analysis of a supersonic two-stage ejector. *Mater. Today Proc.* **2021**, *38*, 2700–2705. [CrossRef]
24. Haida, M.; Smolka, J.; Hafner, A.; Mastrowski, M.; Palacz, M.; Madsen, K.B.; Nowak, A.J.; Banasiak, K. Numerical investigation of heat transfer in a CO2 two-phase ejector. *Energy* **2018**, *163*, 682–698. [CrossRef]
25. Al-Doori, G.; Saleh, K.; Al-Manea, A.; Al-Rbaihat, R.; Altork, Y.; Alahmer, A. A review of axial and radial ejectors: Geometric design, computational analysis, performance, and machine learning approaches. *Appl. Therm. Eng.* **2025**, *266*, 125694. [CrossRef]
26. Liu, G.; Zhao, H.; Deng, J.; Wang, L.; Zhang, H. Performance improvement of CO_2 two-phase ejector by combining CFD modeling, artificial neural network and genetic algorithm. *Int. J. Refrig.* **2023**, *154*, 151–167. [CrossRef]
27. Zhang, K.; Zhang, Z.; Han, Y.; Gu, Y.; Qiu, Q.; Zhu, X. Artificial neural network modeling for steam ejector design. *Appl. Therm. Eng.* **2022**, *204*, 117939. [CrossRef]
28. Ringstad, K.E. CFD Modelling for Improved Components in CO_2 and Ammonia Vapour Compression Systems. Ph.D. Thesis, Norwegian University of Science and Technology, Trondheim, Norway, 2023.
29. Wang, H.; Cai, W.; Wang, Y. Optimization of a hybrid ejector air conditioning system with PSOGA. *Appl. Therm. Eng.* **2017**, *112*, 1474–1486. [CrossRef]
30. Li, Y.; Deng, J.; He, Y. Numerical study on the interaction of geometric parameters of a transcritical CO_2 two-phase ejector using response surface methodology and genetic algorithm. *Appl. Therm. Eng.* **2022**, *214*, 118799. [CrossRef]
31. Bencharif, M.; Croquer, S.; Fang, Y.; Poncet, S.; Nesreddine, H.; Zid, S. Prediction of Performance and Geometrical Parameters of Single-Phase Ejectors Using Artificial Neural Networks. *Thermo* **2023**, *3*, 1–20. [CrossRef]
32. Al-Manea, A.; Buttsworth, D.; Saleh, K.; Malpress, R.; Leis, J. Measurement of turbulent supersonic steam jet flow characteristics using TDLAS. *Flow Meas. Instrum.* **2022**, *87*, 102212. [CrossRef]
33. Dvořák, V.; Kotek, M. PIV measurement of constant area mixing in subsonic air ejector. *EPJ Web Conf.* **2012**, *25*, 01011. [CrossRef]
34. Chen, W.; Chong, D.; Yan, J.; Liu, J. The numerical analysis of the effect of geometrical factors on natural gas ejector performance. *Appl. Therm. Eng.* **2013**, *59*, 21–29. [CrossRef]
35. Pradere, C.; Clerjaud, L.; Batsale, J.C.; Dilhaire, S. High speed heterodyne infrared thermography applied to thermal diffusivity identification. *Rev. Sci. Instrum.* **2011**, *82*, 054901. [CrossRef] [PubMed]
36. Han, Y.; Wang, X.; Sun, H.; Zhang, G.; Guo, L.; Tu, J. CFD simulation on the boundary layer separation in the steam ejector and its influence on the pumping performance. *Energy* **2019**, *167*, 469–483. [CrossRef]
37. Sumeru, K.; Martin, L.; Ani, F.N.; Nasution, H. Energy Savings in Air Conditioning System Using Ejector: An Overview. *Appl. Mech. Mater.* **2014**, *493*, 93–98. [CrossRef]

38. Sumeru; Nasution, H.; Ani, F.N. The application of gas ejector for road transport air conditioning system. *AIP Conf. Proc.* **2012**, *1440*, 1133–1140. [CrossRef]
39. Wang, L.; Cai, W.; Zhao, H.; Lin, C.; Yan, J. Experimentation and cycle performance prediction of hybrid A/C system using automobile exhaust waste heat. *Appl. Therm. Eng.* **2016**, *94*, 314–323. [CrossRef]
40. Suresh, R.; Prakash, R.; Praveen, V.; Datta, S.P. Numerically Optimized Ejector Geometry for Ejector Refrigeration Systems With Low-Global Warming Potential Working Fluids. *J. Energy Resour. Technol.* **2024**, *146*, 101702. [CrossRef]
41. Keeratiyadathanapat, N.; Sriveerakul, T.; Suvarnakuta, N.; Pianthong, K. Experimental and Theoretical Investigation of a Hybrid Compressor and Ejector Refrigeration System for Automotive Air Conditioning Application. *Eng. J.* **2017**, *21*, 105–123. [CrossRef]
42. Takeuchi, H.; Nishijima, H.; Ikemoto, T. *World's First High Efficiency Refrigeration Cycle with Two-Phase Ejector:"EJECTOR CYCLE"*; SAE Technical Papers 2004-01-0916; SAE International: Warrendale, PA, USA, 2004.
43. Oshitani, H.; Gocho, M.; Takano, Y. Ejector-Type Cool Box. *SAE Int. J. Passeng. Cars—Mech. Syst.* **2008**, *1*, 640–645. [CrossRef]
44. Zegenhagen, M.; Ziegler, F. Feasibility analysis of an exhaust gas waste heat driven jet-ejector cooling system for charge air cooling of turbocharged gasoline engines. *Appl. Energy* **2015**, *160*, 221–230. [CrossRef]
45. Zhu, J.; Elbel, S. A New Control Mechanism for Two-Phase Ejector in Vapor Compression Cycles for Automotive Applications Using Adjustable Motive Nozzle Inlet Swirl. *SAE Int. J. Passeng. Cars—Mech. Syst.* **2016**, *9*, 44–51. [CrossRef]
46. Ünal, Ş.; Erdinç, M.T.; Kutlu, Ç. Optimal thermodynamic parameters of two-phase ejector refrigeration system for buses. *Appl. Therm. Eng.* **2017**, *124*, 1354–1367. [CrossRef]
47. Syaukani, M.; Lech, S.; Daniarta, S.; Kolasiński, P. A Systematic Review of Two-Phase Expansion Losses: Challenges, Optimization Opportunities, and Future Research Directions. *Energies* **2025**, *18*, 3504. [CrossRef]
48. Jaruwongwittaya, T.; Chen, G. Application of Two Stage Ejector Cooling System in a Bus. *Energy Procedia* **2012**, *14*, 187–197. [CrossRef]
49. Lawrence, N.; Elbel, S. Experimental and analytical investigation of automotive ejector air-conditioning cycles using low-pressure refrigerants. In Proceedings of the International Refrigeration and Air Conditioning Conference, West Lafayette, IN, USA, 16–19 July 2012.
50. Sukri, M.F.; Musa, M.N.; Senawi, M.Y.; Nasution, H. Achieving a better energy-efficient automotive air-conditioning system: A review of potential technologies and strategies for vapor compression refrigeration cycle. *Energy Effic.* **2015**, *8*, 1201–1229. [CrossRef]
51. Galindo, J.; Gil, A.; Dolz, V.; Ponce-Mora, A. Numerical Optimization of an Ejector for Waste Heat Recovery Used to Cool Down the Intake Air in an Internal Combustion Engine. *J. Therm. Sci. Eng. Appl.* **2020**, *12*, 051024. [CrossRef]
52. Venkataraman, V.; El-Kharouf, A.; Pandya, B.; Amakiri, E.; Steinberger-Wilckens, R. Coupling of engine exhaust and fuel cell exhaust with vapour absorption refrigeration/air conditioning systems for transport applications: A review. *Therm. Sci. Eng. Prog.* **2020**, *18*, 100550. [CrossRef]
53. Pan, M.; Bian, X.; Zhu, Y.; Liang, Y.; Lu, F.; Xiao, G. Thermodynamic analysis of a combined supercritical CO2 and ejector expansion refrigeration cycle for engine waste heat recovery. *Energy Convers. Manag.* **2020**, *224*, 113373. [CrossRef]
54. Chen, Y.; Zou, H.; Dong, J.; Xu, H.; Tian, C.; Butrymowicz, D. Experimental investigation on refrigeration performance of a CO2 system with intermediate cooling for automobiles. *Appl. Therm. Eng.* **2020**, *174*, 115267. [CrossRef]
55. Song, Y.; Cui, C.; Yin, X.; Cao, F. Advanced development and application of transcritical CO_2 refrigeration and heat pump technology—A review. *Energy Rep.* **2022**, *8*, 7840–7869. [CrossRef]
56. Ipakchi, O.; Mosaffa, A.; Farshi, L.G. Ejector based CO_2 transcritical combined cooling and power system utilizing waste heat recovery: A thermoeconomic assessment. *Energy Convers. Manag.* **2019**, *186*, 462–472. [CrossRef]
57. Wu, C.; Wan, Y.; Xu, X.; Liu, C. A transcritical carbon dioxide power cycle enhanced by ejector refrigeration for engine waste heat recovery: Comprehensive analysis and optimization. *Energy Convers. Manag.* **2023**, *292*, 117428. [CrossRef]
58. Xue, H.; Wang, L.; Zhang, H.; Jia, L.; Ren, J. Design and investigation of multi-nozzle ejector for PEMFC hydrogen recirculation. *Int. J. Hydrogen Energy* **2020**, *45*, 14500–14516. [CrossRef]
59. Rogié, B.; Kærn, M.R.; Wen, C.; Rothuizen, E. Numerical optimization of a novel gas-gas ejector for fuelling of hydrogen vehicles. *Int. J. Hydrogen Energy* **2020**, *45*, 21905–21919. [CrossRef]
60. Han, J.; Feng, J.; Chen, P.; Liu, Y.; Peng, X. A review of key components of hydrogen recirculation subsystem for fuel cell vehicles. *Energy Convers. Manag. X* **2022**, *15*, 100265. [CrossRef]
61. Wen, C.; Rogie, B.; Kærn, M.R.; Rothuizen, E. A first study of the potential of integrating an ejector in hydrogen fuelling stations for fuelling high pressure hydrogen vehicles. *Appl. Energy* **2020**, *260*, 113958. [CrossRef]
62. Marsano, F.; Magistri, L.; Massardo, A. Ejector performance influence on a solid oxide fuel cell anodic recirculation system. *J. Power Sources* **2004**, *129*, 216–228. [CrossRef]
63. Kim, Y.Y.; Lee, J.T.; Caton, J.A. The Development of a Dual-Injection Hydrogen-Fueled Engine With High Power and High Efficiency. *J. Eng. Gas Turbines Power* **2005**, *128*, 203–212. [CrossRef]

64. Karnik, A.Y.; Sun, J.; Buckland, J.H. Control analysis of an ejector based fuel cell anode recirculation system. In Proceedings of th 2006 American Control Conference, Minneapolis, MN, USA, 14–16 June 2006.
65. Wee, J.-H. Applications of proton exchange membrane fuel cell systems. *Renew. Sustain. Energy Rev.* **2007**, *11*, 1720–173 [CrossRef]
66. Ahluwalia, R.K.; Wang, X. Fuel cell systems for transportation: Status and trends. *J. Power Sources* **2008**, *177*, 167–176. [CrossRe
67. Zhu, Y.; Li, Y. New theoretical model for convergent nozzle ejector in the proton exchange membrane fuel cell system. *J. Pow Sources* **2009**, *191*, 510–519. [CrossRef]
68. Kuo, J.-K.; Jiang, W.-Z.; Li, C.-H.; Hsu, T.-H. Numerical investigation into hydrogen supply stability and I-V performance of PEM fuel cell system with passive Venturi ejector. *Appl. Therm. Eng.* **2020**, *169*, 114908. [CrossRef]
69. Zhao, Y.; Liu, Y.; Liu, G.; Yang, Q.; Li, L.; Gao, Z. Air and hydrogen supply systems and equipment for PEM fuel cells: A reviev *Int. J. Green Energy* **2022**, *19*, 331–348. [CrossRef]
70. Chen, L.; Xu, K.; Yang, Z.; Yan, Z.; Dong, Z. Optimal Design and Operation of Dual-Ejector PEMFC Hydrogen Supply an Circulation System. *Energies* **2022**, *15*, 5427. [CrossRef]
71. Chen, L.; Xu, K.; Yang, Z.; Yan, Z.; Zhai, C.; Dong, Z. Optimal design of a novel nested-nozzle ejector for PEMFC's hydroge supply and recirculation system. *Int. J. Hydrogen Energy* **2023**, *48*, 27330–27343. [CrossRef]
72. Dadvar, M.; Afshari, E. Analysis of design parameters in anodic recirculation system based on ejector technology for PEM fue cells: A new approach in designing. *Int. J. Hydrogen Energy* **2014**, *39*, 12061–12073. [CrossRef]
73. Kuo, J.-K.; Hsieh, C.-Y. Numerical investigation into effects of ejector geometry and operating conditions on hydrogen recirculatio ratio in 80 kW PEM fuel cell system. *Energy* **2021**, *233*, 121100. [CrossRef]
74. Yin, B.; Li, Z.; Dong, F.; Xu, S.; Ni, H. A novel dual-nozzle ejector for enhancement of hydrogen recirculation applied to proton exchange membrane fuel cell system. *J. Power Sources* **2023**, *580*, 233444. [CrossRef]
75. Antetomaso, C.; Irimescu, A.; Merola, S.; Vaglieco, B.; Di Micco, S.; Jannelli, E. Ejector design for PEM fuel cells and assessment o its scalability. *Int. J. Hydrogen Energy* **2024**, *95*, 1235–1241. [CrossRef]
76. Ding, H.; Zhang, P.; Dong, Y.; Yang, Y. Optimization of hydrogen recirculation ejector for proton-exchange membrane fuel cells (PEMFC) systems considering non-equilibrium condensation. *Renew. Energy* **2024**, *237*, 121748. [CrossRef]
77. Arabbeiki, M.; Mansourkiaei, M.; Ferrero, D.; Santarelli, M. Numerical optimization of ejector for enhanced hydrogen recirculation in proton exchange membrane fuel cells. *J. Power Sources* **2025**, *641*, 236846. [CrossRef]
78. Brunner, D.A.; Marcks, S.; Bajpai, M.; Prasad, A.K.; Advani, S.G. Design and characterization of an electronically controlled variable flow rate ejector for fuel cell applications. *Int. J. Hydrogen Energy* **2012**, *37*, 4457–4466. [CrossRef]
79. Jenssen, D.; Berger, O.; Krewer, U. Improved PEM fuel cell system operation with cascaded stack and ejector-based recirculation *Appl. Energy* **2017**, *195*, 324–333. [CrossRef]
80. Seth, B.; Knecht, S.; Szalai, M.; Haußmann, J. Design of a variable passive ejector for hydrogen recirculation of a PEM fuel cell system. *Int. J. Hydrogen Energy* **2025**, *99*, 956–965. [CrossRef]
81. Song, Y.; Wang, X.; Wang, L.; Pan, F.; Chen, W.; Xi, F. A twin-nozzle ejector for hydrogen recirculation in wide power operation of polymer electrolyte membrane fuel cell system. *Appl. Energy* **2021**, *300*, 117442. [CrossRef]
82. Zhuang, H. Performance Study of Ejector-Based Compressed Air-Powered Vehicles. Beijing University of Technology. 2012. Available online: https://kns.cnki.net/kcms2/article/abstract?v=5ykJdPmCibI5N-v0eYBRcECQf9ijLvfos5Tq8EpxPOh463VuVYqM4Tm7iu2ikIs-l-uvh2VIpXJsqAHV-FKraLKsTEGi6IDTq2wPk2NkGnWklJ0tAqUEwsqBwxH32nL4DNYee-9AUTOzLRLF_LluuzCVsHXrNK0nSfm20nJzYRAcA2tm0-WmQtBLz7QgrbV-&uniplatform=NZKPT&language=CHS (accessed on 23 July 2025).
83. Shi, Y.; Li, F.; Cai, M.; Yu, Q. Literature review: Present state and future trends of air-powered vehicles. *J. Renew. Sustain. Energy* **2016**, *8*, 025704. [CrossRef]
84. Verma, S.S. Air Powered Vehicles. *Open Fuels Energy Sci. J.* **2008**, *1*, 54–56. [CrossRef]
85. Papson, A.; Creutzig, F.; Schipper, L. Compressed Air Vehicles: Drive-Cycle Analysis of Vehicle Performance, Environmental Impacts, and Economic Costs. *Transp. Res. Rec.* **2010**, *2191*, 67–74. [CrossRef]
86. Marvania, D.; Subudhi, S. A comprehensive review on compressed air powered engine. *Renew. Sustain. Energy Rev.* **2017**, *70*, 1119–1130. [CrossRef]
87. Wasbari, F.; Bakar, R.; Gan, L.; Tahir, M.; Yusof, A. A review of compressed-air hybrid technology in vehicle system. *Renew. Sustain. Energy Rev.* **2017**, *67*, 935–953. [CrossRef]
88. Guo, Z.; Deng, G.; Fan, Y.; Chen, G. Performance optimization of adiabatic compressed air energy storage with ejector technology. *Appl. Therm. Eng.* **2016**, *94*, 193–197. [CrossRef]
89. Chen, L.X.; Hu, P.; Zhao, P.P.; Na Xie, M.; Wang, D.X.; Wang, F.X. A novel throttling strategy for adiabatic compressed air energy storage system based on an ejector. *Energy Convers. Manag.* **2018**, *158*, 50–59. [CrossRef]
90. Zhou, S.; He, Y.; Chen, H.; Xu, Y.; Deng, J. Performance analysis of a novel adiabatic compressed air energy system with ejectors enhanced charging process. *Energy* **2020**, *205*, 118050. [CrossRef]

91. Jankowski, M.; Pałac, A.; Sornek, K.; Goryl, W.; Żołądek, M.; Homa, M.; Filipowicz, M. Status and Development Perspectives of the Compressed Air Energy Storage (CAES) Technologies—A Literature Review. *Energies* **2024**, *17*, 2064. [CrossRef]
92. Rabi, A.M.; Radulovic, J.; Buick, J.M. Radulovic and J. Buick, Comprehensive Review of Compressed Air Energy Storage (CAES) Technologies. *Thermo* **2023**, *3*, 104–126. [CrossRef]
93. Liu, T.; Wu, S.; Zhong, L.; Yao, E.; Hu, Y.; Xi, G. Parametric assessment and multi-objective optimization of an ejector-enhanced compressed air energy storage system based on conventional and advanced exergy. *J. Renew. Sustain. Energy* **2024**, *16*, 054101. [CrossRef]
94. Sadeghi, S.; Ahmadi, P. Thermo-economic optimization of a high-performance CCHP system integrated with compressed air energy storage (CAES) and carbon dioxide ejector cooling system. *Sustain. Energy Technol. Assess.* **2021**, *45*, 101112. [CrossRef]
95. Cao, Z.; Zhou, S.; He, Y.; Xu, Y.; Chen, H.; Deng, J. Numerical study on adiabatic compressed air energy storage system with only one ejector alongside final stage compression. *Appl. Therm. Eng.* **2022**, *216*, 119071. [CrossRef]
96. Yang, Y.; Wang, N.; Ren, X.; Li, Y.; Guo, X.; Zhang, K.; Shen, S. Research on similarity characteristics of critical states transferring of adjustable ejector. *Int. J. Refrig.* **2023**, *148*, 64–74. [CrossRef]
97. Ikeda, M.; Yoshida, K.; Suzuki, M. *Flow Control Technique to Modify the Aeroacoustic and Aerodynamic Characteristics of a High-Speed Pantograph by Ejection of Jets*; Institute of Noise Control Engineering: Washington, DC, USA, 2004.
98. Ikeda, M.; Suzuki, M.; Yoshida, K.; ITO, M. Application of Jet Ejection to Control Contact Force of Pantograph for High-speed Trains. In Proceedings of the 6th Symposium on Smart Control of Turbulence, Tokio, Japan, 6–9 March 2005; pp. 311–319.
99. Suzuki, M.; Ikeda, M.; Koyama, T. Flow Control on Pantograph with Air Intake and Outlet. *Q. Rep. RTRI* **2007**, *48*, 236–239. [CrossRef]
100. Huang, S.; Zhang, B.D.; Li, Z.W.; Zhao, J.P.; Peng, W.J.; Lin, J.R. Aerodynamic Characteristics of High-speed Train Pantographs Based on Jet Flow Control. *J. Appl. Fluid Mech.* **2024**, *17*, 1536–1551. [CrossRef]
101. Guo, Z.; Li, X.; Kagawa, T. Sewerage force adjustment technology for energy conservation in vacuum sanitation systems. *Chin. J. Mech. Eng* **2013**, *26*, 334–340. [CrossRef]
102. Fujino, K.; Yamamoto, H.; Yamamoto, N.; Youn, C.; Kagawa, T.; Ito, M.; Xin, L.; Kawashima, K.; Taniguchi, K.; Yamamoto, H. Transient pressure and flow rate measurement of the vacuum toilet system for train. In Proceedings of the 8th JFPS International Symposium on Fluid Power, Okinawa, Japan, 25–28 October 2011.
103. Fujino, K.; Yamamoto, H.; Yamamoto, N.; Youn, C.; Kagawa, T. Study of the energy consumption of a vacuum toilet system for train. In Proceedings of the 9th JFPS International Symposium Fluid Power, Matsue, Japan, 28–31 October 2014.
104. Starcheous, Y.; Danileychenko, A.; Lupikov, K. Development of transport heating systems with cascade transformers of energy. *Teka Kom. Motoryz. I Energetyki Rol.* **2014**, *14*, 275–285.
105. Iranzo, A.; Salva, J.A.; Guerra, J.; Barea, G.; Pino, F.J. Air Ventilation and Comfort in Railway Vehicles Operating in Ventilation Mode. *J. Therm. Sci. Eng. Appl.* **2019**, *11*, 031010. [CrossRef]
106. Schmeling, D.; Zierke, O.; Maier, J.; Dehne, T.; Volkmann, A.; Marggraf-Micheel, C.; Goerke, P. Multi-jet personalized ventilation in passenger trains: Objective and subjective thermal comfort. *Build. Environ.* **2025**, *270*, 112510. [CrossRef]
107. Vasic, G. *New Rail Materials and Coatings*; RRUK Report A(PRUK/A2/1); University of Sheffield: South Yorkshire, UK, 2003.

www.ingramcontent.com/pod-product-compliance
Lightning Source LLC
Chambersburg PA
CBHW061814210926
51850CB00006B/703
9783725849895

MDPI AG
Grosspeteranlage 5
4052 Basel
Switzerland
Tel.: +41 61 683 77 34

Energies Editorial Office
E-mail: energies@mdpi.com
www.mdpi.com/journal/energies